T0231153

RUBIACEAE (Part 3)

B. Verdcourt & D. Bridson

Since the publication of Part 2 of this family (1988) E. Robbrecht has published a new classification of the family — 'Tropical Woody Rubiaceae', Opera Bot. Belg. 1, 272pp. (1988). The value of the presence of rhaphides is given much less prominence and the position of several tribes is altered. As far as the East African genera and tribes are concerned the most important differences compared with the classification followed in this Flora are as follows: *Craterispermum* (*Craterispermeae*), *Pentanisia* and *Paraknoxia* (*Knoxieae*) are removed from the *Rubioideae* and placed in the *Antirheoideae*. *Otiophora* is removed from the *Anthospermeae* and placed in the *Hedyotideae*; *Virectaria* is removed from *Virectarieae* and also placed in *Hedyotideae*. *Vanguerieae* are removed from *Cinchonoideae* and placed in *Antirheoideae*. *Cinchonoideae* and *Ixoroideae* are maintained separate; *Cremasporeae* is placed in *Gardenieae* subtribe *Diplosporinae*.

We do not agree with these changes save for the removal of *Otiophora* from the *Anthospermeae*. We would not, however, agree to placing it in the *Hedyotideae*; rather it would be best as a small tribe between the *Anthospermeae* and the *Spermacoceae*. Some *Pentanisia* are so similar to *Pentas* that only an examination of the ovary will separate them and we cannot believe this is due to convergence over a wide range of characters.

Tribe 22. VANGUERIEAE*

KEY TO GENERA**

1. Spines present 2
   Spines absent (rarely some side branches spine-like in
       Vangueria randii and *Cuviera semseii*) 8
2. Calyx-lobes conspicuous, leafy, 0.7–1.6 cm. long, ± 7 times
       as long as calyx-tube, much exceeding the corolla-
       tube, persistent in fruit; spines supra-axillary or present
       on trunks of saplings, often quite robust; pyrenes 1–5   92. **Lagynias
                                                                                      pallidiflora**

   Calyx-lobes not so conspicuous; other characters not so
       combined 3
3. Spines present on trunks of young trees or restricted to
       young or coppice shoots in which case ternate or less
       often paired from leaf-axils; ovary 2-locular; corolla
       lacking ring of deflexed hairs inside; corolla-lobes not
       apiculate; fruit 1.3–2.4 cm. long and wide . . . .   105. **Canthium**
                                                                                      subgen. *Lycioserissa*

   Spines positioned above lateral branches (sometimes
       reduced) or from leaf-axils of more mature stems; other
       characters not combined as above 4
4. Calyx-lobes linear, 1–3.5 mm. long, persistent in fruit;
       ovary always 2-locular 5
   Calyx-lobes triangular, acute, up to 1 mm. long or even
       shorter; ovary 2- or 3–5-locular 6

* Genera 91–101,103,104 by B. Verdcourt, genera 102, 105–108 by D. Bridson.
** See also multi-access key on p.756.

5. Fruit ± 1.3 cm. long; stigmatic knob not widened at base   100. **Rytigynia bugoyensis**

Fruit ± 2 cm. long; stigmatic knob widened at base (assumed from generic description, flowers of actual species not known) . . . .   101. **Vangueriella**

6. Spines only occasionally arising above reduced cushion-shoots; corolla-tube 2–4 mm. long; lobes shortly or long apiculate; fruit 2–5-locular, seldom 1 cm. across   100. **Rytigynia** (in part)

Spines always arising above reduced cushion-shoots (fig. 130/4); corolla smaller, tube up to 2 mm. long; lobes blunt or apiculate; fruit if 3–5-locular larger, if 2-locular small 7

7. Ovary 4–5-locular; fruit moderately large, ± subglobose, 1.7–2 cm. in diameter; corolla-lobes with apiculum 0.5 mm. long . . . .   104. **Meyna**

Ovary 2-locular; fruit smaller, laterally flattened, indented at apex, ± 1 cm. across; corolla-lobes scarcely apiculate . . . .   105. **Canthium** subgen. **Canthium**

8. Inflorescence umbellate (sometimes 1-flowered), entirely enclosed in bud by paired connate persistent bracts; style attached to base of stigmatic knob (fig. 131/18, p. 752); flowers ♂ or unisexual; leaves usually with tertiary venation obscure; corolla-throat densely congested with hairs . . . .   106. **Pyrostria**

Inflorescence never enclosed by paired bracts; style attached in recess above the base of stigmatic knob (fig. 131/13–17), rarely as above; other characters not so combined 9

9. Corolla-lobes linear-lanceolate, 2–4 cm. long, greatly exceeding tube, erect, tomentose or pubescent outside; fruit 2–4 cm. long with calyx-remains triangular to linear; pyrenes 2 . . . .   95. **Vangueriopsis**

Corolla-lobes not as above; fruit smaller or if large then with more than 2 pyrenes or with calyx-remains cupular 10

10. Climbing plants 11

Trees, shrubs (sometimes subscandent), herbs or subshrubby herbs 12

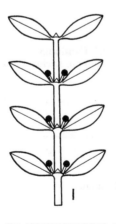  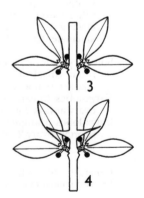

FIG. 130. VANGUERIEAE, diagrammatic representation of inflorescence positions — **1**, *Keetia*-type; **2**, *Vangueria*-type; **3**, *Canthium pseudoverticillatum*-type; **4**, *Meyna*-type. Drawn by Sally Dawson.

11. Leaves coriaceous, drying bright green, glabrous; stipules with a strongly keeled lobe; inflorescences not pedunculate; calyx-limb reduced to a rim, shorter than disk; anthers reflexed; fruit didymous; seeds with endosperm not streaked with granules   . . . . .   107. **Psydrax** subgen. **Phallaria**

    Leaves chartaceous to coriaceous, not drying bright green, glabrous or pubescent; stipules lacking keeled lobe; inflorescence pedunculate; calyx-limb at least equalling disk, often dentate, occasionally lobed; anthers exserted but not reflexed; fruit ± heart-shaped; seed with endosperm streaked with granules, or rarely not . . . .   108. **Keetia**

12. Subshrubby herbs or single-stemmed shrubs from a woody rootstock, up to 2 m. tall 13

    Shrubs or trees (1-)2-20 m. tall 20

13. Ovary (or fruit) 2-locular; stigmatic knob 2-lobed at tip, e.g. fig.131/13-18, p. 752 14

    Ovary (or fruit) 3-8-locular, sometimes with a few 2-locular ovaries as well; stigmatic knob 3-8-lobed at tip, e.g. fig. 131/19-22 16

14. Leaf-blades coriaceous; corolla coriaceous, drying wrinkled; pyrenes very thickly woody, fig. 132/14, p. 754; inflorescences frequently supra-nodal . . .   102. **Multidentia concrescens**

    Leaf-blades not coriaceous; corolla not coriaceous; pyrenes not thickly woody; inflorescences axillary 15

15. Small subshrub 15-30 cm. tall; leaves glabrous or with sparse bristly hairs on both sides; calyx-lobes subfoliaceous, narrowly oblong, elliptic or lanceolate, fig. 131/11, p. 752; fruit 1.5-2 cm. across . . . .   96. **Pygmaeothamnus**

    Larger subshrubs, 0.5 m. or taller, leaves velvety beneath, discolorous; calyx-lobes linear; fruit up to 1 cm. across . . . .   100. **Rytigynia** subgen. **Fadogiopsis**

16. Small subshrub 5-25 cm. tall; fruit 1.5-1.7 cm. across; leaves usually paired, glabrous or with seta-like hairs on both sides; calyx-lobes linear or linear-lanceolate   93. **Pachystigma pygmaeum**

    Larger subshrubs, often with virgate stems; fruit not usually more than 1 cm. across; other characters not combined as above 17

17. Leaves glabrous to pubescent, only occasionally velvety; corolla glabrous or pubescent; calyx-lobes absent to quite well developed, up to 3.5 mm. long; leaves almost invariably in whorls of 3-6 . . . . . .   99. **Fadogia**

    Leaves velvety, entirely covered with silky hairs; corolla velvety; calyx-lobes triangular or linear; leaves paired or sometimes in 3's 18

18. Calyx-lobes 2-4 mm. long, linear or linear-lanceolate; fruits with orange-brown velvety tomentum on drying . . . .   97. **Tapiphyllum** (in part)

    Calyx-lobes short and triangular, 0.5-1 mm. long; fruit almost glabrous 19

19. Corolla-tube 3-5 mm. long, straight . . . .   98. **Fadogiella**

    Corolla-tube 1.6-1.8 cm. long, slightly curved . . . .   99a. **Ancylanthos**

20. Ovary or fruit 2-locular (rarely occasional 3-locular ones as well); stigmatic knob 2-lobed at tip, e.g. fig. 131/13-17, p. 752 21

    Ovary or fruit 3-5-locular (rarely occasional 2-locular ones as well); stigmatic knob 3-5-lobed at tip, e.g. fig. 131/19-22 30

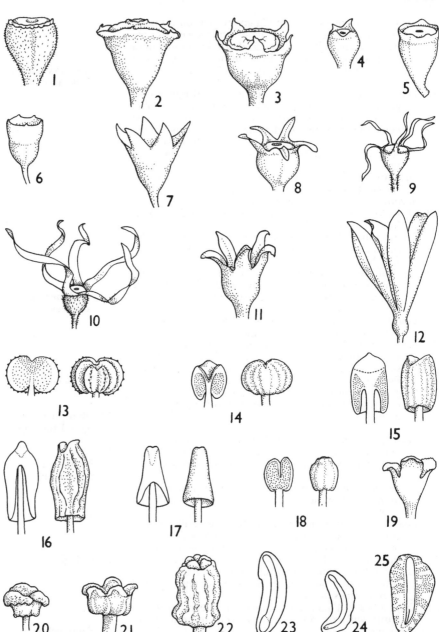

FIG. 131. VANGUERIEAE — 1–12— calyx-types, 13–22 stigmatic knobs (most with longitudinal sections) & 23–25 longitudinal sections of seeds, all × 3; 1, *Canthium lactescens*, × 6; 2, *Temnocalyx nodulosus*, × 5; 3, *Meyna tetraphylla* subsp. *comorensis*, × 8; 4, *Pyrostria lobulata*, × 6; 5, *Canthium kilifiensis*, × 6; 6, *Multidentia exserta* subsp. *exserta*, × 6; 7, *M. kingupirensis*, × 6; 8, *Canthium oligocarpum* subsp. *oligocarpum*, × 6; 9, *Rytigynia bugoyensis* subsp. *bugoyensis*, × 4; 10, *Vangueria apiculata*, × 4; 11, *Pygmaeothamnus zeyheri* var. *zeyheri*, × 6; 12, *Lagynias pallidiflora*, × 2; 13, *Canthium lactescens*, × 10; 14, *C. oligocarpum* subsp. *oligocarpum*, × 10; 15, *Psydrax dicoccos*, × 10; 16, *Keetia gueinzii*, × 10; 17, *Psydrax kraussioides*, × 10; 18, *Pyrostria lobulata*, × 10; 19, *Temnocalyx nodulosus*, × 3; 20, *Rytigynia caudatissima*, × 6; 21, *Meyna tetraphylla* subsp. *comorensis*, × 18; 22, *Vangueria madagascariensis*, × 12; 23 *Canthium coromandelicum*; 24, *Psydrax dicoccos*; 25, *Keetia gueinzii*. 1, 13, from *Troupin* 2751; 2, from *Stolz* 2429; 3, 21 from *Greenway & Kanuri* 14619; 4, 18, from *Ford* 729; 5, from *R.M. Graham* in *F.D.* 1711; 6, from *Anderson* 826; 7, from *Vollesen* in *M.R.C.* 4277; 8, 14, from *Harley* 9567; 9, from *Shabani* 352; 10, from *Carmichael* 535; 11, from *Holmes* 1224; 12, from *Drummond & Hemsley* 3940; 15, from *Cooray* 69042208R; 16, from *Parnell* 2160; 17, from *Torre & Correia* 18674; 19, from *Gillett* 17729; 20, from *D. Thomas* 3781; 22, from *Haarer* 1752; 23, from *Fosberg* 51891; 24, from *Wirwan et al.* 867; 25, from *White* 3196. Drawn by Sally Dawson.

21. Style usually at least twice as long as corolla-tube; stigmatic
     knob ± twice as long as wide, ± cylindric, fig. 131/15–17;
     stipules glabrous within; disk sometimes pubescent;
     inflorescences usually subtended by leaves; seed with
     cotyledons perpendicular to ventral face, fig. 131/24,
     25 22
     Style usually much less than twice as long as corolla-tube
     (save in *Multidentia exserta* which has stipules hairy
     within); stigmatic knob mostly as broad as long, fig.
     131/13,14; stipules hairy or glabrous within; disk
     glabrous; inflorescences subtended by leaves or not;
     seeds with cotyledons parallel to ventral face, fig.
     131/23 23
22. Trees or shrubs, sometimes scandent; leaves typically
     subcoriaceous to coriaceous, drying light green, or
     occasionally chartaceous in deciduous species; if
     scandent, then stipules triangular to truncate at base
     with strongly keeled lobe; calyx-limb a dentate to
     repand rim only occasionally equalling disk, usually
     much smaller; anthers usually reflexed; fruit not or
     scarcely indented at apex except when ± didymous;
     pyrene cartilaginous to woody with shallow apical crest,
     fig. 132/24, p. 754 . . . .   107. **Psydrax**
     Scandent bushes with lateral branches set at right-angles,
     often subtended by modified leaves; leaves
     chartaceous to subcoriaceous, rarely coriaceous;
     stipules lanceolate to ovate or triangular, acuminate;
     anthers usually erect; fruit strongly or slightly indented
     at apex, typically heart-shaped; pyrene woody with lid-
     like area surrounding crest (either positioned on
     ventral face or across apex), fig. 132/25, 26 . . . 108. **Keetia**
23. Calyx-lobes subfoliaceous, 2.5–12 mm. long, persistent on
     fruit; inflorescence clearly branched bearing linear or
     subfoliaceous bracts and bracteoles; corolla-lobes
     acuminate or shortly appendaged . . . .   94. **Cuviera**
                                                          (in part)
     Calyx-lobes if present, not as above; inflorescence
     branched or unbranched but bracts and bracteoles
     usually inconspicuous; corolla-lobes blunt or if
     acuminate or appendaged then inflorescence ±
     unbranched 24
24. Calyx-limb with tubular part cupular, repand or lobed but
     lobes never exceeding it, fig. 131/6, 7; fruit large;
     pyrenes very thickly woody, strongly irregularly ridged
     with lines of dehiscence apparent, fig. 132/4; leaves
     discolorous, mostly with conspicuous network of
     tertiary nerves; corolla-tube with a ring of deflexed
     hairs inside; lobes never apiculate; stigmatic knob
     spherical to elongate-ellipsoid, ribbed . . . . . . 102. **Multidentia**
     Calyx-limb with tube obsolete or less well developed;
     lobes absent or, if present, usually greatly exceeding
     tube; pyrenes not so thickly woody nor ridged as above;
     other characters not combined as above 25
25. Flowers in branched mostly many-flowered pedunculate
     dichasial or complicated branched cymes, occasionally
     subumbellate by reduction; leaves restricted to new
     growth or not strictly so 26
     Flowers solitary or in a few–several-flowered fascicles,
     subumbellate or less often with rudimentary branches,
     peduncles mostly, but not always suppressed; leaves
     well spaced along branches or restricted to
     cushion-shoots 27

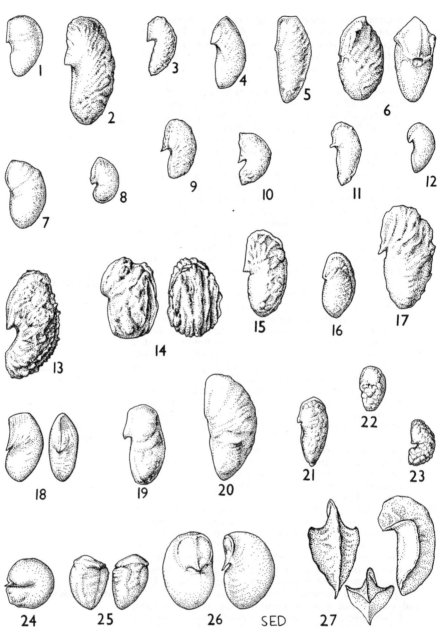

FIG. 132. VANGUERIEAE — pyrene-types; **1**, *Temnocalyx nodulosus*, × 2; **2**, *Lagynias pallidiflora*, × 2; **3**, *L. rufescens* subsp. *angustiloba*, × 1; **4**, *Pachystigma burttii* subsp. *burttii*, × 2; **5**, *Vangueriopsis longiflora*, × 1; **6**, *Pygmaeothamnus zeyheri* var. *zeyheri*, × 2; **7**, *Tapiphyllum burnettii*, × 2; **8**, *Fadogiella stigmatoloba*, × 2; **9**, *Fadogia stenophylla* subsp. *odorata*, × 2; **10**, *Ancylanthos rogersii*, × 2; **11**, *Rytigynia uhligii*, × 2; **12**, *R. monantha* var. *monantha*, × 2; **13**, *Vangueriella rhamnoides*, × 2; **14**, *Multidentia concrescens*, × 1; **15**, *Vangueria madagascariensis*, × 1; **16**, *V. praecox*, × 1; **17**, *Meyna tetraphylla* subsp. *comorensis* × 2; **18**, *Canthium glaucum* subsp. *frangula*, × 2; **19**, *C. kilifiense*, × 2; **20**, *C. oligocarpum* subsp. *oligocarpum*, × 2; **21**, *C. pseudosetiflorum*, × 2; **22**, *Pyrostria bibracteata*, × 2; **23**, *Psydrax schimperiana* subsp. *schimperiana*, × 2; **24**, *P. kraussioides*, × 2; **25**, *Keetia gueinzii*, × 2; **26**, *K. zanzibarica* subsp. *zanzibarica*, × 2; **27**, *Cuviera semseii*, × 1. 1, from *Gillett* 17729; 2, from *Haerdi* 604/0; 3, from *Rodgers & Mwasumbi* 1227; 4, from *Hornby* 617; 5, from *Haerdi* 435/0; 6, from *Brummitt et al.* 13879; 7, from *Haerdi* 445/0; 8, from *Fanshawe* 683; 9, from *Paulo* 235; 10, from *Angus* 847; 11, from *Gardner* 2305; 12, from *Peter* 37611; 13, from *Synnott* 657; 14, from *Milne-Redhead & Taylor* 7879A; 15, from *Richards* 20808; 16, from *Renvoize & Abdallah* 2309; 17, from *Bally* 1636; 18, from *Angus* 1084; 19, from *Rawlins* 856; 20, from *Maitland* s.n.; 21, from *Verdcourt* 766; 22, from *Faulkner* 4262; 23, from *Ruffo* 209; 24, from *Pawek* 2901; 25, from *White* 3196; 26, from *Shabani* 267; 27, from *Bidgood et al.* 1760. Drawn by Sally Dawson, with 27 by Diane Bridson.

26. Stipules seldom sheathing at base when mature, often
     becoming corky outside, if lobed then lobe not
     decurrent and often caducous; leaves strictly restricted
     to new growth; inflorescence with flowers usually
     arranged to one side of ultimate inflorescence-branch;
     calyx-limb ± obsolete; fruit heart-shaped and strongly
     indented at apex or obovate . . . .   105. **Canthium** subgen.
            Afrocanthium
            (in part)

     Stipules sheathing at base, bearing a linear to subulate
     often decurrent lobe; leaves occasionally restricted to
     new growth; inflorescence not as above; calyx-limb
     lobed to base or almost so; fruit slightly indented at
     apex . . . .   105. **Canthium** subgen.
            Lycioserissa

27. Leaves restricted to very short reduced branchlets or
     cushion-shoots giving a pseudo-verticillate appearance,
     fig. 130/3, p. 750; inflorescences always at leafless
     nodes . . . .   105. **Canthium** subgen.
            Afrocanthium
            (in part)

     Leaves not restricted to short branchlets; inflorescences
     in axils of normal leaves, fig. 130/1 28
28. Flowers functionally unisexual, the ♀ inflorescence 1-
     flowered, the ♂ few–many-flowered; stipules not
     pubescent inside; corolla-tube very short, 1–2 mm.
     long; lobes erect . . . .   105. **Canthium** subgen.
            Bullockia

     Flowers not functionally unisexual; stipules pubescent
     to villous inside; corolla-tube usually, but not always,
     exceeding 2 mm. long 29
29. Corolla-tube glabrous or with few hairs within; calyx-limb
     lobed to base, fig. 131/8, p. 752; foliage glabrous; young
     stems scarcely lenticellate; fruit 1.3–2.5 cm. across;
     corolla-lobes not acuminate nor appendaged . . .   105. **Canthium** subgen.
            Lycioserissa

     Corolla-tube with a ring of deflexed hairs within or
     occasionally glabrous; calyx-limb usually shortly
     tubular, dentate; foliage glabrous to hairy; young stems
     usually conspicuously lenticellate; fruit 0.8–1.2 cm.
     across; corolla-lobes blunt, acuminate or appendaged   100. **Rytigynia** (in part)
30. Corolla conspicuous, yellow; tube 2.5 cm. long, slightly
     curved, glabrous within; flowers solitary, long-
     pedicellate; calyx-limb reduced to a ± truncate rim, fig.
     131/2; stigmatic knob not hollow, fig. 131/19, p. 752   91. **Temnocalyx**
     Corolla not as above; flowers not solitary; calyx-limb
     produced into distinct lobes or short teeth; stigmatic
     knob hollow at base 31
31. Calyx-lobes long and ± leafy, 0.8–1.6 cm. long 32
     Calyx-lobes usually shorter, mostly linear or triangular,
     at the most subfoliaceous 33
32. Calyx-lobes 0.7–1.6 cm., ± 7 times as long as the calyx-tube
     and usually equalling or ± exceeding the corolla-tube,
     fig. 131/12; cymes umbel-like with a ± cup-like bract;
     pedicels long; one species glabrous, the other with
     ferruginous red indumentum . . . .   92. **Lagynias**
     Calyx-lobes leafy, ± 8 × 4.7 mm., about 4 times as long as
     calyx-tube, much exceeding the corolla-tube; cymes
     branched and with scattered ± leafy bracts and
     bracteoles; pyrene with pronounced keel, fig. 132/27,
     p. 754 . . . .   94. **Cuviera semseii**

33. Inflorescence a dichasial cyme with flowers scattered
along the arms, usually lax and many-flowered but a
few species with few-flowered inflorescences; fruit
large and globose, 1.4–5 cm. in diameter (when dry)   103. **Vangueria***
Inflorescences varying from solitary flowers to fascicles or
dichasial cymes but mostly condensed; fruits often
small, not over 3 cm. in diameter when dry (not known
for 100a, *Gen. ? nov. A*)   .34
34. Plant velvety-tomentose including the corolla 35
Plant glabrous to hairy, occasionally velvety-tomentose, but
then corolla never velvety   .37
35. Calyx-lobes reduced to triangular teeth; corolla-tube 1.6–
1.8 cm. long, slightly curved   . . . .   99a. **Ancylanthos**
Calyx-lobes well developed; corolla-tube much shorter,
straight 36
36. Calyx-tube ± ⅓ the length of the lobes; lobes ovate to
oblong-triangular, distinctly less pubescent inside than
out; flowers 1–4(–several); leaves oblong-ovate or
almost round   . . . .   93. **Pachystigma
schumannianum**

Calyx-tube much less than ⅓ the length of the lobes;
lobes linear or linear-lanceolate, equally hairy on both
sides; flowers several–many; leaves elliptic, oblong-
elliptic or oblong-ovate   . . . .   97. **Tapiphyllum**
37. Mature leaf-blades 5.5–16.5 cm. long, paired at apex of
branches, absent or immature at time of flowering, fig.
130/2, p. 750; fruit 2.2–2.6 cm. wide   . . . . .   103. **Vangueria**
subgen. **Itigi**

Mature leaf-blades seldom as large, often less than 5 cm.
long, with one to several pairs on each branch, mature,
immature or rarely absent at time of flowering; fruit not
exceeding 1 cm. wide (not known for 100a, *Gen. ?
nov. A)* 38
38. Leaf-blades rounded to obtuse at apex; calyx-lobes ovate,
oblong or oblong-linear; corolla-lobes always distinctly
acuminate; flowers sessile (or subsessile), solitary or in
few-flowered fascicles   . . . .   93. **Pachystigma**
Leaf-blades acuminate or occasionally acute to obtuse at
apex; calyx-lobes reduced, dentate or linear; corolla-
lobes blunt, apiculate or distinctly acuminate; flowers
solitary or in few-flowered pedunculate to subsessile
inflorescences 39
39. Stipules not terminating in a club-shaped colleter   100. **Rytigynia**
(in part)
Stipules terminating in a club-shaped colleter   . . . .   100a. **Gen. ? nov. A**

NON-DICHOTOMOUS GUIDE TO GENERA

Since the delimitation of genera in the Vanguerieae is problematical and the
distinguishing characters are for the most part weak, variable or need to be used in
combination, it is thought useful to provide a non-dichotomous guide to genera. This
guide will prove especially helpful for incomplete material; it will either indicate the
genus or at least greatly restrict the choice between genera.

Genera and subgenera are numbered as follows: bold indicates the generic number,
and non-bold the species number.

*See also 103a, *Gen. ? nov. B.*

Temnocalyx **91** (1)
Lagynias **92** (1–2)
   subgen. *Lagynias* **92.1** (1)
   subgen. *Bembea* **92.2** (2)
Pachystigma **93** (1–6)
Cuviera **94** (1–4)
Vangueriopsis **95** (1–2)
Pygmaeothamnus **96** (1)
Tapiphyllum **97** (1–6)
Fadogiella **98** (1)
Fadogia **99** (1–14)
Anyclanthos **99a** (1)
Rytigynia **100** (1–51)
   subgen. *Rytigynia* **100.1** (1–42, 46–51)
   subgen. *Fadogiopsis* **100.2** (43)
   subgen. *Sali* **100.3** (44, 45)
Gen. ? nov. A **100a**(1)

Vangueriella **101** (1)
Multidentia **102** (1–8)
Vangueria **103** (1–8)
   subgen. *Vangueria* **103.1** (1–7)
   subgen. *Itigi* **103.2** (8)
Gen. ? nov. B **103a** (1)
Meyna **104** (1)
Canthium **105** (1–24)
   subgen. *Canthium* **105.1** (1)
   subgen. *Afrocanthium* **105.2** (2–16)
   subgen. *Lycioserissa* **105.3** (17)
   subgen. *Bullockia* **105.4** (18–24)
Pyrostria **106** (1–9)
Psydrax **107** (1–18)
   subgen. *Psydrax* **107.1** (1–16)
   subgen. *Phallaria* **107.2** (17,18)
Keetia **108** (1–16)

The guide is divided into two parts: the first (Guide 1) lists only *restricted characters*, i.e. characters which do not occur in more than three genera, while the second (Guide 2) lists *widely distributed characters*.

Guide 1 only includes a selection of genera, and is intended to aid quick recognition of the more characteristic ones. Check the list of characters against the specimen and note the generic numbers of any characters it possesses. If more than one number is noted either proceed to Guide 2 or the dichotomous key.

In **Guide 2** each character is considered in two or more states and these are listed against the corresponding generic numbers. In order to use Guide 2 proceed as follow:

1. Select from the list *any* character clearly present on the specimen and list the generic numbers corresponding to the character state present.

2. Next, working in *any order*, proceed to a second character. Refer to the list of generic numbers previously noted and delete any that are not indicated against the list for the new character state. Ignore any additional numbers.

3. Continue until only one number remains or no more can be eliminated. Refer to the dichotomous key and search for couplets that may help the choice between the short-listed genera.

Brackets have been used in the number-lists to indicate the following:

a) the character state is present only in the species indicated, e.g. **99**(14) —occurs in *Fadogia fuchsioides* but not in any other species of *Fadogia*.
b) the character state occurs only infrequently or partially in the genus or species e.g. (**104**) or (**94**(1)).

Guide 1. **Restricted characters**

leaves subcoriaceous to coriaceous, shiny above drying yellowish green — **96, 102** (7, 8), **107**
bracts paired, connate and persistent, completely enclosing the young umbellate inflorescence, fig. 158/C, p. 889 — **106**
bracts linear-lanceolate, conspicuous; inflorescence cymose, figs. 136/1, p. 772 & 137/3, p. 775 — **94, 95**
flowers unisexual — **105.4, 106** (2, 5, 7)
corolla subcoriaceous, wrinkled when dry — **102** (5, 7, 8)
corolla-tube deep red, large — **99** (14)
corolla-tube somewhat curved — (**91**), **99** (12, 13), **99a**
corolla-lobes linear-lanceolate, distinctly longer than tube, fig. 137/4, p. 775 — **92**(1), **95**
stigmatic knob attached at base, fig. 131/18 & 19, p. 752 — **91, 106**
stigmatic knob widened at base, fig. 131/17 — **94**(1), **101***, **107.2**

---

* Corolla and stigmatic knob not known for species in Flora area, data based on other species

pyrenes with lid-like area; endosperm streaked with granules, figs. 131/25, p. 752; 132/25 & 26, p. 754 — **108**
cotyledons positioned parallel to ventral face of seed, fig. 131/24 & 25 — **107, 108**
stipules terminating in club-shaped colleter — **100a**

Guide 2. **Widely distributed characters**

General habit characters
climbers — **101, (104), 107.1** (13,15), **107.2, 108**
subshrubby herbs or shrubs (mostly single-stemmed) from a woody rootstock, up to 1.5(–2) m. tall — **93** (1), **96, 97, 98, 99, 99a, 100.2, 102** (8)
shrubs to small trees, branched, sometimes scandent, (1–)2–10 m. tall — **91, 92, 93** (2–4), **94, 95** (1), **97, 99, 99a, 100, 100a, 101, 102, 103, 104, 105.1, 105.2, 105.3, 105.4, 106, 107.1, (108)**
trees over 10 m. tall — **92, 94** (3, 4), **95** (2), **(100), 100a, 102, (103), 103a, 105.2** (5, 6), **105.3, 106** (4), **107** (1, 2)

spines present — **92.1, (94** (1)), **100, 101, 103** (7), **104, 105.1, 105.3, (106**(8))
spines absent — **91, 92.2, 93, 94, 95, 96, 97, 98, 99, 99a, 100, 100a, (101), 102, 103, 105.2, 105.3, 105.4, 106, 107, 108**

Inflorescence position
inflorescences subtended by mature leaves (occasionally also a few at nodes where leaves have fallen especially in fruiting stage), fig. 130/1, p. 750 — **92, 93, (94), 96, 97** (4, 5), **98, 99, 100, 101, 102** (8), **105.3, 105.4, 106, 107, 108**
inflorescences always borne at nodes from which the leaves have fallen; immature or mature leaves borne at apex of stem, or absent at time of flowering, fig. 130/2 — **91, 92, 93, 94, 95, 96, 97, 98, (99), 99a, 100, 100a, 101, 102, 103, 105.2, (105.3), (106), 107** (3, 4)
inflorescences borne on brachyblasts below immature or mature leaves or leaves absent at time of flowering, fig. 130/3 & 4, p. 750 — **100, 101, 104, 105.1, 105.2** (2–4)

Young stems
conspicuously lenticellate — **(93), (94** (1)), **(95** (1)), **100a, 100.1, 101, 103, 104, 105.1, (105.2)**
lenticels inconspicuous — **91, 92, 93, 94, 95, 96, 97, 98, 99, 99a, 100a, 100.1, 100.2, 100.3, 101, 103, 104, 105.1, 105.2, 105.3, 105.4, 106, 107, 108**

Leaves
always paired — **91, 92, 93, 94, 95, 97, 98, (99), 99a, 100, 100a, 101, 102, 103, 104, 105, 106, 107, 108**
some at least in whorls of 3–6 — **(93), 96, 97, 98, 99, 99a, 100.1** (28), **101, (102** (8)), **105.3, (105.4** (23)), **107** (14)

always glabrous save for domatia — **92.2, 93, 94** (1–3), **99, 100, 100a, 102, 103.1** (1, 3, 4, 7), **103.2, 103a, 104, 105.1, 105.2, 105.3, 105.4, 106, 107, 108**
sparsely to densely pubescent, but leaf-surface visible beneath — **91, 92.2, 93, 94** (3), **95, 96, 99, 100, 101, 102** (7), **103.1** (3, 5, 7), **104, 105.2, 105.4, 107** (3, 4, 10), **108**
densely velvety or felted beneath, leaf-surface entirely obscured — **93** (4), **94** (4), **95, 97, 98, 99, 99a, 100.1** (40), **100.2, 103.1** (2, 6), **105.2** (14, 16)

Stipules
conspicuously hairy within — **91, 93, 94, 95, (99), 99a, 100, 100a, 101, 102, 103, 104, 105.1, 105.2, 105.3**
hairs absent or inconspicuous within — **91, 92, 93, 94, 95, 96, 97, 98, 99, 99a, (100), (102), 103, 103a, 104, 105.1, (105.2), 105.4, 106, 107, 108**

Inflorescence
flowers solitary — **91, 93** (2), **99, 100, 104, 105.2** (3–4), **105.4, 106** (8), **107** (16)
peduncle absent or occasionally very short; inflorescence unbranched; flowers few-many, usually pedicellate — **93** (3, 4), **97** (7), **100, 100a, 104, (105.2** (2, 4)), **105.4, 107.1** (6–15), **107.2**
peduncle short or well developed; inflorescence unbranched or slightly branched, ±

2-10-flowered — **92, 93, (94), 97, 98, 99, 99a, 100, 101, 103, 104, 105.1, 105.2, (2, 4), 105.3, 105.4, 106, 107, (108)**

peduncle short or well developed; inflorescence distinctly branched, many-flowered — **94, 95, 96, (97), 102, 103, 103a, 105.2, 105.3, 107, 108**

Calyx-limb

reduced to a rim shorter than disk, sometimes dentate, fig. 131/1–4, p. 752 — **91, 99, 99a, 100.1, 104, (105.1), 105.2, 106, 107**

tube present, small (at least equalling disk) or well developed, unlobed or with lobes at most equalling tube, fig. 131/5–7, p. 752 — **93** (2–4), **95** (1), **98, 99, 99a, 100.1, 102, (105.1), (105.2), 105.4, (106), 107, 108**

tube ± undeveloped; lobes linear, linear-oblong or, less often, triangular, fig. 131/8–10, p. 752 — **95** (2), **97, 100.1** (40), **100.2, 100.3, 100a, 103, ?103a, 105.3, 106**

tube present (at least 1.5 mm. long); lobes always exceeding tube, triangular, triangular-ovate or large and foliaceous, fig. 131/11, 12, p. 752 — **92, 93, 94, 96, 99, 102** (8), **108** (8)

Corolla-tube

large, 1–2.5 cm. long — **91, 99** (12–14), **99a, 102** (2)

medium, 0.5–1 cm. long — **92, 93, 95, (97), 99, 100, 100a, 101*, 102** (2), **(103), 106** (6, 7), **107** (17), **108** (16)

small, 2–5 mm. long — **93, 94, 95, 96, 97, 98, 99, 100, 101*, 102, 103, 104, 105.2, 105.3, 106, 107, 108**

very small, 1–1.75 mm. long — **99** (11), **100** (1, 28), **105.1, 105.2** (2–4), **105.4, 107, 108** (14)

ring of deflexed hairs present inside — **92, 93, 94, 95** (1), **96, 97, 99, 99a, 100, 100a, 101*, 102, 103, 104, 105.1, 105.4, 107, 108**

ring of deflexed hairs absent inside — **91, 95** (2), **100, 105.2, 105.3, 106, 107** (3–5), **108** (8)

Corolla-lobes

blunt to acute — **91, 95** (2), **(99), 99a, 100, 100a, 102, 103, 105, 106, 107.1, 108**

acuminate to apiculate — **98, 99, 100, 101*, 103, 104, 106, 107.2**

distinctly tailed — **92, 93, 94, 95** (1), **96, 97, 99, 100, 101*, 103**

Disk

glabrous — **91, 92** (1), **93, 94, 95, 96, 97, 98, 99, 99a, 100, 100a, 101, 102, 103, 104, 105, 106, 107**

hairy — **92** (2), **93** (5), **107** (1, 2), **108**

Style

equalling or shortly exceeding corolla-tube — **91, 92, 93, 96, 97, 98, 99, 99a, 100, 100a, 101*, 102, 103, (104), 105, 106**

exceeding corolla-tube by at least twice — **94, 95, 102** (3), **(104), 107, 108**

Stigmatic knob

± as wide as long or somewhat longer than wide, fig. 131/13, 14 & 18–21, p. 752 — **91, 92, 93, 97, 98, 99, 99a, 100, 100a, 102, 103, 104, 105, 106**

distinctly longer than wide, fig. 131/15–17 & 22 — **94, 95, 96, 101*, 102** (7, 8), **107, 108**

Locules

2; stigmatic knob 2-lobed, fig. 131/13–18, p. 752 — **94** (2–4), **95, 96, 100, 101, 102, 105, 106, 107, 108**

(2–)3–5; stigmatic knob (2–)3–5-lobed, fig. 131/19–22 — **91, 92, 93, 94** (1), **97, 98, 99, 99a, 100, 100a, 103, 103a, 104**

Fruit

small to medium, up to 1 cm. across — **91, 93, 97, 98, 99, 99a, 100, 105, 106, 107, 108**

large, 1–5 cm. across — **92, 93, 94**, 95, 96, 99** (13), **100.1** (24), **100.3** (43), **101, 102, 103, 103a, 104, 106** (7)

± globose, often 3–5-lobed when dry — **91, 92** (1), **93, 94** (1), **95, 96, 97, 98, 99, 99a, 100, (102), 103, 103a, 104, (106)**

somewhat flattened, clearly 2-lobed — **?94**, 95, 100, 101, (102), 105, 106, 107, 108**

---

* See footnote, p.757

** Fruit not known in 2-locular species

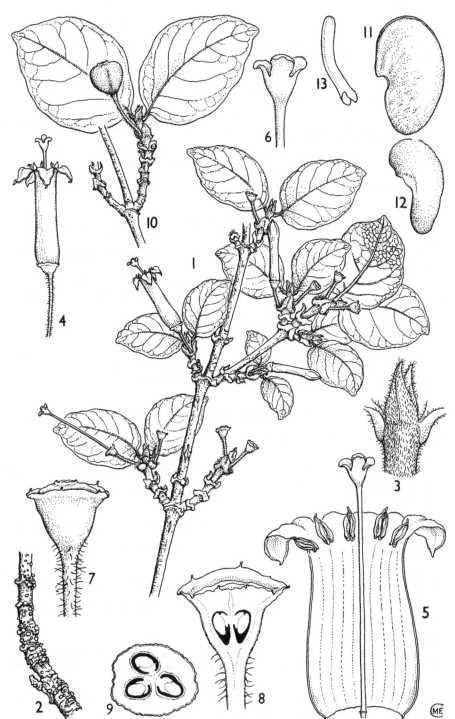

FIG. 133. *TEMNOCALYX NODULOSUS* — **1**, flowering branch, × ⅔; **2**, part of older branch, × ⅔; **3**, stipule, × 4; **4**, flower, × 1; **5**, corolla opened out, × 2; **6**, stigmatic knob, × 3; **7**, calyx, × 5; **8**, longitudinal section through ovary, × 6; **9**, transverse section through ovary, × 8; **10**, fruiting branch, × ⅔; **11**, pyrene, × 4; **12**, seed, × 4; **13**, embryo, × 4. 1, 3, from *Davies* 1; 2, from *Stolz* 2429; 4–9, from *Gillett* 17729; 10–13, from *St. Clair-Thompson* 901. Drawn by Mrs M.E. Church.

91. TEMNOCALYX

Robyns in B.J.B.B. 11: 317 (1928), sensu stricto; Verdc. in K.B. 36: 532 (1981)

Shrubs or small trees. Leaves paired, distinctly shortly petiolate, the blades with hairy domatia beneath; stipules connate into a sheath, broadly triangular or ovate at the base, drawn out into a short acumen. Flowers ♂, axillary, solitary, borne from the upper nodes of the congested leafy shoots, on quite long pedicels; buds obtuse or bluntly apiculate. Calyx-tube obconic, the limb very short, truncate or slightly erose. Corolla-tube ± cylindrical, straight or slightly curved, glabrous within; lobes 5, oblong-triangular, acute and shortly apiculate, papillate inside. Stamens 5, the anthers oblong, partly exserted; filaments very short. Ovary 3–4-locular; style relatively slender; stigmatic club continuous with style (i.e. latter *not* fitting into hollowed base of the club), obconic, prominently 3–4-lobed at the apex. Fruit a subglobose drupe with (2–)3–4 pyrenes developed, often 2–4-lobed when dry.

As restricted by me this is a monospecific genus; the other 4 species formerly included by Robyns have been transferred back to *Fadogia* in which they were originally described. *Temnocalyx* differs from *Fadogia* in its very different habit, opposite leaves with short but more distinct petioles, domatia, lack of hair within corolla-tube and above all the positioning of the stigmatic club (fig. 131/19, p. 752). They differ also in testa-cell characters but these have not been examined for all *Fadogia* species.

T. nodulosus *Robyns* in B.J.B.B. 11: 318 (1928); T.T.C.L.:533 (1949); Verdc. in K.B. 36: 533 (1981) & in K.B. 42, fig 2/H–K (1987). Type: Tanzania, Rungwe District, Mwakaleli, *Stolz* 2296 (K, holo.!, BM, EA, iso.!)

Shrub or small tree (2–)3–6 m. tall, with mostly rather thick gnarled often lichen-covered grey branchlets; leafing shoots with congested nodes, the youngest parts with dense ± ferruginous curled hairs. Leaf-blades ovate-oblong, 3.3–8 cm. long, 2.2–6.5 cm. wide, very shortly acuminate to a blunt apex, rounded at the base, drying slightly discolorous, sparsely hairy above and on the venation beneath, also with tuft-domatia in some of the nerve-axils beneath; petiole 2–3 mm. long, hairy; stipules 4–7 mm. long, with appendage 2–3 mm. long, densely hairy. Pedicels 0.8–2.5 cm. long, shortly hairy. Calyx-tube 2.5 mm. long, glabrous. Corolla yellow, glabrous outside and inside; tube 2.5 cm. long, 0.7 cm. wide near apex, widening to 1–1.5 cm. near the base; lobes 7–8(–10) mm. long, 3.5–4(–5) mm. wide. Stigmatic club ± 4 mm. long. Fruit ± 1 cm. diameter. Figs 131/2, 19, p. 752; 132/1, p. 754 & 133.

TANZANIA. Mbeya District: Poroto [Mporoto] Mt. ridge road, 14 km. E. of junction with Mbeya-Tukuyu road, 11 Nov. 1966, *Gillett* 17729!; Rungwe Mt., 4 Nov. 1931, *R.M. Davies* R1! & same area, 4 Feb. 1914, *Stolz* 2515!
DISTR. **T** 7; not known elsewhere
HAB. Bamboo thicket, margins of *Hagenia*, *Olea* and *Podocarpus* forest; 1200– 2850 m.

NOTE. I have been unable to localize *St. Clair-Thompson* 901, Mchombo R., but the altitude is given as '4000–4500 ft.'. Gillett points out that the plant would be a good ornamental shrub.

92. LAGYNIAS

Robyns in B.J.B.B. 11: 312 (1928)

Glabrous or velvety hairy shrubs or understorey trees or sometimes ± scandent. Leaves opposite, petiolate, mostly discolorous when dry; stipules subulate from a very short triangular base, soon falling. Flowers small, pale-coloured, in few–many-flowered axillary cymes or terminating short lateral shoots; peduncles short and pedicels mostly elongate. Calyx-tube subglobose; lobes 5, much longer than the tube and usually almost as long as or longer than the corolla-tube, elongate-spathulate or narrowly oblong-spathulate, obtuse. Buds elongate with cylindrical tube and clavate limb, 5-apiculate. Corolla-tube narrowly cylindrical, inside glabrous or sparsely hairy at the throat but with a ring of deflexed hairs below middle; lobes 5, triangular-lanceolate, tailed at the apex, usually reflexed. Stamens 5, inserted at the throat, the anthers short, usually completely exserted. Ovary 5-locular; ovules solitary in each locule; style filiform, mostly distinctly exserted; stigmatic club cylindrical, 5-ribbed, the apex very shortly 2-lipped. Fruit varying in shape according to number of pyrenes developed, asymmetrically ellipsoid, pyriform or subglobose; pyrenes 1–5, with woody walls. Fig. 132/2, 3, p. 754.

A small genus of 4 well-defined species restricted to eastern and south-eastern Africa as far as Transvaal and Natal, close to *Pachystigma, Cuviera, Robynsia* and *Vangueria* but with a characteristic facies. One specimen of *L. lasiantha* (Sond.) Bullock from Mozambique, Inhaca I., is definitely spiny. *L. rufescens* belongs to subgen. *Bembea* Verdc., see K.B. 42: 143 (1987).

Glabrous . . . . 1. *L. pallidiflora*
Young stems, leaf-venation beneath, calyces, etc. with striking
    red-brown pubescence . . . . 2. *L. rufescens*

1. **L. pallidiflora** *Bullock* in K.B. 1931: 273 (1931); T.T.C.L.: 503 (1949); K.T.S.: 449 (1961); Verdc. in K.B. 42, fig. 1B (1987). Type: Tanzania, E. Usambara Mts., Sigi, *Zimmermann in Herb. Amani* 3219 (K, holo.!, EA, iso.!)

Glabrous shrub or small tree 3–10(–20) m. tall with smooth light-coloured bark and rather drooping branches; stems with pale peeling epidermis; paired spines to 3 cm. sometimes present. Leaf-blades narrowly elliptic to elliptic or oblong-elliptic or obovate-oblanceolate, 1.4–14.5 cm. long, 0.7–5.5 cm. wide, acuminate or ± rounded at the apex, cuneate at the base, drying very discolorous, pale beneath and dark above, the lateral nerves obscure beneath and the venation not visible; petioles 5–10 mm. long; stipules filiform-subulate, 4–10 mm. long. Inflorescences umbel-like, (1–)2–6-flowered; peduncles 0.2–1.2 cm. long, bearing auriculate bracts at apex; pedicels slender, 0.9–3 cm. long. Calyx-tube subglobose, 1.5–2.5 mm. long; lobes oblong-linear-spathulate, drying densely speckled red-brown inside, cream, green or greenish yellow in life, 0.7–1.6 cm. long 1.5–4 mm. wide, rounded at the apex, somewhat accrescent, ± reflexed. Buds with club-shaped limb, 5-tailed at the apex. Corolla-tube cream, greenish white or greenish yellow, narrow, 0.9–1.2 cm. long, with a ring of deflexed hairs inside; lobes pale yellow, lanceolate, 6–8 mm. long, including the distinctive tail-like appendage up to 3–4 mm. long, becoming reflexed. Anthers 1.3 mm. long, the filaments exserted 0.5 mm. Ovary 5-locular; style exserted 0.5–2 mm.; stigmatic club 0.8 mm. long. Fruit brown, containing (1–)3 pyrenes, asymmetrically ellipsoid to subglobose, at least when dry, lobed according to number of pyrenes, 1.6 cm. long, 1 cm. wide, 5 mm. thick, or ± 1.7 cm. in diameter, dark red tannin granules present in flesh. Pyrenes woody, narrowly reniform, 1.6 cm. long, 8–10 mm. wide, 5.5 mm. thick. Seeds oblong-ellipsoid, 1.3 cm. long, 4 mm. wide, 2.5 mm. thick. Figs. 131/12, p. 752; 132/2, p. 754 & 134.

KENYA. Kwale District: Buda Mafisini Forest, 21 Aug. 1953, *Drummond & Hemsley* 3940! & Shimba Hills, Lango ya [Longo] Mwagandi, 24 Mar. 1968, *Magogo & Glover* 441! & Shimba Hills Game Reserve, road to Giriama Point, 2 Dec. 1972, *Spjut* 2731!
TANZANIA. Lushoto District: E. Usambara Mts., Sigi, 16 Oct. 1940, *Greenway* 6047!; Ulanga District: Mahenge, Liondo, 26 Apr. 1931, *Schlieben* 2135! & near Mahenge, Ujiji, Sept. 1960, *Haerdi* 604/0!; Pemba I., Ngezi Forest, 22 Jan. 1933, *Vaughan* 2065!
DISTR. **K** 7; **T** 3, 6, ?7; **P**; not known elsewhere
HAB. Evergreen forest, thickets; 0–900 m.

SYN. *L. littoralis* Bullock & Greenway in K.B. 1933: 148 (1933). Type: Pemba I., Ngezi Forest, *Greenway* 2697 (K, holo.!, EA, K, iso.!)

2. **L. rufescens** (*E.A. Bruce*) *Verdc.* in K.B. 42: 143 (1987). Type: Tanzania, Uluguru Mts., Tanana, *E.M. Bruce* 765 (K, holo.! & iso.!, EA, iso.!)

Shrub ± 2.4 m. tall, with rigid ascending branches, or tree to 12 m.*; youngest internodes densely covered with ± adpressed rusty somewhat bristly hairs; older parts of branchlets greyish. Leaves opposite; blades oblong to oblong-elliptic or slightly oblong-obovate, 3–9(–13.5*) cm. long, 1–3.8(–6*) cm. broad, acuminate at the apex, rounded at the base, drying markedly discolorous, adpressed ferruginous pubescent on both sides but particularly on the main venation beneath where the red-brown nerves contrast with the green surface (in life as well as when dry *fide* collector); petiole 5–10 mm. long, densely ferruginous hairy; stipules with a triangular-ovate base ± 3 mm. long, with a subulate apex 6 mm. long. Inflorescences 1–2-flowered, densely ferruginous hairy; peduncle 4–9 mm. long; pedicels 2–5 mm. long; bracts ± cupuliform, 2–3 mm. long, often lobed. Calyx-tube semi-globose, 2.5–3 mm. long, densely ferruginous hairy and verrucose; lobes leafy, narrowly obovate-oblong, 1–1.4 cm. long, 2–4.5 mm. wide, obtuse, densely adpressed ferruginous-pubescent, particularly on the back of the midrib, with ± tubercular-based bristly hairs. Buds conspicuously 5-tailed. Corolla* yellowish green; tube 6–7 mm. long, 5 mm. wide at the throat, densely ferruginous-pubescent outside, glabrous inside save for a

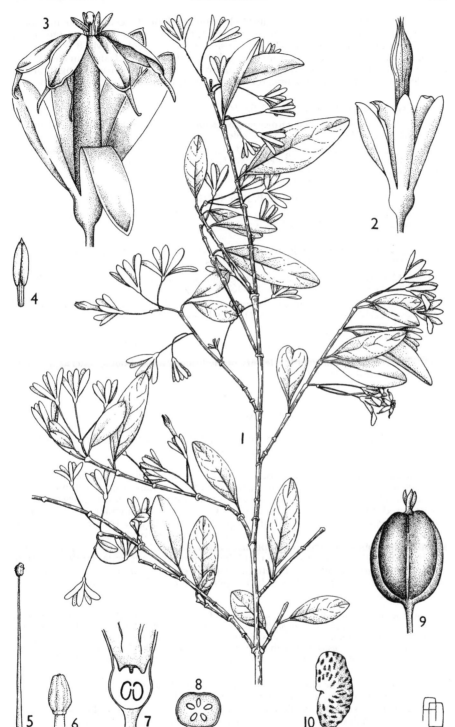

FIG. 134. *LAGYNIAS PALLIDIFLORA* — **1**, flowering branch, × ⅔; **2**, flower bud, × 2 ⅔; **3**, flower, × 4; **4**, stamen, × 10; **5**, style and stigmatic knob, × 4; **6**, stigmatic knob, × 14; **7**, longitudinal section through ovary, × 6; **8**, transverse section through ovary, × 6; **9**, fruit, × 1½; **10**, pyrene, × 1½. 1–8, from *Drummond & Hemsley* 3940; 9, 10, from *Haerdi* 604/0. Drawn by Ann Farrer, with 6 by Sally Dawson.

ring of deflexed hairs 1.5 mm. long just below middle; lobes lanceolate, 5–6 mm. long, 1.5–2.5 mm. wide, caudate-acuminate at the apex, the appendage 2–4 mm. long, ferruginous-pubescent outside, glabrous inside. Ovary 5-locular; style 8 mm. long*; stigmatic club obconic-truncate, 1 mm. long. Fruits* globose, 2.5–3 cm. diameter, with 5 pyrenes, reddish pubescent; pyrenes blackish brown, canoe-shaped, with longitudinally rounded dorsum, 15.5 × 7.5 × 5 mm., very rugose, sharply keeled at upper dehiscent end.

subsp. **rufescens;** Verdc. in K.B. 42: 143, fig. 7/M (1987)

Shrub ± 2.4 m. tall, with indumentum on young twigs and foliage ± adpressed. Calyx-lobes narrowly obovate-oblong, 1–1.4 cm. long, 3.5–4.5 mm. wide.

TANZANIA.   Morogoro District: Uluguru Mts., Tanana, 3 Feb. 1935, *E.M. Bruce* 765!; Nguru Mts., above Mhonda Mission, Turiani, 14 Feb. 1988, *Lovett & Congdon* 3071!
DISTR.   **T** 6; not known elsewhere
HAB.   Evergreen forest; 1350–1400 m.

SYN.   *Ancylanthos rufescens* E.A. Bruce in K.B. 1936: 477 (1936); T.T.C.L.: 482 (1949)

NOTE.   This subspecies is known only from the specimens cited. The reference by both E.A. Bruce and Brenan to peduncles 5–9 cm. long is a slip.

subsp. **angustiloba** *Verdc.* in K.B. 42: 143, fig. 7/A–L (1987). Type: Tanzania, Iringa District, Uzungwa Mts., Sanje, *Lovett* 289 (K, holo.!, BR, DSM, iso.!)

Tree to 12 m. with indumentum on young twigs and foliage more spreading. Calyx-lobes linear-oblong, 1.2–1.3 cm. long, 2–2.5 mm. wide. Fig. 132/3, p. 754.

TANZANIA.   Iringa District: Mwanihana Forest Reserve, near Sanje village, June 1981, *Rodgers & Mwasumbi* 1227! & Sanje, Logger's Camp, 2 Jan. 1981, *Rodgers* 490! & Sanje, near Logger's Camp, 8 Sept. 1984, *Bridson* 649!
DISTR.   **T** 7; not known elsewhere
HAB.   Evergreen forest of *Parinari, Newtonia, Allanblackia,* etc.; 900–1400 m.

NOTE.   The rufous indumentum is a feature of the living plant and not an artefact of drying according to the field notes.

93. **PACHYSTIGMA**

Hochst. in Flora 25: 234 (1842); Robyns in B.J.B.B. 11: 117 (1928); Verdc. in K.B. 36: 541–547 (1981)

Subshrubby herbs or small shrubs, glabrous or less often pubescent. Leaves opposite, rarely in whorls of 3, shortly petiolate; stipules connate, hairy within, often at length deciduous. Flowers fairly small in axillary usually opposite cymes or fascicles, often few-flowered; bracts and bracteoles present. Calyx-lobes 5, triangular or linear, erect, obtuse, ± leafy, sometimes exceeding the corolla-tube. Corolla green or greenish yellow to greenish white; tube cylindrical, with a ring of deflexed hairs inside and the throat usually slightly hairy; lobes reflexed, often apiculate. Stamens inserted at the throat, the anthers exserted. Ovary 3–5-locular with one pendulous ovule per locule; style slender, exserted, the stigmatic knob cylindrical, smooth or ± sulcate, obscurely 5-lobed. Fruit globose or obovoid, with 2–5 pyrenes, crowned by the persistent calyx-limb. Pyrenes ± woody. Fig. 132/4, p. 754.

A small genus of about 10 species in southern and central Africa, mostly originally included in *Vangueria* but differing in habit. *Tapiphyllum* is very close to *Pachystigma* differing in the consistently velvety leaves, pedunculate inflorescences with more numerous flowers, longer calyx-lobes equally hairy on both sides and usually smooth stigmatic knob. Some South African species are certainly intermediate in character, but I have preferred to retain the two for the purposes of this Flora.

1. Perennial subshrubby herb 5–15(–25) cm. tall; leaf-blades
    oblong to oblong-elliptic, elliptic or oblanceolate, 4–
    9.5(–15.5) × 1.6–4.7 cm. (**T** 7, 8) . . . .   1. *P. pygmaeum*
    Shrubs or small trees with smaller leaves   2

* Descriptive portions marked * refer to the subsp. *angustiloba,* both specimens of subsp. *rufescens* being in flower bud only.

2. Flowers solitary; pedicels pubescent, 5–8 mm. long (**T** 5,
    Kondoa area) . . . .   2. *P. solitariiflorum*
   Fascicles 1–several-flowered; pedicels if pubescent
    then up to 5 mm. long 3
3. Leaf-blades sparsely pubescent to tomentose 4
   Leaf-blades entirely glabrous or pubescent at base on
    midrib only 5
4. Leaf-blades sparsely pubescent to pubescent (**K** 5, Meru
    area) . . . .   3. *P. gillettii*
   Leaf-blades tomentose . . . .   4. *P. schumannianum*
5. Leaf-blades entirely glabrous; petiole 1–2 mm. long; lateral
    branches, numerous and held at right-angles; bark
    white or greyish white (predominantly coastal)   6. *P. loranthifolium*
   Leaf-blades pubescent at base on midrib; leaves essentially
    sessile but much narrower at base so there appears to
    be a winged petiole; branching not as above; bark
    brown or pale to dark grey (**T** 5, Dodoma and Mpwapwa
    Districts) . . . .   5. *P. burttii*

1. **P. pygmaeum** (*Schlecht.*) *Robyns* in B.J.B.B. 11: 122, figs. 17, 18 (1928); Verdc. in K.B. 36: 541 (1981). Type: South Africa, Transvaal, near Johannesburg, Eilandsfontein, *Gilfillan* 1416 (B, holo.†)

Perennial subsrubby herb, 5–15(–25) cm. tall, usually with a long rhizome, erect or prostrate; stems at first sparsely to densely setose-hairy but becoming glabrous and covered with brown finely fissured bark. Leaves paired, very rarely in whorls of 3; blades oblong to oblong-elliptic, elliptic or oblanceolate, 4–9.5(–16) cm. long, 1.6–4.7 cm. wide, obtuse, acute or shortly acuminate at the apex, cuneate at the base, glabrous to covered on both surfaces with dense seta-like hairs which do not in any way hide the surface, often quite scabrid; petiole 0–3(–7) mm. long; stipules scarious to thicker, with subulate appendage 2–6 mm. long from a base 1–2 mm. long, densely setose or glabrescent. Inflorescences simple, 3–7(–many)-flowered, often at ground-level, densely setose; peduncles 5–6(–12) mm. long; pedicels 5–8 mm. long; bracts scarious, usually several at base of inflorescence and conspicuous, obovate, 2–3 mm. long and wide, obtuse. Buds filiform-apiculate, sparsely setose; appendages up to 3 mm. long. Calyx-tube 1.5 mm. long, glabrous to sparsely setose; limb-tube 0–2(–3) mm. long; lobes 5 or more, erect, narrowly oblong or linear, 2–8.5 mm. long, usually setose. Corolla-tube white or pale greenish cream, subcampanulate to quite narrowly cylindrical, 4–8 mm. long; lobes green outside, narrowly triangular, 4–6 mm. long including the filiform appendage, 2 mm. wide. Anthers brown, exserted, 1–1.7 mm. long; filaments short, ± 1 mm. long. Ovary 5-locular; styles slender, 5–9 mm. long; stigmatic club pale green, cylindrical or coroniform, 0.75–1 mm. long, slightly 5-lobulate. Fruit yellow-brown to black, subglobose, pear-shaped or oblique, 1.5–1.7 cm. diameter, with (1–)2–5 pyrenes; pyrenes ellipsoid, obtusely subtrigonous, 1.1–1.5 cm. long, 6–8 mm. wide, with sharp point above point of attachment, rugulose. Fig. 135, p. 766.

TANZANIA. Iringa District: Mufindi, ?Nundwe, 16 Feb. 1969, *Paget-Wilkes* 371!; Rungwe District: near Kiwira R., 15 May 1973, *Shabani* 1059!; Songea airfield, 14 Feb. 1956, *Milne-Redhead & Taylor* 8693! & 24 Feb. 1956, *Milne-Redhead & Taylor* 8693a!
DISTR. **T** 7, 8; Malawi, Zambia, Zimbabwe, Swaziland and South Africa (Transvaal)
HAB. Grassland derived from *Brachystegia, Uapaca* woodland; 1020–1950 m.

SYN. *Vangueria pygmaea* Schlect. in J.B. 35: 342 (1897)
    V. setosa Conrath in K.B. 1908: 224 (1908). Type: South Africa, Transvaal, Modderfontein, *Conrath* 339 (K, holo.!)
    V. rhodesiana S. Moore in J.B. 47:130 (1909). Type: Central Zimbabwe, *Rand* 1349 (BM, holo.!)
    Pachystigma rhodesianum (S. Moore) Robyns in B.J.B.B. 11: 119 (1928)

NOTE. Only the 4 sheets cited above have been seen from the Flora area; those from Songea have thinner more acuminate leaves and the indumentum is longer and less scabrid. The species is very variable and material from the north of the range has longer corolla-tubes, perhaps fewer bracts, shorter calyx-lobes, etc. In the present state of our knowledge I am not prepared to subdivide. Although *P. rhodesianum* was made a new subgenus by Robyns it does not differ sufficiently to be kept distinct — the length of the calyx-limb tube and the indumentum are very variable.

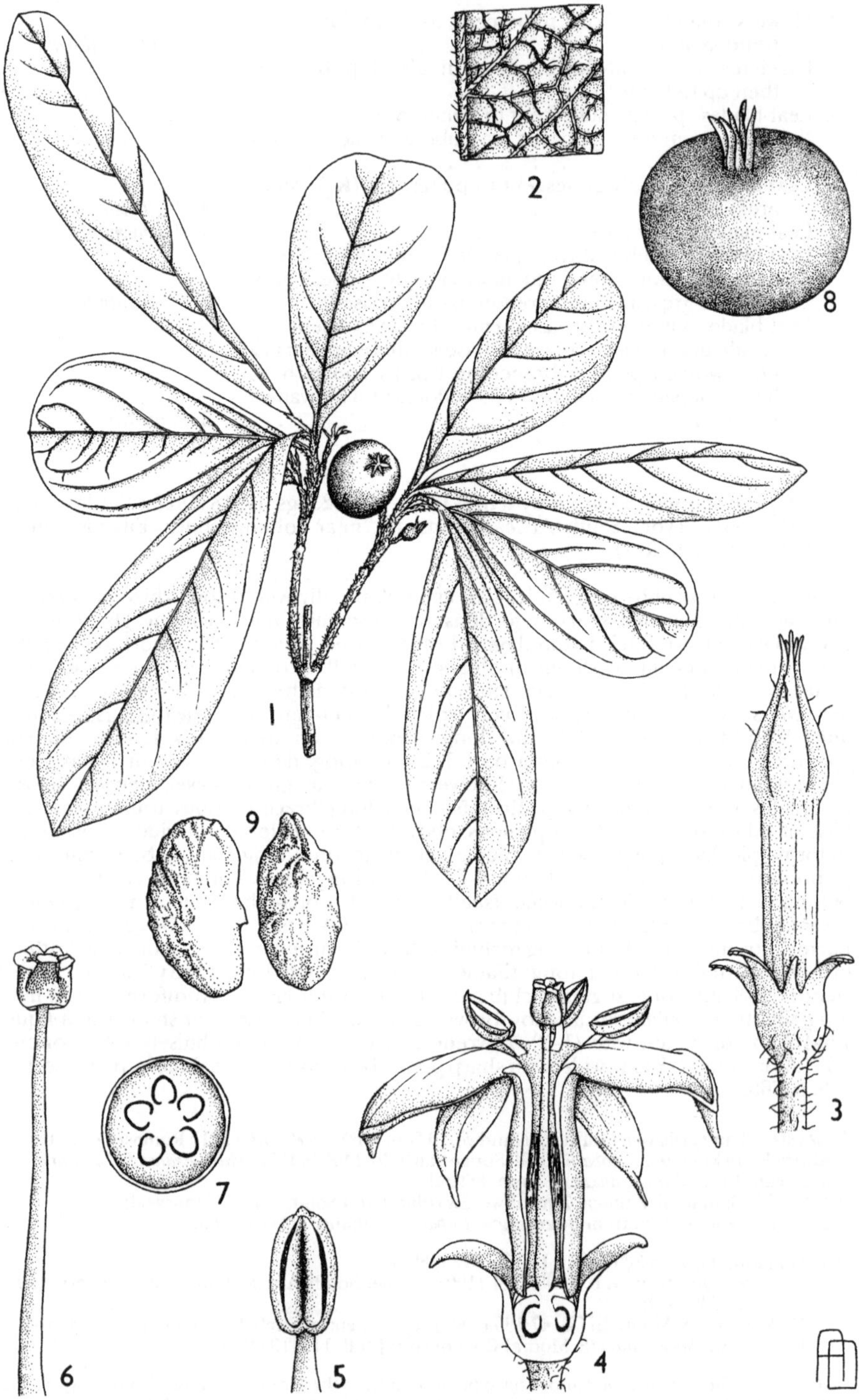

FIG. 135. *PACHYSTIGMA PYGMAEUM* — **1**, habit, × ⅔; **2**, detail from lower surface of leaf, × 2; **3**, flower bud, × 4; **4**, flower with section removed, × 4; **5**, stamen, × 8; **6**, style and stigmatic knob, × 6; **7**, transverse section of ovary, × 10; **8**, fruit, × 1⅓; **9**, pyrenes (2 views), × 2. 1, 2, 8, 9 from *Fanshawe* 10077; 3–7, from *Milne-Redhead & Taylor* 8693. Drawn by Ann Farrer, with 9 by Sally Dawson.

2. **P. solitariiflorum** *Verdc.* in K.B. 36: 541, fig. 9 (1981). Type: Tanzania, Kondoa District, Kolo, *Polhill & Paulo* 1122 (K, holo.!)

Shrub or small tree 2–4.5 m. tall, with branches paired or in whorls of three, the foliage either closely conferted with short internodes or more normally spaced, the branches with close internodes and nodose at leafless nodes; bark smooth, pale; young shoots densely covered with pale ferruginous spreading hairs. Leaves paired, drying a grey-green with midrib and sometimes the margins pale yellowish; blades broadly elliptic, ± round or ovate, 0.7–3 cm. long, 0.5–2.1 cm. wide, rounded or very shortly and obtusely acuminate at the apex, cuneate to rounded at the base, glabrous save for a few pale ferruginous hairs on both sides of the midrib or ± densely pubescent all over; petiole 1–3 mm. long, hairy; stipules triangular, ± 1–2 mm. long, with subulate tip ± 2 mm. long, hairy. Flowers solitary; pedicels 5–8 mm. long, pubescent. Buds hairy in middle, with long tails. Calyx-tube hemispherical, 2 mm. long, pubescent; lobes 5, narrowly triangular to lanceolate, 3–4 mm. long, 1–2 mm. wide, glabrous, with red resinous patches. Corolla-tube pale green, cylindrical, 4.5–5.5 mm. long, ± densely spreading pubescent above, throat hairy but not densely filled, with ring of deflexed hairs inside; lobes cream, yellow towards the throat, narrow triangular-lanceolate, 8–8.5 mm. long including the caudate apex, 2.2–2.5 mm. wide. Anthers narrowly ovate-cordate, 1.5–2 mm. long, shortly exserted. Ovary 5-locular; style pale green, 0.9–1.1 cm. long; stigmatic club coroniform, 2 mm. wide. Fruit subglobose, ± 1 cm. diameter, very sparsely pubescent, with up to 4 pyrenes but often oblique and with only one developed.

TANZANIA. Kondoa District: Salanga Hill, 14 Dec. 1927, native collector for *B.D. Burtt* 1177! & Kinyassi Mt., 2 Jan. 1928, native collector for *B.D. Burtt* 936! & Kolo, 10 Jan. 1962, *Polhill & Paulo* 1122!
DISTR. **T** 2, 5; not known elsewhere
HAB. Mixed dry bushland; 1560 m.

NOTE. The two *Burtt* sheets cited differ in habit; one has shoots with branchlets in whorls of 3 and the leaves spaced and the other opposite irregular branching with nodulate branchlets bearing closely placed nodes. There is, however, no doubt they are the same species. Brenan (T.T.C.L.: 529 (1949)) obliquely mentions the two *Burtt* gatherings above under *Rytigynia monantha* but does not cite them. K. Schumann's description of *R. monantha* does not agree.

3. **P. gillettii** (*Tennant*) *Verdc.* in K.B. 42:140 (1987). Type: Kenya, Northern Frontier Province, Dandu, *Gillett* 12765 (K, holo.!, EA, iso.)

Somewhat straggling shrub or small tree, 2–3 m. tall; bark pale grey, smooth; branches setose-pubescent when young but soon glabrescent and covered with minutely longitudinally fissured grey-brown bark. Leaves borne on very abbreviated side-shoots, broadly elliptic or obovate, 1–2.3 cm. long, 0.7–1.8 cm. wide, rounded at the apex, cuneate at the base, setulose-pubescent on both surfaces; lateral nerves 2–4 on each side; petioles ± 2 mm. long often together with the mid-rib blue-black on drying; stipules narrowly triangular, at length corky, pilose inside. Fascicles ± 5-flowered, borne on the abbreviated side-shoots; pedicels 3–5 mm. long, pubescent, often drying blue-black. Calyx-tube campanulate, 1–2 mm. long, pubescent; lobes triangular-lanceolate, 3.2–4.5 mm. long, 1–1.7 mm. wide, pubescent. Buds distinctly 5-apiculate and setulose-pubescent at the apex. Corolla yellow; tube cylindrical-suburceolate, 3–3.7 mm. long, 3.5 mm. wide, pilose at the throat and with a ring of deflexed hairs inside about the middle, sparsely to densely setulose-pubescent outside; lobes ovate-triangular, ± 3 mm. long, with apicula 1.2–2 mm. long. Anthers 1.5–2 mm. long, exserted, minutely apiculate. Ovary 3–4-locular; style narrowly obclavate, 5.5 mm. long; stigmatic club cylindric, 1.2 mm. long, obscurely sulcate. Only imperfect fruits with one pyrene seen, asymmetric, ellipsoid-subreniform, 9.5 mm. long, 6 mm. wide, 5 mm. thick, sparsely pubescent.

KENYA. Northern Frontier Province: near Ndoto Mts, Nguronit Mission, 31 Oct. 1978, *Gilbert & Gachathi* 5266!; Meru District: Meru National Park, Main Ura Gate road crossing Kiolu R., 26 Dec. 1972, *Ament* 552! & near same locality, Kiolu R. crossing near Mughwango, 28 Nov. 1979, *Hamilton* 593!
DISTR. **K** 1, 4; Somalia
HAB. Dense mixed *Acacia, Commiphora* bushland, steep rocky river banks, among granite rocks and also on deep red sandy loam on plains; 650–1200 m.

SYN. *Rytigynia gillettii* Tennant in K.B. 19: 279 (1965)
     Pachystigma kenyense Verdc. in K.B. 36: 542, fig. 10 (1981). Type: Kenya, Meru National Park, Kiolu R. crossing, *Hamilton* 593 (EA, holo.!)

4. **P. schumannianum** (*Robyns*) *Bridson & Verdc.*, comb. nov. Type: Tanzania, Moshi District, Himo, *Volkens* 1848 (B, holo.†, BR, iso.!, K, fragment)

Much-branched intricate shrub or small tree, 1.2–4.5(–10) m. tall; branchlets rugose and verruculose, nodular, covered with greyish fissured bark, velvety grey pubescent above but eventually glabrescent. Leaves opposite; blades oblong-ovate to almost round, 0.9–4.5 cm. long, 0.8–4 cm. wide, obtuse or slightly retuse at the apex, rounded to subcordate at the base, densely grey velvety hairy on both surfaces with curled hairs; petiole 0.5–3 mm. long; stipules triangular, acuminate, 2.5–4 mm. long, pubescent outside with a few silky hairs inside. Flowers grey-pubescent-tomentose, solitary or in fascicles of 1–4(–several); peduncles obsolete; pedicels up to 2.5–5 mm. long. Calyx grey-pubescent; tube 1.5–2 mm. long; lobes ovate to oblong-triangular, 1–2.5 mm. long, 1–1.2 mm. wide, obtuse, puberulous inside. Buds distinctly apiculate. Corolla green to whitish green or cream, becoming yellow-green or golden yellow; tube cylindrical, 3–3.5(–4) mm. long; lobes (4–)5(–6), elliptic-oblong, 3.5–5 mm. long, 1.5–2 mm. wide, distinctly caudate. Style 4.5–6.5 mm. long; stigmatic club coroniform-cylindric, 1.2 mm. long. Fruits ± globose, 0.8–1.4 cm. diameter, sparsely pubescent when ripe, on pedicels up to 5–7 mm. long, crowned by the persistent calyx, with (1–)3–5 pyrenes, each ± 1–1.1 cm. long.

subsp. **schumannianum**

Leaves smaller, usually ± 1.2–1.7(–2.7) cm. long, 1.2–1.8 cm. wide.

KENYA.   Masai District: W. foot of Ngong Hills, 30 Mar. 1957, *Greenway* 9170! & W. Ngong Hills on circular road, 18 Oct. 1964, *Greenway* 11767! & Narok road, 17 Mar. 1935, *V.G.L. van Someren* in *C.M.* 3635!
TANZANIA.   Moshi District: road between Moshi and Paneta, Jan. 1894, *Volkens* 1748!
DISTR.   **K** 6; **T** 2; not known elsewhere
HAB.   Deciduous bushland and woodland (*Acacia*, *Commiphora* or *Combretum*) often on rocky ground; (900–)1150–1980 m.

SYN.   *Tapiphyllum schumannianum* Robyns in B.J.B.B. 11: 109 (1928); T.T.C.L.: 512 (1949); K.T.S.: 473 (1961); Robyns in B.J.B.B. 32: 137 (1962); Verdc. in K.B. 36: 538 (1981) & in K.B. 42: 144 (1987)

subsp. **mucronulatum** (*Robyns*) *Bridson & Verdc.*, comb. nov. Type: Kenya, N. Kitui, Migwani Location, *Napper* 1590 (BR, holo.!, EA, K, iso.!)

Leaves distinctly larger, up to 4.5 cm. long, 4 cm. wide; indumentum rather more woolly.

KENYA.   Embu District: Embu–Kangonde road, 3.5 km. towards Embu from turnoff to Siakago, 6 June 1974, *R.B. & A.J. Faden* 74/722!; Machakos District: Kilungu, beyond Kikoko Hill, Kavata Nzou [Kavatanzoo] School, 22 Aug. 1971, *Mwangangi* 1667! & Nzaui Hill, 16 Feb. 1969, *Kokwaro* 1884!
DISTR.   **K** 4, 7; not known elsewhere
HAB.   Thicket and bushland, especially *Carissa*, *Securinega*, *Vernonia*, *Rhus*, and *Acacia*, *Grewia*, also grassland with scattered *Acacia* and *Lantana* and *Combretum* woodland; 1140–1900 m.

SYN.   *Vangueria tomentosa* K. Schum. in P.O.A. C: 385 (1895); De Wild. in B.J.B.B. 8: 65 (1922), *non* Hochst. (1842), *nom. illegit.* Type: Kenya, Kitui District, Ikanga, *Hildebrandt* 2835 (B, holo.†)
      Rytigynia tomentosa (K. Schum.) Robyns in B.J.B.B. 11: 150 (1928)
      Tapiphyllum mucronulatum Robyns in Bull. Jard. Bot. Brux. 32: 151 (1962)
      T. schumannianum Robyns subsp. *mucronulatum* (Robyns) Verdc. in K.B. 36: 538 (1981)

NOTE.   No type material of *Vangueria tomentosa* survives it seems, but there is no doubt it is the same plant later described by Robyns as *Tapiphyllum mucronulatum*. He separated *T. mucronulatum* from *T. schumannianum* by having larger leaves, up to 5-flowered inflorescences, sepals puberulous within and anthers distinctly mucronulate. There is no doubt that it is not specifically distinct from the extensive Kenya material labelled as *T. schumannianum*. Unfortunately only a fragment of the type of the latter is extant which lacks complete flowers. The calyx-lobes of this are undoubtedly not glabrous within. The Kenya material varies in leaf-size completely over the range of claimed differences and the inflorescences vary from 1- to 5-flowered. The only other Tanzanian specimen collected since 1894 is unfortunately only in fruit. On the evidence of the Kew type fragment I am not prepared to keep them more than subspecifically separate and the fact that Robyns has named (in 1962), *Bally* 792 (4.8 km. from Machakos) *T. cf. schumannianum* indicates his inability to distinguish the supposed taxa. Volkens is, however, the only person to mention a height of 10 m. and more material from the Moshi area is required if still extant.
      This species has previously been placed in the genus *Tapiphyllum* and it indeed bears a strong resemblance to *T. parvifolium* (Sond.) Robyns from Botswana and South Africa. However, the fewer-flowered inflorescences, shorter, broader calyx-lobes less densely pubescent inside than outside, and the smaller broader leaves are a better fit with *Pachystigma*. One specimen (*Greenway & Napper* 13555 from Ngong Hills) bore an enlarged fruit which proved to be galled by the fungus *Aecidium vangueriae* Cooke.

5. **P. burttii** *Verdc.* in K.B. 36: 545, fig. 11 (1981) & in K.B. 42: fig.1A (1987). Type: Tanzania, Dodoma District, 11 km. from Dodoma on Kongwa road, *Leippert* 5510 (EA, holo.!, K, iso.!)

Shrub or small tree 2–4.5 m. tall; branches slender with brown or pale to dark grey bark with faint impressed pattern, sometimes peeling; younger parts densely covered with spreading hairs. Leaf-blades obovate-spathulate, 1.3–4.2 cm. long, 0.6–2 cm. wide, very shortly rounded-acuminate or rounded at the apex, gradually narrowed at the base into a short petiole 1 mm. long, glabrous or with few scattered hairs on margins and midrib; stipule-base 1 mm. tall with subulate apex 2 mm. long. Inflorescences sessile fascicles of 5–10 flowers borne on very short ± obsolete shoots; pedicels 1.5 mm. long, slightly pubescent. Calyx-tube globose or slightly urceolate, 1.5 mm. diameter, glabrous to densely shortly pubescent; limb-tube very short, 0.2 mm. long; lobes linear-lanceolate, (1–)2.5–3.5 mm. long. Buds conspicuously apiculate, with 5 diverging tails. Corolla yellow or greenish yellow; tube 2.5–3 mm. long, pubescent or with very sparse hairs at the top and a ring of deflexed hairs within; lobes 5–6, elliptic-oblong, tapering into a long subulate tip, 4 mm. long overall, the actual tip ± 2 mm. long. Ovary 5-locular; style 4.5–5 mm. long; stigmatic club truncate-obconic, 0.7 mm. long. Fruit at first subglobose, 5 mm. diameter, later if with 1 pyrene eccentric-ellipsoid, 1 cm. long, 8 mm. wide, ± 4 mm. thick, if with 2 then 1–2 cm. long and wide; pyrenes straw-coloured, subreniform, 9.5–10 mm. long, 5–5.5 mm. wide, 3.5–4.5 mm. thick, rugulose and pitted, notched near middle.

subsp. **burttii**

Calyx and corolla glabrous or with very few hairs. Calyx-lobes longer and relatively narrower, 2.5–3.5 mm. long. Fig. 132/4, p. 754.

TANZANIA.   Dodoma District: 11 km. from Dodoma on Kongwa road, 27 Jan. 1965, *Leippert* 5510!; Mpwapwa District: Godegode, 8 Feb. 1933, *B.D. Burtt* 4569! & Mpwapwa, 30 Mar. 1937, *Hornby* 617!
DISTR.   T 5; not known elsewhere
HAB.   Deciduous thicket and bushland, particularly *Commiphora, Cordyla* and *Delonix, Albizia, Strophanthus* associations; 900–1200 m.

subsp. **hirtiflorum** *Verdc.*, subsp. nov. a subsp. *burttii* tubo calycis dense patenter breviter pubescenti, lobis brevioribus 1–1.5 mm. longis, tubo corollae extra pubescenti differt. Type: Tanzania, Morogoro District, Ruaha valley, *Lovett* 1130 (K, holo.!, MO, iso.)

Calyx-tube densely spreading pubescent; lobes shorter and relatively broader, 1–1.5mm. long. Corolla-tube pubescent outside.

TANZANIA.   Morogoro District: Ruaha valley, 20 Dec. 1986, *Lovett* 1130!; Iringa District: road outside Ruaha National Park boundary, 5km. from Idodi, 2 Mar. 1987, *Lovett* 1629!
DISTR.   T 6,7; not known elsewhere
HAB.   *Acacia, Commiphora, Combretum* bushland with *Adansonia*, etc.; 700–850m.

6. **P. loranthifolium** (*K. Schum.*) *Verdc.* in K.B. 42: 140 (1987). Type: Tanzania, Tanga District, Doda Creek, *Holst* 2935 (B, holo.†, K, iso.!)

Laxly branched almost completely glabrous shrub or small tree 3–6 m. tall, with several main branches from the base, with greyish white slightly fissured bark; ultimate branchlets mostly numerous and held at right-angles, decussate or all in one plane, up to 1.5–9 cm. long, sometimes almost spine-like. Leaf-blades elliptic or obovate, (0.7–)1.4–4 cm. long, 0.4–2 cm. wide, rounded at the apex, cuneate at the base, somewhat coriaceous; petiole 1–2 mm. long; stipules rounded, ± 1.5 mm. long and produced in the centre into an apiculum ± 1 mm. long, the youngest with a few short stiff bristly adpressed hairs, ultimately deciduous. Cymes 1–3-flowered or peduncle suppressed and 2–3 flowers apparently borne separately at each node; peduncle 0–3 mm. long; pedicels (1–)7–9 mm. long, with a minute bracteole at base; bracts ovate, ± 1.5 mm. long. Flowers (4–)5-merous. Calyx-tube campanulate, 1.5–2 mm. long; limb-tube 0.8–1.5 mm. long; lobes broadly triangular, 1–2.5 mm. long. Buds acuminate. Corolla green when first opening, turning yellow or orange; tube 3–4 mm. long, with a ring of deflexed hairs at the middle inside; lobes oblong-lanceolate, 5.5 mm. long, 2.2 mm. wide including the narrow acute 1–1.5 mm. long apical appendage. Anthers ± 1 mm. long, minutely apiculate, just exserted. Ovary 5-locular; style green, exserted ± 2.5 mm.; stigmatic club subcylindric, ± 1 mm. long, constricted in the middle, sulcate. Fruit globose, ± 1.5 cm. diameter, crowned by the persistent calyx-limb; pyrenes 4–5, oblong-ellipsoid, widened in the upper third, 11.5 mm. long, 6 mm. wide, densely filled with dark red tanniniferous inclusions, keeled along the upper rounded margin.

subsp. **loranthifolium**

Leaf-blades 1.4–4 cm. long. Pedicels 7–9 mm. long.

KENYA. Mombasa District: Nyali Beach, 28 May 1934, *Napier* 3305 in *C.M.* 6281!; Kilifi District: below Jilore Forest Station, 26 Nov. 1972, *Spjut & Ensor* 2671!; Tana River District: Kurawa, 13 Oct. 1961, *Polhill & Paulo* 652!

TANZANIA. Handeni District: Kideleko, 22 Apr. 1954, *Faulkner* 1425! & 19 km. on Mkata–Kwamsisi road, Mar. 1965, *Procter* 2919!; Tanga District: N. of Kwale, 13 July 1937, *Greenway* 4967!

DISTR. **K** 7; **T** 3; not known elsewhere

HAB. *Cynometra* forest and derived thicket, *Acacia nigrescens* and *Terminalia*, *Acacia*, *Euphorbia* woodland and bushland, grassland and scattered bushland of *Zanthoxylum*, *Dobera*, *Strychnos*, *Commiphora*, *Sideroxylon*, etc., on sandy shores and sandy soil with black clay on coral, also margins of granite outcrops; 0–600 m.

SYN. *Vangueria loranthifolia* K. Schum. in P.O.A. C: 385 (1895); De Wild. in B.J.B.B. 8: 58(1922)
     Rytigynia loranthifolia (K.Schum.) Robyns in B.J.B.B. 11: 204 (1928); T.T.C.L.: 531 (1949); K.T.S.: 472 (1961)

subsp. **salaense** *Verdc.* in K.B. 42: 140 (1987). Type: Kenya, Teita District, Sala, *Hucks* 932 (EA, holo.!)

Leaf-blades ± 7 mm. long, 4 mm. wide. Pedicels 1–2 mm. long.

KENYA. Teita District: Sala area, 24 Dec. 1966, *Hucks* 932!

DISTR. **K** 7; not known elsewhere

HAB. Open bushland on red sandy soil; ± 300 m.

NOTE. From a single specimen it is impossible to decide on the status of this but despite its very different appearance, I think it is only from an isolated population of *P. loranthifolium* showing distinctive characters; but more material may indicate merely an abnormal variant.

94. CUVIERA

DC. in Ann. Mus. Hist. Nat. Paris 9: 222, t. 15 (1807); N. Hallé in Bull. Soc. Bot. Fr. 106: 342 (1960), *nom. conserv.*

Small trees or shrubs, unarmed or sometimes spinose. Leaves opposite, often large, mostly ± oblong or elliptic, petiolate, usually coriaceous and persistent; stipules small, basally ± connate, often acuminate, deciduous. Flowers ♂ or sometimes infertile (polygamo-dioecious *fide* Benth. & Hook.f., G.P.) in subsessile or pedunculate many-flowered axillary cymes; bracts and bracteoles linear to lanceolate or elliptic, often leafy and accrescent. Calyx-tube obconic or turbinate, sometimes 3–4-angled; lobes 3–6, linear to ovate, often leafy and accrescent, mostly longer than the corolla, persistent. Corolla funnel-shaped, campanulate or barrel-shaped, retrorsely hairy or bristly inside but throat glabrous, glabrous or pilose outside; lobes 5–6, spreading or reflexed, elongate, mostly caudate-acuminate, sometimes very markedly so. Stamens 5, inserted in the throat, the anthers exserted. Disk depressed, lobed. Ovary (1–)2–5-locular; ovules solitary in each locule, pendulous; style thick, narrowed at both ends, stiffly pubescent to glabrous, sometimes with a conspicuous globular swelling near the base (some W. African species only); stigmatic club cylindrical, mitriform or peltate, 2–10-grooved. Fruit drupaceous, often large, ovoid or subglobose, sometimes obscurely angled, with 1–5, 1-seeded pyrenes.

A genus of about a score of species in tropical Africa; most occur in western tropical Africa including a small subgenus *Globulostylis* (Wernham) Verdc. but 4, belonging to subgen. *Cuviera*, occur in the extreme south of the Flora area and are still imperfectly known.

1. Leaf-blades velvety pubescent on both surfaces; ovary 2-
     locular . . . . 4. *C. tomentosa*
   Leaf-blades glabrous save for hairy domatia in leaf-axils
     beneath or with sparse long setae beneath; ovary 2- or
     5-locular 2
2. Calyx-lobes 2.3–2.6 mm. long . . . . 3. *C. schliebenii*
   Calyx-lobes exceeding 6 mm. long 3
3. Calyx-lobes ovate to lanceolate, 6–15 × (2–)4.7–6 mm.; ovary
     5-locular . . . . 1. *C. semseii*
   Calyx-lobes lanceolate, 12 × 2–2.5 mm.; ovary 2-locular 2. *C. migeodii*

1. **C. semseii** *Verdc.* in K.B. 11: 449 (1957). Type: Tanzania, Lindi District, Rondo Plateau, Mchinjiri, *Semsei* 620 (EA, holo.!, BR, K, PRE, iso.!)

Small tree or shrub 2–6 m. tall, with grey-brown rugulose glabrous lenticellate branchlets; lateral branches sometimes reduced to spines 2.5 cm. long. Leaf-blades elliptic, 4.4–8.5 cm. long, 1.6–4 cm. wide, acuminate at the apex, the actual tip sometimes narrowly rounded, cuneate at the base, thin, glabrous above, with hairy domatia in the main nerve-axils and sparse long setae beneath; petioles 0.5–1.3 cm. long, glabrous or sparsely setose; stipules linear-triangular to oblong, 5 mm. long, 1 mm. wide, rather thick, ± acute, the nodes bristly within. Inflorescences axillary, opposite, several–many-flowered, the branches ± bifariously shortly pubescent; peduncles 5 mm. long; secondary peduncles up to 1 cm. long; bracts and bracteoles oblong-oblanceolate, 2–5 mm. long, 0.5–1.5 mm. wide; true pedicels 1–2 mm. long. Calyx-tube subglobose to turbinate, 2 mm. long and wide; lobes 4–5, ovate to lanceolate, 0.6–1.5 cm. long, (2–)4.7–6 mm. wide, acuminate, narrrowed and often with colleters at the base, leafy, nervose, ± pubescent and ciliate (densely so in S. Malawi). Corolla yellow or greenish yellow; tube 4.5 mm. long, 3 mm. wide at base and apex, 3.75 mm. wide at the middle, glabrous outside, with a ring of deflexed bristles 1 mm. long inside; lobes 5, linear-triangular, 5–6 mm. long, 1.5 mm. wide, acuminate, sparsely pilose outside. Ovary 5-locular; style 7.5–9 mm. long, narrowed towards the apex; stigmatic club cylindric, 1.1 mm. long. Fruits subglobose, ± 3–3.5 cm. long, 2.5 cm. wide, glabrous and a little shining, angular and wrinkled when dry; pyrenes of extraordinary characteristic shape, 2.3 cm. long, 1.4 cm. wide and 1.3 cm. thick, the outer part with radial faces forming conspicuous curved wings which form a rough V-shape in transverse section, either symmetrically 3-winged with two laterals forming a V-shape and shorter than the dorsal mid-wing or asymmetrical, one or both laterals much reduced, sometimes to a mere incurved ridge. Figs. 132/27, p. 754; 136, p. 772.

TANZANIA. Lindi District: Lake Lutamba, 30 Aug. 1934, *Schlieben* 5202! & 16 Jan 1935, *Schlieben* 5878!; Newala District: SE. Kitangari, *Gillman* 1332! & Selous Game Reserve, Nunga, 26 Feb. 1971, *Ludanga* 1280! & same locality, 27 Feb. 1972, *Ludanga* 1372!
DISTR.   T 8; S. Malawi, Mozambique
HAB.   Swamp forest, woodland and in thickets; 240–810 m.

NOTE.   This species is poorly known and the sheets cited are by no means identical, the type having 5 ovate calyx-lobes and *Gillman* 1332 4 lanceolate lobes although I do not doubt that they belong to the same species. A suggestion on the folders at Kew that this is probably synonymous with the W. African *Cuviera nigrescens* (Oliv.) Wernham is not correct. The single specimen seen from S. Malawi has much more pubescent flowers and shorter corolla lobes.

2. **C. migeodii** *Verdc.* sp. nov. affinis *C. semseii* Verdc. sed bracteis inflorescentiae calycis lobisque angustioribus apice anguste attenuatis, pedicellis et corollae lobis glabris differt. Typus: Tanzania, Lindi District, Tendaguru, *Migeod* 1052 (BM, holo.!)

Slender shrub probably almost identical in habit with *C. semseii* but less leafy; stems slender, dull purplish brown, longitudinally striate and with peeling fissured epidermis, glabrous save for dense white bristle-like hairs within the stipule-base remnants. Leaves borne on very short lateral shoots, oblong to lanceolate, up to 3.5 cm. long, 1.3 cm. wide (fide collector), ± obtuse to narrowly attenuate, lanceolate at the apex, cuneate at the base, glabrous; petiole 2–3 mm. long. Cymes axillary 1–2-flowered; peduncle and pedicels up to 4 mm. long; bracts linear-lanceolate 1 cm. long, 1.5 mm. wide, attenuate at apex. Calyx-lobes lanceolate, 1.2 cm. long, 2–2.5 mm. wide, attenuate at apex. Corolla (in bud only) yellow, ± 8 mm. long including 1 mm. long apical tails. Ovary 2-locular, attenuate into pedicel. Only fruit seen with only one locule developed.

TANZANIA.   Lindi District: Tendaguru, 18 Dec. 1930, *Migeod* 1052!
DISTR.   T8; not known elsewhere
HAB.   Wooded grassland; 180 m.

NOTE.   Migeod notes that the leaves and flowers are scarcely distinguishable.

3. **C. schliebenii** *Verdc.* in K.B. 33: 497 (1979). Type: Tanzania, Lindi District, Lake Lutamba, *Schlieben* 5576 (B, holo.!, BR, EA, K, LISC, iso.!)

Shrub or small tree 2–18 m. tall; youngest shoots drying black, sparsely pubescent; older shoots covered with a pale brown rather corky bark, irregularly fissured. Leaf-blades elliptic to oblong-elliptic, 5.2–15 cm. long, 1.8–7.4 cm. wide, acuminate at the apex,

FIG. 136. *CUVIERA SEMSEII* — **1**, flowering branch, × ⅔; **2**, branch with reduced shoots, × ⅔; **3**, domatium, × 9; **4**, stipule, × 6; **5**, bud with one calyx-lobe folded back, × 2; **6**, flower, × 3; **7**, corolla opened out, × 3; **8**, stigmatic knob, × 9; **9**, section through ovary, × 6; **10**, fruit (immature), × 1. 1, 3, from *Vollesen* in *M.R.C.* 1889; 2, 5–9, from *Schlieben* 5202; 4, from *Vollesen* in *M.R.C.* 3702; 10, from *Ludanga* in *M.R.C.* 1280. Drawn by Sally Dawson.

cuneate at the base, drying dark, glabrous save for very small hairy domatia in the nerve axils beneath; petiole 2–3 mm. long; stipules linear-lanceolate, up to 1.5 cm. long from broad bases which are slightly connate at the base. Inflorescences appearing with or without leaves, cymose, much branched, 3–5 cm. long, the ultimate parts spreading pubescent; peduncle 2–3 cm. long, pedicels 1–3 mm. long, pubescent; bracts and bracteoles linear-lanceolate, ± 3 mm. long, 0.8 mm. wide. Calyx-tube top-shaped, 1.2 mm. long, densely covered with short spreading hairs; lobes oblong or linear-oblong or slightly elliptic, 2.3–2.6 mm. long, 0.7–0.9 mm. wide, ± glabrous; buds with 5 minute apicula at the apex. Corolla yellow, probably slightly fleshy; tube shortly cylindric, 3 mm. long, glabrous outside, inside with a ring of white strongly deflexed hairs 1.3 mm. long affixed at ± the upper ⅓ of the tube; lobes triangular-lanceolate, 2.2 mm. long, 1.2 mm. wide, with a slender apical appendage ± 1 mm. long. Anthers inserted at the throat, oblong, 1 mm. long, ± acute, just exserted. Ovary 2-locular; style slightly thickened at the base, 4.8 mm. long, glabrous; stigmatic club cylindric-clavate, 12-ridged, slightly bifid at the apex. Disk annular, thick. Fruit oblong, compressed, 2.3 cm. long, 1.3 cm. wide, 9 mm. thick, with 1–2 pyrenes; pyrenes fusiform-canoe-shaped, 1.9 cm. long, ± 5 mm. wide and thick, the hilar notch close to top.

TANZANIA.   Lindi District: 40 km. W. of Lindi, Lake Lutamba, 1 Nov. 1934, *Schlieben* 5576! & same locality, 5 Oct. 1934, *Schlieben* 5438! & near W. edge of Rondo Forest Reserve, 8 Feb. 1991, *Bidgood et al.* 1416!
DISTR.   T 8; not known elsewhere
HAB.   Deciduous woodland on hillsides and forest edge; 240–750 m.

4. **C. tomentosa** *Verdc.* in K.B. 36: 557 (1981). Type: Tanzania, Kilwa District, 15 km. SW. of Kingupira, *Vollesen* in M.R.C. 3122 (K, holo.!, EA, K, MRC, WAG, iso.)

Shrub to 3 m. or tree 10–15 m. high; stems of young leafy shoots densely covered with spreading grey hairs; older stems glabrous or obscurely pubescent, with grey-brown finely flaking bark, minutely punctate with scars of hair-bases; branchlets spreading at right-angles. Leaves opposite; blades elliptic or oblong-elliptic, up to ± 10 cm. long 6 cm. wide, narrowly rounded to acuminate at the apex, subcordate at the base, densely velvety on both surfaces with short ± spreading hairs; petioles short, up to 2 mm. long; stipules with ± triangular base 1.5 mm. long with linear-lanceolate apex 8–11 mm. long. Flowers fairly numerous in pubescent branched dichasial cymes; peduncle 1–2 cm. long; secondary branches 0.3–1 cm. long; pedicels 1–3 mm. long; bracts leafy, oblong-subspathulate, 4–5 mm. long, 1–2 mm. wide, glabrous save for base and margins. Calyx-tube 1 mm. long, densely spreading pubescent; lobes narrowly oblong-spathulate, 3–4 mm. long, 0.8–1.2 mm. wide, glabrous inside, ciliate at margins and pubescent outside. Buds yellowish green, acute and shortly 5-tailed, finely rather sparsely spreading puberulous. Corolla olive-green; tube 2–3 mm. long, with ring of long deflexed hairs inside but throat glabrous; lobes narrowly triangular, 2.5–3.5 mm. long including the tails. Ovary 2-locular; style exserted 1.5 mm.; stigmatic club 0.6 mm. long. Fruit not known.

TANZANIA.   Kilwa District: 15 km. SW. of Kingupira, 21 Dec. 1975, *Vollesen* in M.R.C. 3122!
DISTR.   T 8; N. Mozambique
HAB.   *Brachystegia*, *Erythrophleum*, etc. mixed woodland; 175 m.

95. VANGUERIOPSIS

Robyns in B.J.B.B. 11: 248 (1928), pro parte; Verdc. in K.B. 42: 187 (1987)

Shrubs or small trees with mostly stiff thick branches. Leaves opposite, somewhat coriaceous, hairy or velvety tomentose; stipules thick, triangular, ± joined into a sheath at the base, long-caudate at apex. Flowers large and conspicuous, ± thick, in opposite simple to much-branched many-flowered axillary cymes; bracts fairly conspicuous. Calyx-lobes erect, triangular to linear-lanceolate or linear-oblong. Buds elongate-lanceolate, beaked. Corolla-tube cylindrical, ⅔–⅛ the length of the linear-lanceolate lobes, with a ring of deflexed hairs or glabrous within save at base; throat glabrous. Stamens inserted at the throat; anthers exserted, linear. Ovary 2-locular (5-locular in *V. gossweileri*), each locule with a solitary pendulous ovule; style slender, long-exserted; stigmatic club cylindrical. Fruit fairly large, didymous when 2 pyrenes developed but unilateral and oblique when 1 is aborted, smooth or irregularly ribbed; presumably ± globose in *V. gossweileri*.

Robyns does not indicate a type for this genus but I have accepted *Rostranthus* Robyns as the type section of the genus (i.e. section *Vangueriopsis*) and *V. lanciflora* as the type species. I have restricted the genus to the four species which fall within *Rostranthus* Robyns; two occur in the Flora area, which are not at all closely related, one *V. longiflora* being referable to subgen. *Guranivea* Verdc.

Calyx-lobes triangular to lanceolate, 1.5–5.5 mm. long; corolla-
    lobes linear-lanceolate, 2–2.5 cm. long; bark flaking to
    expose a brownish pink to rusty red under-bark; fruit
    probably smooth in life . . . .   1. *V. lanciflora*
Calyx-lobes linear to linear-lanceolate, 4.2–7.5(–10) mm. long;
    corolla-lobes linear, 3.4–4 cm. long; bark not flaking to
    expose such a layer; fruit ribbed . . . .   2. *V. longiflora*

1. **V. lanciflora** (*Hiern*) *Robyns* in B.J.B.B. 11: 252 (1928); T.T.C.L.: 538 (1949); F.F.N.R.: 425, fig. 68B,C (1962); Launert in Prodr. Fl. SW. Afr. 115: 27 (1966); Palmer & Pitman, Trees S. Afr. 3: 2086 (1972); Palmer, Field Guide Trees S. Afr.: 44 (1977); Coates Palgrave, Trees S. Afr.: 876 (1977); Verdc. in K.B. 42: 187, fig. 3C (1987). Type: Zambia, near Victoria Falls, *Kirk* (K, holo.!)

Deciduous shrub or much-branched small tree (0.9–)1.5–6(–13) m. tall; branches thick and stiff; bark grey, flaking off to expose a brownish pink or rusty red under-bark; young branches densely grey-pubescent. Leaves opposite; blades elliptic or oblong-elliptic, (1.5–)5.3–21.5 cm. long, (0.5–)1.6–12.5 cm. wide, ± rounded or subacute at the apex, rounded to cuneate at the base, markedly discolorous, rather scabrid pubescent above, velvety grey or yellowish tomentose or pubescent beneath; petiole 0.5–1.5 cm. long; stipules thick, triangular, up to 9 mm. long, joined to form a sheath at the base, grey-tomentose outside, densely hairy inside, with a thick obtuse subulate apex 0.8–1.2 cm. long, eventually deciduous. Inflorescences densely yellowish velvety pubescent, simply cymose or branched, axillary on the leafless nodes of older branches; cymes several-flowered; bracts lanceolate or subulate, 4–6 mm. long; peduncles and secondary branches 1–2 cm. long; pedicels very short or up to 8 mm. long. Buds elongate, beaked, grey-tomentose, up to 2–3 cm. long, with divaricate tails at apex. Calyx-tube subcampanulate, 4 mm. long, 4 mm. wide, densely tomentose; lobes erect, narrowly lanceolate to triangular or oblong-lanceolate, 1.5–5.5 mm. long, densely tomentose; corolla whitish or yellow-green; tube cylindric, 5 mm. long, with a ring of deflexed hairs inside; lobes reflexed, linear-lanceolate, 2–2.5 cm. long, glabrous inside, tomentose outside, apiculate at the apex. Anthers 2.5–3 mm. long, exserted for 5 mm. Style green, rather stout, exserted for up to 1.5 cm., constricted near the apex; stigmatic club cylindric, 2–2.5 mm. long, smooth, 2-lobed at the apex. Fruit rather like a medlar, rounded, compressed, didymous or oblique, 2–4 cm. long, 1.7–2 cm. wide, or the size of a peach (ex Zimbabwe label), crowned by the calyx-limb, sparsely pubescent, with 1–2 pyrenes; pyrenes narrowly ellipsoid, up to 2.1 cm. long, 9 mm. wide, with woody wall 1.7 mm. thick.

TANZANIA. Tabora District: Kakoma, *C.H.N. Jackson* 116!; Mbeya District: Magangwe, 1 Dec. 1970, *Greenway & Kanuri* 14750! & same locality, 9 Apr. 1970, *Greenway & Kanuri* 14303! & Magangwe [Nanganjwa], 26 Aug. 1969, *Greenway & Kanuri* 13808!
DISTR. **T** 4, 7; Zaire, Malawi, Zambia, Zimbabwe, Angola, Botswana, Namibia and South Africa (Transvaal)
HAB. *Brachystegia*, *Combretum*, etc. woodland, inter-zones between *Brachystegia* and flood pans; 1140–1350 m.

SYN. *Canthium lanciflorum* Hiern in F.T.A. 3: 146 (1877); Oliv. in Hook., Ic. Pl. 23, t. 2252 (1893)
    C. platyphyllum Hiern, Cat. Afr. Pl. Welw. 1: 479 (1898). Type: Angola, Huila, Monino, *Welwitsch* 2583 (LISU, holo., BM, iso.)
    Vangueria lateritia Dinter in F.R. 24: 367 (1928). Type: Namibia, Gaub, *Dinter* 2421 (B, holo.†)

NOTE. The fruit is edible but has an acid after taste. Launert (Prodr. Fl. SW. Afr.) gives the authority as Robyns ex Good in J.B. 64, Suppl. 2: 22 (1926) but the genus *Vangueriopsis* is not described and the combination has no validity.

2. **V. longiflora** *Verdc.* in Kirkia 5: 275 (1966) & in K.B. 42: 188, fig. 3D, 6D (1987). Type: Tanzania, Ulanga District, near Ifakara, Itula, *Haerdi* 435/0 (EA, holo.!, BR, K, iso.!)

Tree 10–13 m. tall, with elongated unbranched trunk below; branches ferruginous or grey-brown pubescent or shortly hairy. Leaves opposite; blades large, oblong-ovate to broadly elliptic, 10–21 cm. long, 5.5–14.3 cm. wide, shortly acuminate at the apex, rounded

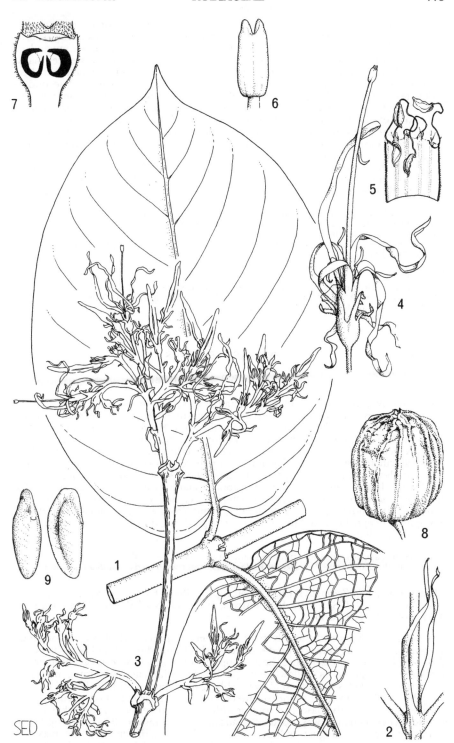

FIG. 137. *VANGUERIOPSIS LONGIFLORA* — **1**, branch with leaves, × ⅔; **2**, stipule, × 2; **3**, flowering branch, × ⅔; **4**, flower × 2; **5**, corolla-tube opened out, × 3; **6**, stigmatic knob, × 9; **7**, longitudinal section through ovary, × 6; **8**, fruit, × 1; **9**, pyrene (2 views), × 1. All from *Haerdi* 435/0. Drawn by Sally Dawson.

or subcordate at the base, discolorous, pubescent above particularly on the main nerves, velvety ferruginous-tomentose beneath, rather thin; petiole 0.7–1.3 cm. long, shortly hairy; stipules with subtriangular base up to 1 cm. wide and a long upper appendage 1.5–2.5 cm. long, 0.5–1 mm. wide, hairy inside, soon deciduous. Inflorescences many-flowered, borne at the ends of ± leafless branches but actually axillary, much branched, 3–5 cm. long; bracts linear, 4–5 mm. long, ± 1 mm. wide; peduncles and secondary peduncles up to 2 cm. long; pedicels 0.6–0.8(–1.3) cm. long. Calyx-tube campanulate, 2.5–3.5 mm. long, 3.2 mm. wide, obtusely 5-ribbed, velvety tomentose; lobes 5, subequal, linear to linear-lanceolate, 4.2–7.5(–10) mm. long, 0.8–1.2 mm. wide, tomentose all over. Corolla green; tube cylindric, 3–5 mm. long, 3.5 mm. wide, tomentellous outside, glabrous inside save at base; lobes linear, 3.4–4 cm. long, 2.3 mm. wide, shortly pubescent. Anthers exserted 2.5–3.5 mm., apiculate. Ovary 2-locular; style coarsely filiform, ± 3.5 cm. long, narrowing above; stigmatic club cylindric, 2.5 mm. long. Fruit obovoid or ellipsoid, 2.3–3.2(–4) cm. long, 1.5–2.6(–3.5) cm. wide, 1.1–1.3 cm. thick, 1–2-locular, much compressed, irregularly ribbed, ferruginous pubescent. Figs. 132/5, p. 754; 137, p. 775.

TANZANIA. Ulanga District: near Ifakara, Itula, 1 Feb. 1960, *Haerdi* 435/0!; Iringa District: Mwanihana Forest Reserve, 5 Jan. 1980, *Rodgers & Bulstrode* 124!
DISTR. **T** 6, 7; not known elsewhere
HAB. Wetter parts of forests and thickets on hills, *Erythrophleum, Parkia, Milicia, Newtonia, Dialium* forest; 500–700 m.

SYN. *V. sp. nov.* sensu Haerdi in Acta Trop., suppl. 8: 148 (1964)

NOTE. Two sterile specimens, *Bridson* s.n. (Iringa District, Sanje) and *Rodgers & Hall* 4526 (Iringa District, Luhomero), described as trees 10–20 m. tall with leaves up to 27 × 14.5 cm., possibly belong to this species, but the indumentum is coarser and sparse.

96. **PYGMAEOTHAMNUS**

Robyns in B.J.B.B. 11: 29 (1928)

Subshrubby herbs mostly under 30 cm. tall, with several–many stems from a woody rhizome. Leaves opposite or in whorls of 3–4, mostly fairly shortly petiolate; stipules triangular, usually shortly subulate-caudate at the apex. Flowers in simple or branched cymes or fascicles, sometimes borne at lower often leafless nodes; bracts small. Calyx-lobes 5, erect, linear-oblong, oblong, lanceolate or ± triangular, obtuse or subacute. Buds shortly apiculate, usually 5-tailed, glabrous or pubescent. Corolla-tube cylindric, with a ring of deflexed hairs within, sparsely to densely pilose at the throat; lobes reflexed, usually shortly apiculate, glabrous within, about equalling the tube. Stamens inserted at the throat, the anthers oblong or lanceolate, just exserted. Ovary 2-locular, each locule with a solitary pendulous ovule; style slender, slightly exserted; stigmatic club cylindric or subcapitate, deeply bilobed. Fruit globose or obpyriform, didymous or asymmmetrical, containing (1–)2 pyrenes.

A genus of 4 species mostly confined to southern Africa but 2 extending to Central Africa and 1 occurring in the Flora area. *P. concrescens* Bullock has been transferred to *Multidentia* Gilli.

P. zeyheri (*Sond.*) *Robyns* in B.J.B.B. 11: 30, fig. 7, 8 (1928); F.F.N.R.: 418 (1962); Verdc. in K.B. 42:131, fig. 2c (1987). Type: South Africa, Transvaal, N. of Magaliesberg, *Zeyher* 766 (MEL, holo., K!, TCD, iso.)

Subshrub with 3–6 slender woody stems 17–30 cm. tall from a fairly slender ± horizontal rhizome; glabrous or stems covered with dense short spreading rather bristly hairs and also some longer adpressed ones. Leaves opposite or mostly in whorls of 3 or 4, drying dark brownish green or yellow-green (as in aluminium accumulators), usually ± subcoriaceous; blades obovate to obovate-oblong or elliptic, 4.2–12 cm. long, 1.8–5 cm. wide, those of lowest whorls rounded, others obtusely to subacutely shortly acuminate at the apex, cuneate at the base, glabrous or with rather sparse yellowish bristly hairs on both surfaces which in no way obscure the surface; petioles distinct, 0.5–1 cm. long, bristly pubescent; stipules triangular at base, sometimes 3-cusped, drawn out at the apex, 4–5 mm. long, glabrous or bristly pubescent. Inflorescences borne in the lower axils including the lowest from which the leaves have fallen or never fully developed, few- to ± 40-flowered, much branched, up to 4 cm. long; peduncles (0.5–)1–2 cm. long; secondary

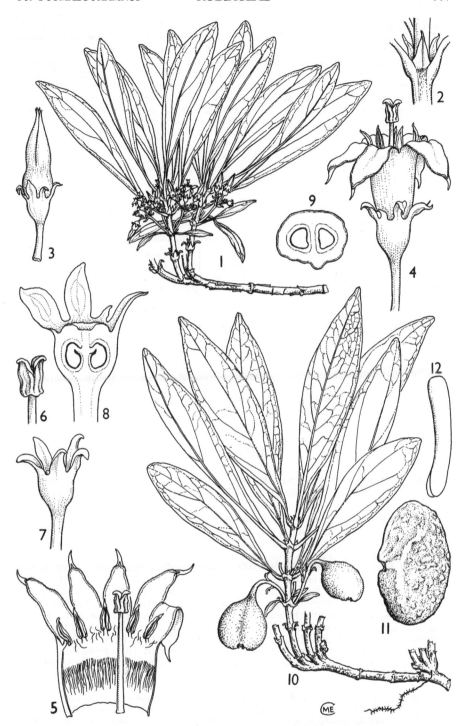

FIG. 138. *PYGMAEOTHAMNUS ZEYHERI* var. *ZEYHERI* — **1**, habit (flowering), × ⅖; **2**, stipule, × 2; **3**, flower bud, × 4; **4**, flower, × 6; **5**, corolla opened out, with style and stigmatic knob, × 6; **6**, stigmatic knob, × 10; **7**, calyx, × 6; **8**, longitudinal section of ovary, × 10; **9**, transverse section of ovary, × 12; **10**, habit (fruiting), × ⅖; **11**, pyrene, × 3; **12**, embryo, × 3. 1–9, from *Holmes* 1224; 10, from *Milne-Redhead* 3458; 11, 12, from *Brummitt et al.* 13879. Drawn by Mrs M.E. Church.

cymose branches sometimes elongate up to 3 cm. long with the flowers spaced; pedicels 3–7 mm. long; bracts and bracteoles narrowly elliptic-lanceolate, 3–7 mm. long, 0.5 mm. wide, acuminate. Calyx-tube globose, ± 1 mm. diameter, bristly pubescent; lobes leafy, narrowly oblong, elliptic or lanceolate, 0.8–2.5(–5) mm. long, 0.3–0.8 mm. wide, glabrous or sparsely bristly pubescent. Buds slender, acuminate, bristly pubescent or glabrous. Corolla red or greenish; tube 3–4 mm. long; lobes oblong-lanceolate, 4–5 mm. long, reflexed or sometimes probably erect. Ovary 2-locular, densely filled with reddish inclusions. Fruits oblong, subglobose or pyriform, ± 2.5–3 cm. long, 1.5–2 cm. wide, glabrous, containing (1–)2 pyrenes, each 1.2 cm. long, 6 mm. wide. Figs. 131/11, p. 752; 132/6, p. 754; 138, p. 777.

SYN. *Pachystigma zeyheri* Sond. in Linnaea 23: 56 (1850)
    Vangueria zeyheri (Sond.) Sond. in Fl. Cap. 3: 15 (1865); De Wild. in B.J.B.B. 8: 66 (1922)
    Fadogia zeyheri (Sond.) Hiern in F.T.A. 3: 153 (1877)

var. **rogersii** *Robyns* in B.J.B.B. 11: 34 (1928). Type: Zimbabwe, Victoria Falls, *F.A. Rogers* 5431 (PRE, holo., BOL, K, iso.!)

Stems, calyces, buds, etc. sparsely to quite densely spreading hairy.

TANZANIA.   16 km. E of Biharamulo, Aug. 1958, *Procter* 977!; Tabora District: Ichemba Increment Plot, 21 Sept. 1967, *Shabani* 50! & Tabora golf course, 10 Oct. 1949, *Bally* 7647!
DISTR.   T1,4; Burundi, Zaire, Zambia, Zimbabwe and South Africa (Transvaal)
HAB.   Wooded grassland with *Parinari*, etc, and derived grassland and plots; 1200–1440 m.

SYN. *Fadogia livingstoniana* S. Moore in J.B. 57: 88 (1919). Type: Zambia, Livingstone, *Rogers* 7466 (BM, holo.!, K, iso.!)
    Pygmaeothamnus zeyheri (Sond.) Robyns var. *livingstonianus* Robyns in B.J.B.B. 11: 34 (1928). Type: Zambia, Livingstone, *Rogers* 7466 (?PRE, holo., BM, K, iso.!)

NOTE.   Robyns' name '*livingstonianus*', is not a new combination but the same epithet based on the same gathering. He appears to have overlooked S. Moore's name altogether and does not mention it under *Fadogia* in his revision. Variously hairy variants of *P. zeyheri* occur throughout the range of the species but are mostly infrequent. I had considered that the material from the Flora area belonged to a distinct species but it is best considered an extreme form of the hairy variety.

97. TAPIPHYLLUM

Robyns in B.J.B.B. 11: 117 (1928) & B.J.B.B. 32: 133 (1962); Verdc. in K.B. 36: 533–539 (1981) & K.B. 42: 143–145 (1987); Havard & Verdc. in K.B. 42: 605 (1987)

Herbs with stems from a woody rhizome, subshrubs, shrubs or even small trees, the foliage, inflorescences and young shoots characteristically covered with dense velvety tomentum. Leaves opposite or in whorls of 3–4, subsessile or shortly petiolate, usually markedly discolorous; stipules connate below, drawn out to a linear apex, at length deciduous. Flowers fairly small in subsessile to shortly pedunculate axillary dense usually opposite few–many-flowered ± globose cymes or 1–few-flowered fascicles; bracts and bracteoles present, often fairly conspicuous. Calyx-lobes 5, mostly erect and linear to linear-lanceolate, less often ovate to oblong-triangular. Buds apiculate. Corolla green, yellow or white; tube cylindrical, glabrous, less often hairy at the throat, with a ring of reflexed hairs within; lobes reflexed, usually apiculate. Stamens inserted at the throat, the anthers ± exserted. Ovary 4–5(–6)-locular, with one pendulous ovule per locule; style slender, usually exserted; stigmatic club cylindric or coroniform, sulcate and 4–5-lobulate. Fruit subglobose, with 3–5 pyrenes, or oblique and reduced to 1 pyrene, mostly crowned by persistent calyx.

An ill-defined genus characterised by little but the velvety tomentum and mostly elongate calyx-lobes but certainly having a recognisable facies; the species are equally poorly defined; Robyns in his 1962 paper recognised 29 species but this could probably be reduced by a third. A more wholesale reduction with most of the taxa reduced to variants of a very few species could certainly be made but scarcely be useful. As with so many savanna and *Brachystegia* woodland genera the variation defies normal taxonomic treatment; I have recognised 6 species from the Flora area but further study with fresh material will probably reduce this to 5. The key refers only to East African material.

1. Corolla-tube elongate, 1.1–1.3 cm. long, very slender . . . 6. *T. burnettii*
   Corolla-tube shorter, (2.5–)4.5–6 mm. long, relatively wider   2
2. Leaves in whorls of 3 or 4; herbs or small shrubs 3
   Leaves paired 5
3. Stems persistently velutinous only at the apex, pubescent or
     glabrescent below but not hiding epidermis; leaf-blades
     3.5–5 × 1–1.8 cm.; calyx-lobes 4–5 mm. long; stems slender
     (T 4) . . . .   2. *T. cinerascens* var.
                                                         inaequale (in part)

   Stems persistently velvety for 15–40 cm. below the shoot
     apex hiding the dark epidermis 4
4. Leaves in whorls of 3–4, the blades 2–2.8 cm. wide, thinner;
     calyx-lobes linear, ± 5 mm. long (T 1) . . . . . .   5. *T. verticillatum*
   Leaves in whorls of (2–)3, the blades more ovate, 1.5–6.5(–
     7.5) cm. wide, thicker; calyx-lobes linear-lanceolate to
     triangular, 3–5 mm. long (T 1, 4) . . . .   4. *T. discolor*
5. Inflorescences lax, the peduncles ± equalling the distinct
     secondary peduncles, both 0.5–1.3 cm. long, separated by
     a pair of oblong-cucullate bracts 4 mm. long; corolla-
     lobes much exceeding the corolla-tube; small tree (T 8)   1. *T. schliebenii*
   Inflorescences denser, usually very dense 6
6. Leaf-blades with wrinkled margins; calyx-lobes distinctly
     more leafy, up to 1.2 mm. wide; inflorescence-bracts up to
     5 mm. wide; plant of the Great Central Thicket area (T 5,
     7) . . . .   3. *T. obtusifolium*
   Leaf-blades with straight margins; calyx-lobes linear or
     subulate, under 1 mm. wide; inflorescence-bracts mostly
     narrower; not from the Great Central Thicket area   7
7. Subshrubby herb with erect stems, probably always several
     from one rootstock; leaf-blades usually elliptic-ovate or
     broadly elliptic-oblong, 5–10 × 2–6.5 cm., rounded to
     subcordate at the base; petioles distinct, 5–6 mm. long.   4. *T. discolor* (forms
                                                         with paired leaves).

   Branched shrubs or subshrubs; leaf-blades oblong-elliptic
     to ovate- or linear-lanceolate, 2–10.5 × 0.5–3.3(–4.6)cm., ±
     cuneate at the base; petioles often shorter   . . . .   2. *T. cinerascens*

1. **T. schliebenii** *Verdc.* in K.B. 36:534 (1981). Type: Tanzania, Lindi District, Lake Lutamba, *Schlieben* 5854 (BR, holo.!, K, iso.!)

Small tree 8–10 m. tall; youngest parts of new stems densely ochraceous-velvety; older parts blackish purple, practically glabrous, slightly ridged. Leaves paired; blades oblong, oblong-elliptic or oblong-ovate, 6.2–10.5 cm. long, 2.2–5.6 cm. wide, subacute at the apex, the actual tip narrowly rounded, cuneate to rounded at the base, sometimes unequally so, densely persistently grey-brown velvety on both surfaces; petioles 5–7 mm. long; stipules with base triangular, 4–5 mm. long, the subulate part up to 1–2 cm. long, densely velvety on the newest parts but less pubescent and ± same colour as the stem on the older parts, eventually deciduous but leaving a triangular patch of hairs on the otherwise dark stem. Inflorescences cymose, 2–3.5 cm. long including the 0.5–1.3 cm. long peduncle; secondary peduncles 5–10 mm. long; pedicels 5–8 mm. long; all parts densely grey-yellow pubescent; paired bracts at the top of the peduncle, oblong-cucullate, 4 mm. long, densely tomentose. Calyx-tube subglobose-hemispherical, ± 1.2 mm. tall; limb-tube 0.5 mm. long; lobes linear, 4–7 mm. long, all parts densely pubescent-tomentose. Buds elongate, tapering, the tails connivent, 1.2 cm. long, densely hairy outside. Corolla yellow; tube 4.5 mm. long; lobes narrowly linear-oblong, 8.5 mm. long, 1.3 mm. wide, inclusive of 2 mm. long apical appendages, reflexed. Ovary 5-locular; style exserted ± 5 mm.; stigmatic club cylindric, 1 mm. long. Fruits not known.

TANZANIA. Lindi District: Lake Lutamba, 10 Jan. 1935, *Schlieben* 5854!
DISTR. **T** 8; not known elsewhere
HAB. Deciduous woodland; 240 m.

2. **T. cinerascens** (*Hiern*) *Robyns* in B.J.B.B. 11: 107 (1928) & B.J.B.B. 32: 136 (1962); Verdc. in K.B. 36:534 (1981) & 42:144 (1987). Type: Angola, Pungo Andongo, Mutollo, *Welwitsch* 3162 (LISU, holo., BM, K, iso.!)

Erect shrub (0.15–)0.9–3 m. tall or small tree, with ± spreading slender branches, or subshrub with 1–several shoots; rootstock woody, often large; branchlets slender, densely velvety with adpressed pale ferruginous hairs when young, later with sparser hairs or glabrescent. Leaves paired or occasionally in whorls of 3; blades oblong-elliptic to ovate-lanceolate, or less often linear-lanceolate, 2–10.5 cm. long, 0.5–3.3(–4.6) cm . wide, acute or subacute at the apex, cuneate to subcordate at the base, occasionally somewhat unequal, slightly to distinctly discolorous, somewhat scabrid-pubescent above, densely velvety tomentose beneath with adpressed pale ferruginous or grey hairs; petiole 3–6 mm. long; stipules joined into a sheath 1–2 mm. long at base and with linear apices 2–4 mm. long. Flowers in shortly pedunculate subglobose dense velvety hairy several–15-flowered cymes; peduncles 2–7 mm. long; pedicels 1–3 mm. long; bracts elliptic to lanceolate, 4–7 mm. long, 1–5 mm. wide. Calyx densely pale ferruginous hairy; tube 1–2 mm. long; lobes linear to linear-lanceolate, 1.5–8(–10) mm. long. Buds with distinct to long tails of up to 2 mm., densely ferruginous hairy or tomentose. Corolla pale yellow or greenish white; tube ± cylindric, 4–5 mm. long; lobes elliptic to lanceolate, 4–5 mm. long (including caudate appendages ± 2 mm. long), 1.2–2 mm. wide. Ovary 3–5-locular; style exserted ± 1.5 mm.; stigmatic club subcylindric or coroniform, 0.8 mm. long, sulcate. Fruits yellow or orange-brown to reddish when ripe, subglobose, 8–9 mm. diameter, ferruginous velvety tomentose and with longer hairs, with 1–5 pyrenes, crowned with the persistent calyx.

KEY TO INFRASPECIFIC VARIANTS

Calyx-lobes short, 2(–3) mm. long; leaf-blades mostly small (**T** ?5,7,8)   .   .   .   .   c. var. **laevius**
Calyx-lobes longer, up to 6(–8) mm. long; leaf-blades ± 8 × 3 cm. or larger:
    Calyx-lobes 3–5(–6) mm. long, mostly ± 4 mm.; leaf-blades elliptic (**T** 4)   .   .   .   .   a. var. **cinerascens**
    Calyx-lobes usually distinctly longer, (3–)5–6(–8) mm. long; leaf-blades mostly elliptic but sometimes quite narrow, linear-lanceolate to oblong-elliptic (**T** 4)   .   .   .   b. var. **inaequale**

a. var. **cinerascens**

Leaf-blades larger, mostly ± 8 × 3 cm. Calyx-lobes 3–5(–6) mm. long, mostly ± 4 mm.

TANZANIA.   Tabora District: Beekeeping Headquarters area, 20 Dec. 1954, *Akiley* 5017! & Tabora, 3 Dec. 1957, *Friend* 15! & Simbo Forest Reserve, Dec. 1977, *Shabani* 1242!
DISTR.   **T** 4, 5; Zaire, Burundi, Zambia, Malawi and Angola
HAB.   *Brachystegia* and other secondary woodland; 1080–1320 m.

SYN.   *Ancylanthos cinerascens* Hiern in F.T.A. 3: 159 (1877) & in Cat. Afr. Pl. Welw. 1: 484 (1898)
NOTE.   Fruits said to be edible.

b. var. **inaequale** (*Robyns*) *Verdc.* in K.B. 42: 144 (1987). Type: Tanzania, Buha District, Gombe Stream Reserve, *Morris-Goodall* 110 (BR, holo.!, EA, iso.!)

Leaf-blades 3.5–9.5 × 1–2.7 cm. Calyx-lobes 3–8 mm. long, mostly ± 5–6 mm.

TANZANIA.   Buha District: Gombe Stream National Park, Kakombe valley, 10 Jan. 1964, *Pirozynski* 235!; Buha/Kigoma Districts: Gombe Stream Reserve, Nov. 1960, *Morris-Goodall* 110! & Gombe Stream National Park, High Mtumba Watu Path, 15 Feb. 1970, *Clutton-Brock* 419; Kigoma District: near Kigoma, Jan. 1929, *Humbert* 7201!.
DISTR.   **T** 4, 7; Burundi and Zaire
HAB.   "Upper woodland", grassland with scattered trees; 750–1370 m.

SYN.   *Vangueria kaessneri* S. Moore in J.B. 48: 221(1910); De Wild. in B.J.B.B. 8:57 (1922). Type: Zaire, Shaba, Katenina and Kundelungu, *Kassner* 2189 (BM, holo.!)
    Tapiphyllum kaessneri (S. Moore) Robyns in B.J.B.B. 11: 106 (1928); T.T.C.L.: 532 (1949); Robyns in B.J.B.B. 32: 135 (1962)
    T. inaequale Robyns in B.J.B.B. 32: 145 (1962)
    T. cinerascens (Hiern) Robyns subsp. *inaequale* (Robyns) Verdc. in K.B. 36: 535 (1981)

NOTE.   Although I have little doubt that *T. kaessneri* and *T. inaequale* both represent the same variant of *T. cinerascens* I have preferred to use the later name based on a type from the Flora area as is permissible at a lower rank. A specimen from Zambia, *Kassner* 2140, with leaves 7.5 × 3.5 cm., but cited by Robyns as *T. kaessneri* and by S. Moore (J.B.48:222 (1910)) as a form of *T. kaessneri*, seems better placed in var. *cinerascens*. Some plants with leaves to 9.5 × 5.5 cm. and calyx-lobes to 1 cm. seem better placed as another variant of *cinerascens*. The Humbert specimen cited has branchlets and leaves in whorls of 3. The citing of *Kassner* 2723 and 3052 as well as 2189 in the original reference is an error — they are actually *Virectaria*.

c. var. **laevius** (*K. Schum.*) *Verdc.* in K.B. 42:144 (1987): Type: Tanzania, Iringa District, Mgololo, *Goetze* 765 (B, holo.†, BR, iso.!)

Leaf-blades small, up to ±3.5 × 1.7 cm. Calyx-lobes 1.5–3.5 mm. long, mostly ± 2 mm.

TANZANIA.   Iringa District: Mgololo, Mar. 1899, *Goetze* 765!; Songea District: 16 km. W. of Songea, 2 Jan. 1956, *Milne-Redhead & Taylor* 8107! & 1.6 km. S. of Gumbiro, 24 Jan. 1956, *Milne-Redhead & Taylor* 8107B! & about 9.5 km. W. of Songea, 4 Jan. 1956, *Milne-Redhead & Taylor* 8107A! & SE. Songea District, Jan. 1901, *Busse* 967!
DISTR.   T ?4 or 5,7,8; Mozambique
HAB.   *Brachystegia* woodland on red loam, "rolling plateau of red laterite"; 990–1400 m.

SYN.   *Vangueria velutina* Hiern var. *laevior* K. Schum. in E.J. 28: 494 (1900); De Wild. in B.J.B.B. 8: 66 (1922)
    Tapiphyllum velutinum (Hiern) Robyns var. *laevius* (K. Schum.) Robyns in B.J.B.B. 11: 111 (1928); T.T.C.L.: 532 (1949)
    T. tanganyikense Robyns in B.J.B.B. 32: 149 (1962). Type: Tanzania, ? Mpanda or Kilosa District, Uguwe (?Ugwe), *Swynnerton* 2052 (BM, holo.!)
    T. cinerascens (Hiern) Robyns subsp. *laevius* (Robyns) Verdc. in K.B. 36: 536 (1981)

NOTE.   Robyns did not see the type when writing his original monograph, the duplicate now at BR not being donated until 1936.

NOTE.   (on species as a whole). The varieties recognised are of a diffuse and unsatisfactory nature. Many other "species" recognised by Robyns from the Flora Zambesiaca area will probably be better treated in a similar manner.

3. **T. obtusifolium** (*K. Schum.*) *Robyns* in B.J.B.B. 11: 115 (1928); T.T.C.L.: 532 (1949); Verdc. in K.B. 36: 536 (1981). Type: Tanzania, Iringa District, Muhinde Steppe, *Goetze* 513 (B, holo. †, BR, K, fragments!)

Shrub 1.8–5 m. tall, much branched, with dark grey or black bark and pale sage-green woolly foliage (in life); branchlets held at right-angles; young parts with fairly persistent very dense velvety tomentum of spreading pale ferruginous hairs, eventually glabrous and with blackish or dark purplish peeling bark. Leaves paired; blades elliptic-oblong to oblong-obovate, 1.8– 8.5(–10) cm. long, 1–4.5 cm. wide, rounded to subacute at the apex, rounded, truncate or ± cuneate or rarely subcordate at the base, at first not discolorous and very densely velvety floccose tomentose all over with yellowish felted hairs, later paler beneath when indumentum on upper surface has become less dense, margins revolute and undulate when young; petioles 3–7 mm. long, densely hairy; stipules with 1.5–3 mm. long bases connate into a sheath, with linear apex 5–7.5 mm. long, densely pilose. Flowers in globose tomentose (10–)15–20-flowered condensed cymes, the opposite inflorescences sometimes coalescing to form nodal balls of flowers ± 2–3 cm. in diameter, sometimes at leafless nodes; subsessile or peduncles ± 5–7 mm. long; bracts ovate, 9 mm. long, 5 mm. wide, acute; pedicels obsolete or up to 2 mm. long. Calyx yellow-green (in life), densely pale ferruginous hairy; tube hemispherical, 1.5–2.5 mm. long; lobes erect, linear or linear-lanceolate, (2–)5–7.5 mm. long, (0.5–)1–1.2 mm. wide. Buds apiculate, pilose outside. Corolla green or creamy yellow; tube cylindric, 2.5–4 mm. long; lobes oblong-ovate, 3–6 mm. long overall, 1–1.7 mm. wide, with a linear-subulate appendage at the apex. Ovary 5-locular; style 5 mm. long; stigmatic club coroniform, 1 mm. long, 5-sulcate and 5-lobulate. Fruits sage-green, oblong-subglobose, 1.3 cm. long, 1 cm. wide, densely adpressed ferruginous tomentose, grooved between the 2–3 pyrenes or oblique and reduced to 1 pyrene.

TANZANIA.   Singida, 16 Dec. 1933, *Michelmore* 836!; Kondoa Irangi, 22 Dec. 1970, *Greenway & Kanuri* 14819!; Dodoma District: between Mukodowa and Dodoma, 11 Dec. 1965, *Richards* 20930!; Mpwapwa, 14 Feb. 1935, *Mr & Mrs Hornby* 632!; Iringa, 19 Mar. 1932, *Lynes* Ig 234!
DISTR.   T 5, 7; not known elsewhere
HAB.   *Brachystegia* and *Isoberlinia* woodland, *Acacia* etc. bushland, Itigi thicket in rocky places; 900–1590 m.

SYN. *Vangueria obtusifolia* K. Schum. in E.J. 28: 493 (1900); De Wild. in B.J.B.B. 8: 60 (1922)
   Tapiphyllum floribundum Bullock in K.B. 1933: 472 (1933); T.T.C.L.: 532 (1949); Robyns in
   B.J.B.B. 32: 137 (1962). Type: Tanzania, Dodoma District, Manyoni, *B.D. Burtt* 3458 (K, holo.!,
   BR, EA, iso.!)

NOTE. A characteristic species of the Great Itigi Thicket. Bullock mentions *T. obtusifolium* when
   describing *T. floribundum* claiming that the latter can be distinguished by the characteristic long
   persistent ('long' of time not length) indumentum of the otherwise smooth branchlets. The
   abundant material now available including some from Iringa enables one to be certain the
   fragment of type at Kew represents the same species.

4. **T. discolor** (*De Wild.*) *Robyns* in B.J.B.B. 11: 105 (1928) & 32: 138 (1962). Type: Zaire,
Shaba, Biano, *Homblé* 873 (BR, lecto.)

Shrub or subshrubby herb with erect stems, probably several from one root, 0.3–0.6(–
2.4) m. tall; stems densely ferruginous velvety, becoming pubescent or even glabrescent
and purplish brown beneath the indumentum. Leaves paired or mostly in whorls of 3;
blades elliptic to oblong-ovate, 3–11(–14) cm. long, 1.5–6.5(–7.5) cm. wide, subacute to
shortly acuminate at the apex, rounded to slightly subcordate at the base, drying
discolorous, densely softly velvety on both surfaces with pale ferruginous hairs but said to
be flannelly white beneath (in life); petiole 5–6 mm. long, velvety; stipules subtriangular
and connate at base, ± 4 mm. long with subulate apex ± 8 mm. long. Flowers in dense short
velvety several–many-flowered cymes, 1.5–2.5 cm. long, including the 0.5–1.2 cm. long
peduncle; bracts linear-lanceolate, 6–7 mm. long, 1.5–2 mm. wide; pedicels 0–2 mm. long.
Calyx densely pale ferruginous-pubescent; tube 2 mm. long; lobes linear-lanceolate to
triangular, 3–5 mm. long, 1 mm. wide. Buds velvety outside, apiculate. Corolla yellow to
green; tube cylindrical, 4–5 mm. long, glabrous at base outside; lobes 3.5–5 mm. long,
1.3–2 mm. wide. Style 7 mm. long; stigmatic club coroniform, 1 mm. long. Fruits green
turning yellow but drying ferruginous, subglobose, ± 1–1.3 cm. diameter, wrinkled in the
dry state, ferruginous velvety and with sparser longer hairs, crowned with the persistent
calyx, grooved between the 3–5 pyrenes in dry state; pyrenes rounded segment-shaped, 8
mm. long, 5 mm. wide.

TANZANIA. Bukoba District: Kakoma, 12 Feb. 1936, *Lloyd* 20!; Ngara District: Bugufi, Murgwanza, 2
   Dec. 1960, *Tanner* 5396!; Buha District: 6.4 km. from Kibondo, 15 Nov. 1962, *Verdcourt* 3368!
DISTR. **T** 1, 4; Zaire, Burundi and Zambia
HAB. Grassland (including seasonally flooded valleys), woodland (including *Brachystegia*), wooded
   grassland, often on stony hillsides; 1140–1770 m.

SYN. *Fadogia discolor* De Wild. in F.R. 13: 138 (1914) & Notes Fl. Kat. 4: 88 (1914) & Contr. Fl. Katanga:
   213 (1921)
   Tapiphyllum confertiflorum Robyns in B.J.B.B. 11: 103 (1928). Type: Zambia, Batoka District,
   Mangoye, *F.A. Rogers* 8943 (K, holo.! & iso.!)
   T. herbaceum Robyns in B.J.B.B. 11: 104 (1928). Type: Zambia, Mumbwa, *Macaulay* 1080 (K,
   holo.!)
   T. fadogia Bullock in K.B. 1933: 147 (1933); Verdc. in K.B. 36: 537 (1981). Type: Tanzania,
   Bukoba District, Muhutwe, *Haarer* 2033 (K, holo.!, BR, EA, K, iso.!)
   T. grandiflorum Bullock in K.B. 1933: 148 (1933). Type: Zambia, Mumbwa, *Macaulay* 922 (K,
   holo.!)
   T. oblongifolium Robyns in B.J.B.B. 32: 137 (1962). Type: Zambia, Ndola, Lake Ishiku, *Fanshawe*
   426 (K, holo.!)

NOTE. *Procter* 3764 (Mwanza District, Geita, Samina Forest plots, Nov. 1967) has paired thinner
   leaves and is somewhat intermediate with *T. cinerascens; Hornby* 1007 (Bukoba District, Kakoma, 6
   Mar. 1927) is a very similar form. In Tabora District *T. discolor* merges with *T. cinerascens*, there
   being populations of specimens where the leaves are paired and cuneate at the base although
   retaining the facies of the former. The two seem very different in habit.

5. **T. verticillatum** *Robyns* in B.J.B.B. 32: 141 (1962); Verdc. in K.B. 36: 537 (1981). Type:
Tanzania, Mwanza District, Nyamonge, *Carmichael* 486 (K, holo.!, EA, iso., BR, fragment
and photo.!)

Subshrubby herb 45 cm. tall with slender stems (several or ?single) from a woody
rhizome; stems slender, densely velvety with greyish to pale ferruginous hairs. Leaves in
whorls of 3–4; blades elliptic to oblong-elliptic, 4–7 cm. long, 2–2.8 cm. wide, very shortly
acuminate at the apex, ± rounded at the base, discolorous, adpressed pilose above,
densely velvety grey hairy beneath; petiole 4–5 mm. long; stipules with subtriangular base
1.5 mm. long, connate, with subulate apex 5–6 mm. long, densely grey pubescent. Flowers

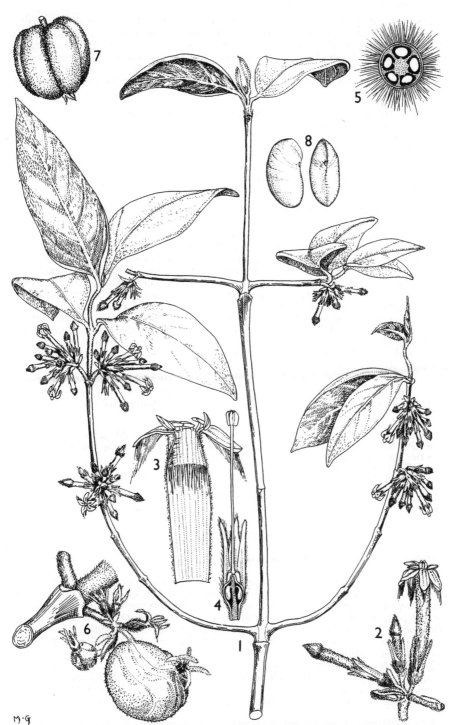

FIG. 139. *TAPIPHYLLUM BURNETTII* — **1**, flowering branch, × 1; **2**, flower and flower buds, × 2; **3**, section through corolla, × 3; **4**, section through ovary with calyx-lobes, style and stigmatic knob, × 3; **5**, transverse section of ovary, × 6; **6**, fruits attached to branch, × 2; **7**, mature fruit, × 2, **8**, pyrene (2 views), × 2. 1–5, from *Richards* 7407; 6–8, from *Haerdi* 445/0. Drawing by Miss M. Grierson, with 8 by Sally Dawson.

in dense shortly pedunculate several-flowered cymes, densely pale ferruginous hairy; peduncles 5–6(–10) mm. long; bracts lanceolate, up to 7 mm. long, 1.5 mm. wide. Calyx densely pale adpressed ferruginous-pubescent; tube 2 mm. long; lobes linear, slightly wider at base, 4–5 mm. long. Buds densely adpressed pubescent, apiculate. Corolla yellowish green; tube subcylindric, 5 mm. long; lobes lanceolate, 4 mm. long, 1 mm. wide, apiculate at the apex. Ovary 5-locular; style 6 mm. long; stigmatic club coroniform, sulcate. Fruit not seen.

TANZANIA. Mwanza District: Geita, Nyamonge, Nov. 1954, *Carmichael* 486!
DISTR. **T** 1; not known elsewhere
HAB. Not known, probably woodland; 1200 m.

NOTE. Perhaps no more than a variety of *T. discolor*.

6. **T. burnettii** *Tennant* in K.B. 19: 280, fig. 1 (1965); Verdc. in K.B. 36: 537 (1981). Type: Tanzania, Ulanga District, Ifakara, Funga, *Haerdi* 445/0 (K, holo.!, BR, EA, iso.!)

Shrub or small tree, (0.6*–)1.5–5 m. tall; branches slender, densely pale ferruginous hairy on young parts but soon glabrescent or glabrous and revealing fissured ± purplish brown bark. Leaves paired; blades oblong to ovate-lanceolate, 1.5–9.5 cm. long, 1–4.4 cm. wide, rounded to acute at the apex, ± rounded at the base, discolorous, green with adpressed ± ferruginous hairs above, densely velvety beneath with silvery to pale ferruginous hairs, the main nervation usually slighter darker ferruginous; petiole 1–2 mm. long; stipules linear-lanceolate, 8 mm. long, deciduous. Flowers in 3–several-flowered cymes, ferruginous pubescent; peduncles 4 mm. long; lower bracts small, upper bracts ± leafy, lanceolate, up to 7 mm. long, 2 mm. wide; pedicels obsolete. Calyx densely pale ferruginous pilose; tube 1.2 mm. long; lobes linear to lanceolate, sometimes slightly spathulate, 2.5–6 mm. long, 0.4–1 mm. wide. Buds capitate, densely ferruginous hairy, acute or apiculate at the apex. Corolla pale green, creamy green or white; tube slender, cylindrical, 1.1–1.25 cm. long; lobes ovate-lanceolate, 3–4 mm. long, 1–2 mm. wide, shortly apiculate. Anthers 1.1–1.5 mm. long, shortly exserted. Style 1.1–1.3 cm. long; stigmatic club coroniform, 1.2 mm. long. Fruits orange-yellow*, subglobose, 1–1.3 cm. diameter, with 2–5 pyrenes, sulcate in dry state, densely ferruginous tomentose with both short and longer hairs; pyrenes reddish brown, up to 9 mm. long, 5.5 mm wide, rugose. Figs. 132/7, p. 754; 139, p. 783.

TANZANIA. Ufipa District: Kawa Falls, 30 Dec. 1956, *Richards* 7404! & Milepa, Dec. 1948, *Burnett* 48/30! & Rukwa Escarpment, Muse Gap, 29 Dec. 1961, *Robinson* 1795!; Rufiji Distict: Utete, Kibiti, 18 Dec. 1968, *Shabani* 247!
DISTR. **T** 4, 5*, 6, 8; Zambia
HAB. Ravines, streams and rocky places, in thickets and woodland; (200–)1250–1500 m.

SYN. *T.sp.* sensu Haerdi in Acta Trop., suppl. 8.: 146(1964)

98. **FADOGIELLA**

Robyns in B.J.B.B. 11: 94 (1928)

Subshrubby herbs or shrubs, completely softly tomentose, mostly much branched. Leaves opposite or less often in whorls of 3, shortly petioled; stipules with triangular or sheathing connate bases and subulate appendages. Flowers fairly small in subsessile or shortly pedunculate, dense, many-flowered cymes. Buds thick, densely tomentose, cylindric or obovoid, obtuse or shortly acuminate at the apex. Calyx-tube globose with a short ± subundulate limb or teeth short, triangular. Corolla mostly yellow or greenish; tube cylindric, glabrous or hairy at the throat, with a ring of deflexed hairs inside; lobes reflexed, obtuse or shortly apiculate. Stamens inserted at the throat, anthers exserted. Ovary (3–)4–5-locular, each locule with a pendulous ovule; style slender, exserted; stigmatic club cylindric or coroniform, sulcate, (3–)4–5-lobed at the apex. Fruit globose, crowned with the calyx-limb, sometimes ribbed, containing (3–)4–5 pyrenes, ± glabrous.

A genus of 2 species mostly in Zaire and Flora Zambesiaca area, one extending into Tanzania, the other known only from the type now destroyed.

* Information from specimens in fruit only and needing confirmation from correlated flowering material.

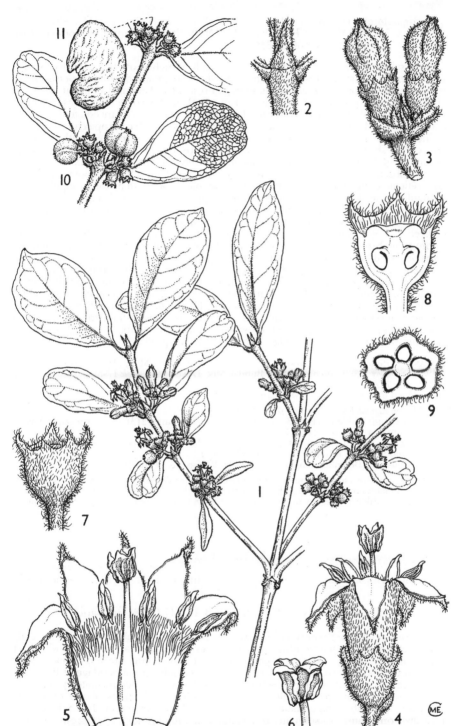

FIG. 140. *FADOGIELLA STIGMATOLOBA* — **1**, flowering branch, × ⅔; **2**, stipule, × 2; **3**, young inflorescence, × 4; **4**, flower, × 5; **5**, corolla opened out, with style and stigmatic knob, × 6; **6**, stigmatic knob, × 10; **7**, calyx, × 6; **8**, longitudinal section through ovary, × 8; **9**, transverse section through ovary, × 8; **10**, fruiting branch, × 1; **11**, pyrene, × 4. 1–9, from *Bullock* 3485; 10, 11, from *Fanshawe* 689. Drawn by Mrs M.E. Church.

F. stigmatoloba (*K. Schum.*) *Robyns* in B.J.B.B. 11: 96 (1928); T.T.C.L.: 494 (1949); F.F.N.R.: 407 (1962); Verdc. in K.B. 42: 125, fig. 1N (1987). Type: Tanzania, Mbeya District, Mbozi Mts., *Goetze* 1428 (B, holo.†, BM, iso.!)

Erect subshrub, 0.5–1.2(–1.8) m. tall, with much-branched often 3–4-angled usually single stem or 2–3 stems from a thick horizontal or vertical woody stock; shoots slender, velvety yellowish tomentose above but soon glabrescent and ultimately dull purplish and glabrous. Leaves predominantly paired or in whorls of 3 at some few nodes bearing 3 side-branches; blades narrowly to broadly elliptic or obovate, or lanceolate, 1.7–8(–10) cm. long, 0.5–4.4(–5.8) cm. wide, narrowed to a rounded or subacute apex, rarely shortly acuneate, cuneate at the base, mostly very discolorous, densely covered with yellowish grey curled short hairs above which do not obscure the surface, ultimately sometimes becoming glabrous, white or yellowish grey velvety tomentose beneath, the surface entirely obscured save in old leaves, venation impressed and ± bullate above and raised and reticulate beneath in mature leaves; petiole 2–3(–6) mm. long; stipules with triangular base, 1–2 mm. long, and subulate appendage, 2.5–3(–6) mm. long, at first tomentose but becoming glabrous. Inflorescences dense and subsessile, the axillary pairs forming dense nodal clusters; peduncles 0–2.5 mm. long, rarely to 1.5 cm. in fruit; pedicels 1–1.5(–5) mm. long, all densely tomentose; bracts and bracteoles scarious, turning dark brown, ovate or narrowly lanceolate, ± 1 mm. long. Buds densely tomentose with adpressed hairs, with 5 short but distinct appendages. Calyx-tube 1–1.5 mm. long, the teeth short, ± triangular, ± scarious, brown, 0.5–1 mm. long, ± glabrous or pubescent. Corolla green or yellow; tube 3–5 mm. long, densely tomentose outside, shortly hairy at throat; lobes linear-triangular, 4–5 mm. long, ± 2 mm. wide, apiculate, with similar indumentum to tube outside. Anthers exserted, 1.5–2 mm. long. Style 6–6.5 mm. long; stigmatic club pale green, subcylindrical, 0.75–1.5 mm. long, 3–5-lobed at apex. Fruits globose, 0.8–1.2(–1.5) cm. diameter, crowned by tomentose limb, shiny, glabrescent or glabrous, with (1–)3–5 pyrenes. Figs. 132/8, p. 754; 140, p. 785.

TANZANIA.   Ufipa District: Lake Kwela, 9 Nov. 1950, *Bullock* 3485!; Mbeya District: Mbozi, 29 Aug. 1933, *Greenway* 3633!; Rungwe District: Bundali Mts., 17 Nov. 1912, *Stolz* 1795!
DISTR.   T 4, 7; Zaire, Malawi and Zambia
HAB.   Open *Brachystegia, Uapaca* woodland; 1530–1700 m.

SYN.   *Fadogia stigmatoloba* K. Schum. in E.J. 30: 414 (1901)
         F. manikensis De Wild. in F.R. 13: 139 (1914); Contr. Fl. Katanga: 213 (1921). Type: Zaire, Shaba, Katentania. *Homblé* 722 (BR, holo.!)
         Fadogiella verticillata Robyns in B.J.B.B. 11: 95 (1928). Type: Zaire, Shaba, Moba [Baudouinville], Kirungu, *De Beerst* 4 (BR, holo.!)
         F. manikensis (De Wild.) Robyns in B.J.B.B. 11: 97, figs. 13, 14 (1928)

NOTE.   The nodes on some specimens of *Stolz* 1795 and other sheets have three axillary branchlets and presumably bore 3 leaves in a whorl but Robyns uses this character to separate off *F. verticillata* the type of which has leaves in 3's with blades 8 × 5.8 cm.; it is not distinguishable from *F. stigmatoloba*.

99. FADOGIA

Schweinf., Reliq. Kotschy.: 47, t. 32 (1868); Robyns in B.J.B.B. 11: 41 (1928); Verdc. in K.B. 36: 515 (1981)

Temnocalyx sensu Robyns in B.J.B.B. 11: 317 (1928), pro parte

Glabrous or hairy herbs or subshrubs; rarely small shrubs, shrubs or small trees usually with several stems from a woody rootstock. Leaves paired or nearly always in whorls of 3–5(–6), mostly subsessile or shortly petiolate; stipules small, connate into a sheath, hairy inside at the base, often persistent, produced into a subulate appendage at the apex. Inflorescences axillary, simple, few-flowered, usually shortly pedunculate cymes or in some species flowers solitary; flowers usually rather small; buds obtuse or shortly apiculate. Calyx-tube campanulate or subglobose, the limb truncate, toothed or lobed, the 5–10 divisions obtuse to subulate. Corolla-tube usually cylindrical, glabrous or hairy outside, with a ring of deflexed hairs inside, pubescent to densely hairy above it and densely hairy or less often glabrous at the throat; lobes 5(–9), acute to apiculate at the apex, papillate on inner surface. Anthers usually oblong, shortly exserted; filaments short. Ovary 3–4(–9)-locular; ovules solitary in each locule, pendulous; style slender or rather thick, usually thicker towards the base, not or scarcely exceeding the anthers; stigmatic club coroniform or rarely cylindrical, sulcate, 3–5(–9)-lobed. Disk usually depressed. Fruits small, often bearing the remains of the calyx-limb, glabrous, containing 1–3(–9) pyrenes.

An African genus of about 40–50 closely related species, 14 of which occur in the Flora area. It is rather well characterised in habit but I have had to merge part of Robyns' *Temnocalyx* with it. There is too much variation in the curvature of the corolla for this to be a useful character; nevertheless the species I have chosen as lectotype of Robyns' genus, *T. nodulosus*, differs markedly from the others he included in it by its distinctive habit, nodose stems, uniformly paired leaves, characteristic corolla lacking deflexed hairs inside and structure of the stigmatic club. The genus *Temnocalyx* has been maintained for this species. The species with larger red flowers are probably bird-pollinated. Several are said to have edible fruits.

1. Leaves densely velvety or papillate beneath, the indumentum ± hiding the lower surface which is markedly discolorous 2
   Leaves glabrous to pubescent beneath but not so densely velvety or papillate as to cover or almost cover the entire lower surface and render blades markedly discolorous; any different colour due entirely to the surfaces themselves (rarely flowering when leafless) 5
2. Indumentum of lower leaf-surface entirely consisting of very small dense white papillae; upper leaf-surface green and glabrous; corolla-tube 3–4 mm. long; calyx-lobes triangular-lanceolate, 2–3.5 mm. long . . . 2. *F. homblei*
   Indumentum of hairs or a mixture of hairs and papillae 3
3. Indumentum of undersides of leaves consisting of a mixture of coarse hairs and an understorey of papillae*; corolla-tube 2.5–3.5 mm. long; calyx-lobes triangular-subulate, 1–2.5 mm. long . . . . . . 1. *F. cienkowskii*
   Indumentum of undersides of leaves consisting of hairs only or with only very few papillae 4
4. Stems short and slender, 15–25 cm. tall; leaves with very fine greyish white indumentum beneath ± hiding surface; corolla-tube 3.5–6 mm. long; calyx-lobes 0.5–1 mm. long . . . . 5. *F. arenicola*
   Stems longer and more robust, to 90 cm. tall; leaves with coarser yellow-grey to ferruginous woolly-velvety indumentum beneath; corolla-tube 4 mm. long; calyx-lobes 1–2 mm. long . . . . 3. *F. elskensii*
5. Calyx-limb truncate or with minute teeth under 0.5 mm. long 6
   Calyx-limb with distinct teeth 14
6. Corolla-tube longer, 0.9–2.8 cm. long, sometimes slightly to distinctly curved 7
   Corolla-tube shorter, straight 9
7. Corolla-tube straight, robust, 1.4–2.7 cm. long, 4–8 mm. wide; leaf-blades 4.5–15 × 1.8–9 cm.; stems and leaves glabrous; flowers usually 6–9-merous . . . . . 14. *F. fuchsioides*
   Corolla-tube ± curved, often more slender; flowers 5-merous 8
8. Corolla-tube short, 0.9–1.4 cm. long, slightly curved; stems pubescent; leaves ± glabrous to densely pubescent   12. *F. verdcourtii*
   Corolla-tube longer 1.8–2.8 cm. long, distinctly curved   13. *F. ancylantha*
9. Plant glabrous save for insides of stipules and inside of corolla-tube 10
   Plant with stems and leaves ± hairy 12

* Best seen under low power (two-thirds) of an ordinary student's compound microscope, particularly after scraping the surface in cases where hairs are also present.

10. Inflorescences borne at the lower nodes of shoots mostly
with reduced leaves; calyx-limb mostly erose or with
short lobes to 0.7 mm. long, but sometimes ± truncate;
corolla-tube 3–4 mm. long; usually short geophyte 6–40
cm. tall with numerous shoots from woody stock; leaves
mostly rounded at apex (**T** 4, 7, 8) . . . .   7. *F. stenophylla*
    Inflorescences more evenly spaced with some from at least
    middle and often the upper nodes bearing normal
    leaves; calyx-limb truncate; leaves mostly acute to
    acuminate at the apex 11
11. Plants from W. Nile District of Uganda with leaves drying
distinctly discolorous, the venation paler above
contrasting with the dark surface and darker beneath
contrasting with the ± glaucous brownish lower
surface; corolla-tube 2.5–4.5 mm. long; plants probably
with few shoots from rootstock . . . .   8. *F. glaberrima*
    Plants from Tanzania; leaves usually not so discolorous;
    corolla-tube often usually distinctly longer, 3–6 mm.
    long; plants with many shoots from a woody base   9. *F. triphylla*
12. Lower inflorescences at least much exceeding the ovate
densely hairy leaves . . . .   10. *F. sp. A**
    Inflorescences shorter than the leaves 13
13. Corolla-tube 1.5–2 mm. long; buds densely spreading
hairy; subshrub ± 30 cm. tall . . . .   11. *F. vollesenii*
    Corolla-tube longer, up to 6 mm. long; buds glabrous or
    nearly so . . . .   9. *F. triphylla*
                                                        (variants)

14. Corolla-tube 2.5 cm. long; calyx-lobes oblong-triangular,
7–8(–10) mm. long; shrub or small tree with paired
leaves, nodose stems and corolla lacking internal ring
of hairs . . . .   *Temnocalyx nodulosus*
                                                        (see p.761)
    Corolla-tube under 1 cm. long 15
15. Inflorescences borne at the lower nodes . . . . .   7. *F. stenophylla*
    Inflorescences not restricted to lower nodes, usually some
    from middle and often upper nodes 16
16. Plant with short shoots 20 cm. long, ± glabrous save for
lower stems and stipules; underside of leaf-blades
without characteristic pattern of stomata**; calyx-lobes
0.7–1 mm. long . . . .   6. *F. sp. B*
    Plant with longer shoots or if short (in one variant of *F.
    elskensii*) then leaf-blades ± densely hairy beneath; leaf-
    blades ± densely hairy or glabrous to sparsely
    pubescent beneath but then with characteristic pattern
    of stomata** 17
17. Leaf-blades ± densely hairy beneath . . . .   3. *F. elskensii*
    Leaf-blades glabrous and ± glaucous beneath or with very
    sparse hairs, the surface with a characteristic pattern of
    stomata . . . .   4. *F. tetraquetra*

1. **F. cienkowskii** *Schweinf.*, Reliq. Kotschy.: 47, t. 32 (1868); Hiern in F.T.A. 3: 154 (1877);
Robyns in B.J.B.B. 11: 79 (1928); Aubrév., Fl. For. Soud.-Guin.: 480, t. 106/4–5 (1950);
Hepper in F.W.T.A., ed. 2, 2: 178 (1963); Verdc. in K.B. 36: 517 (1981) & in K.B. 42: 125, fig.
1/L, fig. 5/B (1987). Type: Ethiopia, Fesoglu, near Fadoga, *Cienkowski* 159 (W, holo.)

Subshrubby herb with few to several(–15) unbranched reddish brown or yellowish
green stems 0.3–1.2 m. tall from the apical parts of a mostly branched woody rhizome;
stems covered with dense pale ferruginous hairs or less often glabrous. Leaves in whorls

* The single specimen seen is pathological and even the length of the inflorescences may be
abnormal.
** Observable under two-thirds objective of student's compound microscope.

of 3–4; blades narrowly to broadly elliptic or oblong-elliptic, to narrowly lanceolate, 2–8.5 cm. long, 0.5–4.5 cm. wide, ± rounded to acute or very shortly acuminate at the apex, broadly to narrowly cuneate or ± rounded at the base, markedly discolorous, often ± bullate, with sparse to fairly dense hairs above which do not obscure the ± glossy surface, very densely continuously velvety tomentose beneath with grey to pale ferruginous matted hairs beneath which is a dense covering of short white papilla-like hairs; petioles short and stout, 1–1.5 mm. long, densely pubescent; stipules 6–9 mm. long, shortly joined at the base, subulate from an oblong-triangular base, pubescent. Inflorescences 2–6-flowered or flowers sometimes solitary; peduncles 3–9(–12) mm. long; pedicels 1.5–9(–12) mm. long, pubescent. Calyx-tube 1–1.5 mm. long; limb with 7–10 triangular to triangular-lanceolate or subulate often membranous usually pubescent teeth (1–)1.5–2.5 mm. long. Buds distinctly apiculate, with as many subulate tails as corolla-lobes, usually densely rather coarsely pubescent. Corolla yellow or cream; tube cylindrical, 2.5–3.5 mm. long, glabrescent to pubescent outside; lobes narrowly lanceolate, ± 3.5–5.5 mm. long including the subulate appendage (0.5–)1.5–2 mm. long. Ovary 3–4-locular; style up to 5–7.5 mm. long; stigma yellow-green or whitish, coroniform, 0.75 mm. long, sulcate, 3–4-lobed. Fruit dark green becoming black and shining, subglobose, up to 1 cm. diameter, with 1–3 pyrenes, when reduced to 1 then oblique, crowned with calyx-lobes.

var. **cienkowskii**; Verdc. in K.B. 36: 517 (1981)

Leaf-blades up to 4.5 cm. wide.

UGANDA. W. Nile District: Logiri, Mar. 1935, *Eggeling* 1870!; Acholi District: Imatong Mts., Langia, Apr. 1943, *Purseglove* 1396!; Karamoja District: Lonyili Mt., Apr. 1960, *J. Wilson* 892!
KENYA. W. Suk District: Kacheliba Escarpment, May 1932, *Napier* 2005!; Trans-Nzoia District: Kitale, Apr. 1964, *Tweedie* 2794! & SE. Cherangani Hills, 12 May 1957, *Symes* 50!
TANZANIA. Kahama District: 64 km. S. of Nungwe, 5 Jan. 1937, *Morgan* 80!
DISTR. U 1; K 2, 3; T 4; W. Africa from Mali to Nigeria, Cameroon, Zaire, Sudan, Ethiopia, Zambia, Malawi, Zimbabwe and Angola, ? South Africa (Transvaal)
HAB. Grassland including upland grassland, seasonally wet grassland, grassland subject to regular burning, wooded grassland particularly with *Protea* or *Combretum, Uapaca* woodland and open *Julbernardia, Brachystegia* woodland, also on rocky slopes and in bamboo zone on some mountains in U 1; 1050–2100 m.

SYN. *Vangueria tristis* K. Schum. in De Wild., Etudes Fl. Katanga 1: 227 (1903) & Contr. Fl. Katanga: 211 (1921). Type: Zaire, Shaba, Pweto, *Verdick* (BR, holo.!)
Fadogia katangensis De Wild. in F.R. 12: 295 (1913); Robyns in B.J.B.B. 11: 77 (1928). Type: Zaire, Shaba, Lubumbashi, *Hock* (BR, holo.!)
F. tristis (K. Schum.) Robyns in B.J.B.B. 11: 76 (1928)
Canthium cienkowskii (Schweinf.) Roberty in Bull. I.F.A.N., A, 16: 61 (1954)

NOTE. *F. cienkowskii* has been much misidentified particularly in Tanzania and further south; almost any plant with thick indumentum has been called by this name but true *F. cienkowskii* has a thick layer of papilla-like hairs beneath the long woolly layer. Some specimens particularly from Cameroon have very few long hairs or hairs restricted to the venation and are intermediate with *F. pobeguinii* Pobég., which basically is *F. cienkowskii* lacking the thick indumentum.

var. **lanceolata** *Robyns* in B.J.B.B. 11: 81 (1928); Verdc. in K.B. 36: 518 (1981). Types: Cameroon, Bosum near Balawala, *Tessmann* 2281 & between Babessi and Balrongo, Babingo, *Ledermann* 5800 & without exact locality, *Elbert* 472 & Zaire, Kimbundo, *Pogge* 406 (B, syn.†)

Leaf-blades very narrow up to ± 1 cm. wide.

TANZANIA. Iringa District: Mufindi, Sao Hill, Sudi Shabani, Nov. 1984, *Procter* 2722!
DISTR. T 7; Zaire and Cameroon
HAB. *Brachystegia* woodland; 1740 m.

NOTE. I have seen no authentic material of this variety and the determination needs confirmation.

2. **F. homblei** *De Wild.* in F.R. 12: 295 (1913) & Notes Fl. Kat. 3: 27 (1913) & Contr. Fl. Katanga: 213 (1921); Robyns in B.J.B.B. 11: 72 (1928); Verdc. in K.B. 36: 518 (1981). Type: Zaire, Shaba, Shinsenda, *Homblé* 555 (BR, holo.!)

Erect subshrubby herb 0.3–1.2 m. tall, with 3–4–many stems from a slender rhizome; stems often reddish, 3–4-angled, glabrous or rarely very sparsely pubescent. Leaves in whorls of 3–5; blades narrowly elliptic or oblong-lanceolate to lanceolate, 2.2–12 cm. long, 0.3–3.8 cm. wide, tapering to a fine acute apex, cuneate at the base, markedly discolorous, green and glabrous above, drying whitish tomentose beneath due to a very dense covering of minute papilla-like hairs, pale green in life; petiole 1–2.5 mm. long; stipules joined to form a sheath 1.5–3 mm. long with 3 filiform appendages 5–8 mm. long.

Inflorescences 2–5-flowered, glabrous; peduncle 3–4 mm. long; pedicels 1.5–6 mm. long. Calyx-tube 1.2–1.5 mm. long, the limb divided into triangular to lanceolate teeth 1–3.5 mm. long. Buds distinctly 5–6-apiculate at the apex, the appendages diverging. Corolla yellowish or bright yellow; tube 3–4(–5) mm. long, with ring of deflexed hairs inside; lobes narrowly triangular or oblong-triangular, 4–5.5 mm. long, 2 mm. wide, drawn out at the apex into a subulate appendage which forms 1–1.5 mm. of the total length. Ovary 3–4-locular; style rather stout, ± 7 mm. long; stigmatic club coroniform, 1–1.5 mm. long, 3–4-lobed. Fruit black, often oblique, 6–10(–13) mm. long, usually with 2–3 pyrenes, crowned with the persistent calyx-limb.

TANZANIA. Mpanda District: 64 km. on Ikola–Mpanda road, 8 Nov. 1959, *Richards* 11736!; Dodoma District: Rungwa Game Reserve, Sulangi, 9 km. W. of Bagamoyo, near Issaua R., 28. Jan. 1969, *Yohannes* in CAWM 5037!; Iringa District: Sao Hill, 8 Jan. 1975, *Brummitt & Polhill* 13625!
DISTR. T 4, 5, 7; Zaire, Mozambique (fide A.E. Gonçalves), Malawi, Zambia, Zimbabwe, Angola and South Africa (Transvaal)
HAB. Grassland, grassland tree clump margins, *Brachystegia* woodland; 1050–2100 m.

SYN. *F. monticola* Robyns in B.J.B.B. 11: 70, fig. 11, 12 (1928). Type: South Africa, Transvaal, N. Magaliesberg, *Lincke* (BR, holo.!)

NOTE. Reference to a shrub 4 m. tall (*Fewdays* s.n. from Zambia, Livingstone) is presumably an error for 4 ft. *F. oleoides* Robyns described from the Transvaal and having longer peduncles is merely a variant.

3. **F. elskensii** *De Wild.*, Pl. Bequaert. 3: 201 (1925); Verdc. in K.B. 36: 518 (1981). Type: Burundi, Kitega, *Elskens* 237 (BR, holo.!)

Subshrubby herb with several unbranched stems 25–90 cm. tall from a woody rootstock; stems ± 4-angled, densely hairy, particularly above. Leaves in whorls of 3–4; blades elliptic, oblong-elliptic or, less often, oblong-lanceolate, 4–9 cm. long, 1.2–4.6 cm. wide, rounded to acute, or even retuse at the apex, usually apiculate, rounded to broadly cuneate at the base, venation impressed above, almost bullate, and raised beneath, pubescent above but not in any way obscuring the surface, densely woolly velvety hairy beneath with grey, whitish, ochraceous or ± ferruginous hairs usually completely obscuring the surface but sometimes thinner and not obscuring it, very discolorous, completely without an understorey of papillae but with a pattern of stomata; petiole ± 2 mm. long; stipules with base ± 2 mm. long, with linear appendage 4–11 mm. long. Inflorescences sessile or pedunculate 1–5-flowered fascicles; peduncles 1–5 mm. long; pedicels 1–6 mm. long. Calyx-tube glabrous to pubescent, 2 mm. long; limb-tube up to 0.5 mm. long; lobes ovate to linear-lanceolate, 1–2.5 mm. long. Buds apiculate, hairy above. Corolla yellow, greenish yellow or cream; tube ± 4 mm. long, glabrous outside; lobes triangular, 3–5 mm. long, 1.5–2.5 mm. wide, apiculate, ± hairy outside above, the apicula up to 2 mm. long. Style pale green, exserted 2 mm.; stigmatic club whitish, oblong or obconic, 1.5 mm. long. Fruit black, glossy, subglobose, ± 1 cm. diameter, with up to 5 pyrenes; pyrenes segment-shaped, 8 mm. long, 4.5 mm. wide, reticulate-rugose.

var. **elskensii**; Verdc. in K.B. 36: 518 (1981)

Leaves with dense velvety indumentum beneath obscuring surface; shoots to 90 cm. tall. Fig. 141.

TANZANIA. Mbulu District: Pienaar's Heights, 8 Jan. 1962, *Polhill & Paulo* 1108!; Buha District: Kalinzi, 22 Nov. 1962, *Verdcourt* 3414!; Songea, 23 Dec. 1955, *Milne-Redhead & Taylor* 7875!
DISTR. T 1, 2, 4, 6–8; Zaire, Burundi and Malawi
HAB. *Brachystegia*, *Isoberlinia* and *Uapaca* woodland, *Protea* wooded grassland, grassland subject to seasonal burning; 1050–1800 m.

SYN. *Vangueria katangensis* K. Schum. in De Wild., Etudes Fl. Katanga 1: 227 (1903), *non Fadogia katangensis* De Wild. Type: Zaire, Shaba [Katanga], *Verdick* (BR, holo.!)
Fadogia cienkowskii sensu Robyns in B.J.B.B. 11: 80 (1928), pro parte

NOTE. This has been always identified as *F. cienkowskii* and is closely related to that species but the lack of an understorey of papillae renders it easily recognisable; a few intermediate specimens exist outside the Flora area. It is closer to *F. tetraquetra* having similar stomatal pattern.

var. **ufipaensis** *Verdc.* in K.B. 36: 519 (1981). Type: Tanzania, Ufipa District, Mbizi Forest, *Napper* 1030 (K, holo.!, EA, iso.!)

Shoots to 25 cm. tall, distinctly angular. Leaves elliptic to oblong, cuneate to broadly rounded at base, shortly bristly pubescent above and on venation beneath but most of the surface clearly visible; venation raised beneath and strongly closely reticulate.

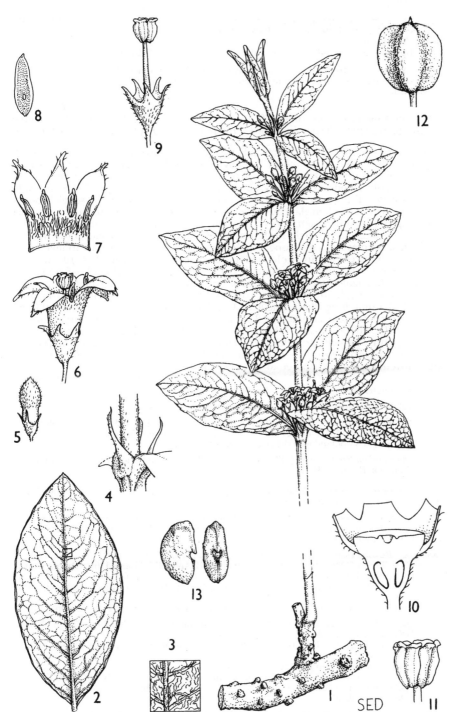

FIG. 141. *FADOGIA ELSKENSII* var. *ELSKENSII* — **1**, habit, × ⅔; **2**, leaf showing lower surface, × ⅔; **3**, detail of lower surface of leaf, × 4; **4** stipule, × 2; **5**, flower bud, × 2; **6**, flower, × 3; **7**, part of corolla opened out, × 4, **8**, dorsal view of anther, × 8; **9**, calyx with style and stigma, × 4; **10**, longitudinal section through ovary, × 8; **11**, stigmatic knob, × 8; **12**, fruit, × 2; **13**, pyrene (2 views), × 2. 1, 7–9, 11, from *Newbold & Jefford* 1707; 2–6, 10, 13, from *Milne-Redhead & Taylor* 7875; 12, from *Lovett* 1269. Drawn by Sally Dawson.

TANZANIA.   Ufipa District: Sumbawanga, 27 Nov. 1954, *Richards* 3443! & Mbizi Forest, 21 Nov. 1958, *Napper* 1030! & Mbizi, 19 Nov. 1949, *Bullock* 1848!
DISTR.   **T** 4; Zaire
HAB.   Upland grassland, ridge grassland and in rocky places; 2100–2400 m.

NOTE.   *Napper* 1030a from the same locality as 1030 has the leaf reticulation less marked and the collector thought it might be a hybrid with another species but I think it is only a minor form of var. *ufipaensis.*

4. **F. tetraquetra** *K. Krause* in E.J. 39: 544 (1907); Robyns in B.J.B.B. 11: 68 (1928); Verdc. in K.B. 36: 519 (1981). Type: Zimbabwe, Mutare [Umtali], *Engler* (1905) 3139 (B, holo.†)

Subshrubby herb 20–90(–120) cm. tall, with up to 7 brown glabrous or very sparsely pubescent to densely hairy mostly 3–4-angled basically unbranched stems from the crowns of a fairly stout rhizome. Leaves in whorls of 3–5(–6), drying distinctly discolorous; blades broadly elliptic to narrowly oblong-elliptic, 4–11 cm. long, 1.5–4.8 cm. wide, acuminate to a fine point at the apex, cuneate to rounded at the base, glabrous to pubescent, paler and often glaucous beneath, the actual surface with distinct appearance under high powers due to denseness of stomata, the main nerves often rather crinkly and undulate perhaps due to slight fleshiness of leaf in life, slightly bullate above in some specimens; petioles 1–2.5 mm. long; stipules connate at the base, the base triangular 3–5 mm. long, pubescent within, prolonged into a subulate appendage (2–)4–10 mm. long. Inflorescences 3–9-flowered; peduncles 2–5(–16) mm. long; pedicels 1.5–5 mm. long; bracts and bracteoles subulate, 3–4 mm. long. Calyx-tube 1.3–1.5 mm. long, often wrinkled when dry; limb with tubular part 0.5–1 mm. long, bearing narrowly triangular teeth 1.5–3.5 mm. long. Buds apiculate. Corolla greenish yellow or honey-coloured, glabrous to ± pubescent outside; tube 2.5–4 mm. long, densely hairy at the throat and with a row of deflexed hairs inside; lobes oblong-triangular, apiculate at the apex, 3.5–5(–7) mm. long. Ovary 3–4-locular; style exserted for up to 1 mm.; stigmatic club coroniform, up to 1.5 mm. long, ribbed and 3–4-lobed. Fruit subglobose or oblique, 6–7 mm. long, usually with 3 pyrenes but often reduced to 1 by abortion.

var. **tetraquetra**; Verdc. in K.B. 36: 521 (1981)

Stems glabrous or sparsely hairy; leaves glabrous. Buds glabrous.

TANZANIA.   Kahama District: 29 km. along the Ushirombo road, 1 Jan. 1936, *B.D. Burtt* 5466!
DISTR.   **T** 4; Zaire, Zambia, Zimbabwe, South Africa (Transvaal) and Swaziland
HAB.   *Brachystegia* woodland with dense grass cover; 1200 m.

SYN.   *F. mucronulata* Robyns in B.J.B.B. 11: 67 (1928). Type: South Africa, Transvaal, Lydenburg, Vaalhoek, *Rogers* 25016 (K, holo.!)

var. **grandiflora** (*Robyns*) *Verdc.* in K.B. 36: 520 (1981). Type: Tanzania, Bukoba, Niamutu–Kusja–Msera, *Braun* in Herb. Amani 5499 (B, holo.†, EA, K, iso.!)

Stems covered with dense crisped hairs. Leaf-blades with scattered short seta-like hairs above and midrib beneath with dense hairs similar to those on the stems or with pubescence ± all over venation beneath. Buds glabrous or hairy.

TANZANIA.   Bukoba District: Kishanda, Oct. 1935, *Gillman* 524!; Biharamulo District: Lusahanga, 15 Oct. 1960, *Tanner* 5322!; Mwanza District: Mbarika, 12 Apr. 1953, *Tanner* 1369!; Tabora District: 36.8 km. from Tabora towards Dodoma, Kigwa, 11 Oct. 1969, *Batty* 676!
DISTR.   **T** 1, 4, 7; Guinée, Ghana, Nigeria, Cameroon, Zambia, Zimbabwe, Angola and South Africa (Transvaal)
HAB.   Deciduous woodland ("light forest"), grassland with scattered trees; 1140–1350 m.

SYN.   *F. glauca* Robyns in B.J.B.B. 11: 62 (1928). Type: Angola, Golungo Alto, Banza, Sobato Bumba, *Welwitsch* 2566 (K, holo.!, BM, iso., BR, fragment!)
        F. variabilis Robyns in B.J.B.B. 11: 63 (1928). Type: Zimbabwe, Chirinda, *Swynnerton* 6502 (K, holo.!, BM, iso.!)
        F. grandiflora Robyns in B.J.B.B. 11: 65 (1928)
        F. dalzielii Robyns in B.J.B.B. 11: 66 (1928). Type: Nigeria, Kontagora, *Dalziel* 218B (K, holo.!)
        [*F. pobeguinii* sensu Hutch. & Dalz., F.W.T.A. 2: 110 (1931) & Hepper in F.W.T.A., ed. 2, 2: 178 (1963), *non* Pobég.]

NOTE.   This variety merges with *F. elskensii* of which it could be considered a variant.

5. **F. arenicola** *K. Schum. & K. Krause* in E.J. 39: 544 (1907); Robyns in B.J.B.B. 11: 84 (1928); Verdc. in K.B. 36: 520 (1981). Type: Tanzania, Songea District, Mironji Camp, *Busse* 980 (B, holo.†, EA, iso.!)

Herb ± 10–25 cm. tall, with several rather slender unbranched stems from a creeping slender woody rhizome; stems pale to dark reddish brown, densely pubescent with short white spreading hairs. Leaves in whorls of 3; blades elliptic to oblong-obovate, 1.7–6 cm. long, 0.8–2.5 cm. wide, obtuse to bluntly acute or even slightly retuse at the apex, cuneate at the base, with dense grey pubescence above which does not hide the leaf-surface or only on venation, usually densely grey velvety beneath rendering the leaves markedly discolorous; lateral nerves 4–6; petioles 1.5–3.5 mm. long; stipules ± 1.5–2 mm. long. Inflorescences 1–4-flowered; peduncle short, up to 2 mm. long; pedicels 2–6 mm. long, densely grey-pubescent, mostly reflexed in fruit. Calyx-tube and limb together ± 1.5 mm. long, pubescent; teeth triangular, 0.5–1 mm. long. Buds rather densely pilose above, obtuse. Corolla-tube greenish yellow or yellow, ± 3.5–6 mm. long, hairy outside near apex, with a ring of deflexed hairs inside; lobes bright yellow, narrowly oblong, spreading, ± 4–5 mm. long, 1.3–2 mm. wide, acute, densely hairy outside. Ovary 3-locular; style 5–8 mm. long; stigmatic club green, coroniform, 2–3-lobed, the lobes 4–6-ribbed, the ribs with transverse grooves. Fruits globose, 5–7 mm. diameter, those seen with only 1 pyrene developed, pubescent.

TANZANIA. Kigoma, Livingstone College, 21 Nov. 1962, *Verdcourt* 3389!; Songea District: Mironji Camp, 5 Feb. 1901, *Busse* 980! & Songea airfield, 2 Feb. 1956, *Milne-Redhead & Taylor* 8582! & 8582A!
DISTR. **T** 4, 8; Zaire, Malawi and Zambia
HAB. Cleared *Brachystegia, Uapaca* woodland on sand, coarse grassland with *Diplorhynchus* on stony sandy hillsides; 750–1020 m.
NOTE. In Zaire (Shaba) specimens occur with the leaf-blades glabrous or glabrescent beneath which seem to be only a variety of this species. *F. schumanniana* Robyns (type: Zaire, Shaba, Lukafu, *Verdick* 350 (BR, holo.!); = *Vangueria verdickii* K. Schum. in De Wild., 1903, *non Fadogia verdickii* De Wild. & Th. Dur.) is probably the same species differing only in more pointed buds due to longer corolla-lobe appendages.

6. **F. sp. B**; Verdc. in K.B. 36: 521 (1981)

Subshrubby herb with several slender flexuous yellowish mostly glabrous or sparsely spreading hairy stems to 20 cm. tall. Leaves in whorls of 3, markedly discolorous on drying; lowest nodes with scale- or very reduced leaves; blades elliptic or obovate-elliptic, 1.5–5 cm. long, 1.5–2.4 cm. wide, obtuse to usually sharply acute or acuminate at the apex, cuneate at the base, drying much paler beneath, glabrous save for a very few scattered seta-like hairs on the venation beneath; petiole 1–2 mm. long. Flowers in 1(–2)-flowered cymes; peduncle 2–3 mm. long; pedicels 3–4 mm. long. Calyx glabrous; tube 1.5 mm. long; limb 0.5 mm. long; lobes linear-triangular to subulate, 0.7–1 mm. long. Corolla colour not stated, probably white; tube 5.5 mm. long; lobes oblong, 5 mm. long, 1.5–2 mm. wide, apiculate. Style 7.5 mm. long; stigmatic club 1.5 mm. long. Fruit not known.

TANZANIA. Tabora District: Ngulu, Goweko–Igalula, 16 Jan. 1926, *Peter* 35073!
DISTR. **T** 4; not known elsewhere
HAB. Not stated, presumably open woodland; 1180 m.
NOTE. This has been confused with *F. triphylla* var. *peteri* but differs in the more discolorous almost glabrous leaves and lobed calyx.

7. **F. stenophylla** *Hiern* in F.T.A. 3: 155 (1877) & Cat. Afr. Pl. Welw. 1: 483 (1898); K. Schum. in Warb., Kunene-Sambesi-Exped.: 390 (1903); Robyns in B.J.B.B. 11: 60 (1928); Verdc. in K.B. 36: 521 (1981). Type: Angola, Huila, R. Lopollo, right bank between Lopollo and Catumba, *Welwitsch* 2570 (LISU, holo., BM, K, iso.!)

Pyrophytic usually glabrous subshrubby herb, with single to mostly several rather short unbranched stems 6–40 cm. tall from a slender creeping woody rhizome; stems glabrous or with rather sparse bristly hairs in a very few specimens; lowest nodes usually leafless. Leaves in whorls of 3–4; blades rounded-elliptic to narrowly elliptic, narrowly oblong or oblanceolate, 1–6.5 cm. long, 0.4–2.7 cm. wide, rounded to bluntly acute at the apex, cuneate to almost rounded at the base, glabrous, usually drying pale olive-green; petioles 1–2 mm. long; stipules oblong-triangular, with a shortly to distinctly produced apex, together 3–4 mm. long, the basal parts connate. Inflorescences (1–)3-flowered, sweet-scented, borne at lower nodes; sometimes flowering in leafless state; peduncles 0.5–2(–3) cm. long and pedicels 1.5–6.5(–10) mm. long, both glabrous or sparsely pubescent, or solitary flowers on stalks up to 1.6 cm. long. Calyx-tube 2–2.5 mm. long, the narrow limb almost membranous, truncate, erose or slightly to distinctly narrowly toothed, teeth 0.7–1.2 mm. long. Buds green, obtuse, glabrous. Corolla white or pale to bright yellow or greenish yellow; tube cylindrical-campanulate, 3–4 mm. long, with ring of deflexed hairs inside; lobes 4–6(–7), triangular-oblong or ovate, 3–5 mm. long, 2.5–2.8 mm. wide, cucullate but scarcely apiculate at the apex, glabrous save for papillate margins and inner surface. Ovary 3–4-locular; style 5.5–6 mm. long; stigma green, coroniform, 3–4-lobed. Fruit red at maturity, globose or often drying very distinctly 2–3-lobed, 7–9 mm. long, 9–10 mm. wide.

subsp. **odorata** (*K. Krause*) *Verdc.* in K.B. 36: 521 (1981). Type: Angola, Humpata, *Fritsch* 249 (B, holo.†)

Leaf-blades rounded-elliptic, narrowly elliptic, narrowly oblong or oblanceolate, up to 2.7 cm. wide. Figs 132/9, p. 754 & 142.

TANZANIA. Buha District: just outside Kibondo on Biharamulo road, 16 July 1960, *Verdcourt* 2870!; Ufipa District: Nsanga Forest, 8 Aug. 1960, *Richards* 13033!; Iringa District: E. of Sao Hill, 2 Aug. 1933, *Greenway* 3455!; Songea District: near Mpapa, Ndenga [Ndengo], 29 Sept. 1956, *Semsei* 2495!
DISTR. T 4, 7, 8; Zaire, Burundi, Mozambique, Malawi, Zambia, Zimbabwe and Angola
HAB. Grassland, particularly on stony hillsides, submontane grassland, open *Brachystegia* woodland, particularly associations which suffer severe burning; 1350–2400 m.

SYN. *F. stenophylla* Hiern var. *rhodesiana* S. Moore in J.B. 40: 253 (1902). Type: Zimbabwe, Harare [Salisbury], *Rand* 629 (BM, holo.!)
F. odorata K. Krause in E.J. 43: 144 (1909); Robyns in B.J.B.B. 11: 52 (1928)
F. stolzii K. Krause in E.J. 48: 415 (1912); Robyns in B.J.B.B. 11: 51 (1928). Type: Tanzania, Rungwe District, Bundali Mts., *Stolz* 108 (B, holo.†, BM, K, iso.!)

NOTE. There is no doubt that S. Moore was correct in considering the narrow-leaved and wide-leaved plants as variants of one species but there does seem to be a geographical basis, the narrow-leaved ones being confined to Angola. Robyns' bud distinctions do not work. There is considerable variation in the development of the calyx-lobes; the most developed seen have been in *Batty* 828 (112 km. from Iringa on the Mbeya road, 6 Nov. 1969) in which the lobes attain 1.2 mm. in length. Specimens with pubescent stems and leaves have been seen from outside the Flora area.

8. **F. glaberrima** *Hiern* in F.T.A. 3: 155 (1877); A. Chev., Etudes Fl. Afr. Centr. Fr.: 157 (1913); Robyns in B.J.B.B. 11: 50 (1928); F.P.S. 2: 434 (1952); Verdc. in K.B. 36: 521 (1981). Type: Sudan, Seriba Ghattas, *Schweinfurth* 1342 (K, holo.!, BM, iso.!)

Glabrous herb or subshrubby herb with ± angular branched pale stems 0.3–1.8 m. tall, presumably several from a woody stock. Leaves in whorls of 3; blades distinctly but not very markedly discolorous on drying, elliptic, oblong-elliptic, elliptic-lanceolate or somewhat obovate-elliptic, 3–10(–11.5) cm. long, 0.7–2.8(–3.5) cm. wide, narrowly rounded at the apex, cuneate at the base, drying brown above with the venation, particularly the midrib, pale and raised, somewhat glaucous beneath with the venation usually darker and contrasting; petiole 1–2 mm. long; stipule-bases broad, connate, 1–1.5 mm. long, with narrow subulate apex about equal twice the length of the base. Inflorescences glaucous, 3–6-flowered; peduncle 0.5–1.6(–3.5) cm. long; pedicels 0.2–1.1(–2) cm. long. Calyx-tube 1.2–2.5 mm. long, the limb reduced to a ± truncate rim. Buds obtuse to acuminate. Corolla greenish white to cream; tube cylindrical-campanulate, 2.5–4.5 mm. long; lobes oblong to triangular, 3 mm. long, 1.8 mm. wide, papillate-puberulous along the margins, acute to apiculate. Ovary 2–4-locular; style 5–6 mm. long; stigmatic club cylindric or coroniform, 1 mm. long, 2–3-lobed. Fruits globose or obliquely ellipsoid, 6–9 mm. diameter, or 8 mm. long, 6.5 mm. wide, with 3 pyrenes or, very often, only one developed.

UGANDA. W. Nile District: Terego, Mar. 1938, *Hazel* 453! & N. of Arua, 16 Mar. 1945, *Greenway & Eggeling* 7215! & E. Madi, Pakelle [Pakelli], May 1932, *Hancock* in F.D. 762! (see note)
DISTR. U 1; Cameroon, Sudan and ? Central African Republic (no specimens seen)
HAB. Open woodland, particularly *Terminalia, Lophira* "savanna"; 600–1350 m.

SYN. *F. ledermannii* K. Krause in E.J. 48: 416 (1912); Robyns in B.J.B.B. 11: 59 (1928); Hepper in F.W.T.A., ed. 2, 2: 178 (1963). Lectotype, see Robyns (1928): N. Cameroon, Pass Tchape, *Ledermann* 2661 (B, lecto.†)

NOTE. The Ugandan specimens have usually been shared between the names *F. glaberrima* and *F. ledermannii*. Robyns kept these quite widely separated because of supposed differences in bud apiculation (and thus corolla-lobes) but this seems very variable. The type of *F. glaberrima* differs from all other material seen in having relatively shorter more broadly elliptic leaves but must be conspecific with *Greenway & Eggeling* 7215. Ugandan material formerly referred to *F. ledermannii*

FIG. 142. *FADOGIA STENOPHYLLA* subsp. *ODORATA* — **1**, habit, × ⅔; **2**, stipule, × 4; **3**, inflorescence, × 3; **4**, corolla opened out, with style and stigmatic knob, × 6; **5**, stigmatic knob, × 8; **6**, calyx, × 6; **7**, longitudinal section of ovary, × 8; **8**, transverse section of ovary, × 8; **9**, fruits, × 1; **10**, pyrene, × 4. 1, 2, from *Stolz* 108; 3–8, from *Watermeyer* 138; 9, 10, from *Paulo* 235. Drawn by Mrs M.E. Church.

certainly matches at least some Cameroon material named by Krause himself (e.g. *Tessmann* 2268!) but other Cameroon material has wider leaves which dry a rich rusty-brown beneath. A few Cameroon specimens have the stems hairy near the base. *Hancock* in *F.D.* 762 is stated to be a "small tree" but there must be some error, possibly of labelling; the leaves in this specimen attain 11.5 × 3.5 cm., the peduncles 3.5 cm. and the pedicels 1–2 cm. Material formerly called *F. ledermannii* all has rather small leaves, e.g. W. Nile District, Mt. Kei Forest Reserve, 25 Feb. 1955, *Dale* U866! & Arua Hill, Mar. 1938, *Hazel* 452! & Koboko [Kobboko], Feb. 1934, *Eggeling* 1494!; Acholi District, W. of Kitgum on Gulu road, Mar. 1934, *Tothill* 2487! Study in the field is needed to confirm my conclusion that there is but one variable species.

9. **F. triphylla** *Baker* in K.B. 1895: 68 (1895); Robyns in B.J.B.B. 11: 57 (1928); Verdc. in K.B. 36: 522 (1981). Type: Zambia, Fwambo, *Carson* 43 (K, holo.!)

Very variable herb or subshrub (10–)22–45(–120) cm. tall, with several branched or unbranched usually erect stems from a thick or fairly slender usually horizontal woody stock; stems 3–4-angled, grooved, glabrous to densely shortly pubescent. Leaves in whorls of 3–4 or opposite above; blades broadly to narrowly elliptic or ovate to ± obovate, (1.7–)3–8.5 cm. long, (0.6–)1.2–4.8(–5.7) cm. wide, acute to obtuse at the apex or very shortly acuminate, cuneate, not markedly discolorous, but usually paler beneath and often drying with a reticulate darker pattern, glabrous to shortly densely pubescent; lowermost leaves often very reduced, 1 × 0.6 cm.; petioles obsolete or very short; stipules ± 2.5–3.5(–4.5) mm. long, the subulate cusp (1.5–)2–3 mm. long from a broad base. Inflorescences (1–)3–7-flowered; peduncles (0.2–)0.5–1.8(–3.2) cm. long; pedicels 0.2–1.4 cm. long, glabrous or shortly pubescent. Flowers fragrant. Calyx oblong-ovoid or obconic, 2.5–4 mm. long including the narrow truncate or almost truncate limb; teeth if present minute, glabrous or shortly pubescent. Buds apiculate and usually tailed, glabrous. Corolla white to greenish yellow or yellow outside, creamy green inside; tube 5–6 mm. long, with ring of deflexed hairs inside; lobes oblong-triangular, 4.5–7 mm. long, 1.5–2(–3.5) mm. wide, papillate inside, appendaged. Anthers half-exserted, with papillate apical appendage, purple, brown or orange. Style green or white, thickened below, 7–8 mm. long; stigmatic club green, coroniform or truncate-obconic, 1–1.5 mm. long. Fruits black, ± 1 cm. in diameter, usually with 3 pyrenes.

KEY TO INFRASPECIFIC VARIANTS

1. Flowers mostly solitary or inflorescence 2–3-flowered, the peduncle usually short or suppressed; shoots mostly short, with only 2–4 leafy nodes, 10–40 cm. tall; stems and leaves glabrescent to shortly pubescent   .   .   .   e. var. **peteri**
   Flowers usually in pedunculate 3–7-flowered umbel-like cymes; shoots often much longer, with up to 9 or more leafy nodes, 10–120 cm. tall   2
2. Plant glabrous save for inside of corolla and stipule-sheath   3
   Stems and/or leaves pubescent   4
3. Leaves broadly to narrowly elliptic or ovate to ± obovate, up to 8.5 cm. long, 5.7 cm. wide, usually over 2 cm. wide   a. var. **triphylla**
   Leaves narrowly elliptic, mostly smaller and narrower, 1.7–5 × 0.8–1.8 cm.   .   .   .   .   c. var. **gracilifolia**
4. Stems, leaves and inflorescences shortly pubescent   b. var. **giorgii**
   Stems and sometimes inflorescences sparsely to densely pubescent; leaves glabrous   .   .   .   .   d. var. **pubicaulis**

a. var. **triphylla**; Verdc. in K.B. 36: 522 (1981)

Plant glabrous save for inside of corolla-tube and stipule-sheath. Leaves often larger, up to 8.5 × 5.7 cm. Stems mostly with over 4 leafy nodes. Inflorescences usually pedunculate several-flowered cymes.

TANZANIA. Iringa District: Sao Hill, 8 Jan. 1975, *Brummitt & Polhill* 13636!; Njombe, 29 Nov. 1931, *Lynes* D 52!; Songea District: ± 1.5 km. S. of Gumbiro, 24 Jan. 1956, *Milne-Redhead & Taylor* 8518!
DISTR. T 1, 7, 8; Cameroon, Zaire, Malawi, Zambia, Mozambique and Angola
HAB. Grassland, *Brachystegia* woodland; 910–2100 m.

SYN. *F. kaessneri* S. Moore in J.B. 49: 152 (1911); Robyns in B.J.B.B. 11: 57 (1928). Type: Zambia, R. Malanguala, *Kassner* 2064 (BM, holo.!)

F. hockii De Wild. in F.R. 11: 535 (1913) & Etudes Fl. Katanga 2: 152 (1913) & Contr. Fl. Katanga:
   213 (1921) & Pl. Bequaert. 3: 204 (1925). Type: Zaire, Shaba, Lubumbashi, *Hock* (BR, holo.!)
F. hockii De Wild. var. *rotundifolia* De Wild. in F.R. 11: 535 (1913). Type: Zaire, Shaba,
   Lubumbashi, *Hock* (BR, holo.!)
F. buarica Robyns in B.J.B.B. 11: 48 (1928). Type: Cameroon, near Buar, *Mildbraed* 9439 (K, holo.!)
F. coriacea Robyns in B.J.B.B. 11: 49 (1928). Type: Zaire, Shaba, Shinsenda, *F.A. Rogers* 10108 (K,
   holo.!)
F. kaessneri S. Moore var. *rotundifolia* (De Wild.) Robyns in B.J.B.B. 11: 58 (1928)

NOTE.   I have included a wide range of habit and leaf variation under this name; more study of the
   habit is needed in the field. Certain specimens are undoubtedly branched shrubs over 1 m. tall and
   seem very different from the pyrophytes with several stems from a woody rootstock. Some may be
   due to protection from burning, e.g. *Milne-Redhead & Taylor* 9091 from Songea District collected in
   seasonally flooded ground by the R. Mkurira; this has many paired leaves and is described as
   much branched from the base. Several specimens from Zambia in *Brachystegia* woodland are
   similar but scarcely protected from burning. A specimen mentioned by Verdcourt & Trump,
   Common Poisonous Pl. E. Afr.: 127 (1969) as *Temnocalyx obovatus* (N.E.Br.) Robyns is in fact a
   shrubby form of *F. triphylla*; it has been used as an antidote to *Acokanthera* arrow-poison.

   b. var. **giorgii** (*De Wild.*) *Verdc.* in K.B. 36: 523 (1981). Type: Zaire, Shaba, Lubumbashi, *De Giorgi*
3102 (BR, holo.!)

Plant with stems, leaves, inflorescences, etc., shortly pubescent.

TANZANIA.   Ufipa District: Lake Tanganyika, Kala Bay, 31 Dec. 1963, *Richards* 18723! & Kalambo
   Falls, 11 Jan. 1975, *Brummitt & Polhill* 13730!; Singida District: Iramba Plateau, 15 Jan. 1949,
   Hammond 181!
DISTR.   T 1, 4, 5, 7; Zaire, Zambia and Malawi
HAB.   *Brachystegia* woodland, chipya woodland, grassland; 775–1590 m.

SYN.   *Vangueria brachytricha* K. Schum. in De Wild., Etudes Fl. Katanga 1: 227 (1903); Th. Dur., Syll. Fl.
   Congol.: 269 (1909); De Wild., Contr. Fl. Katanga: 211 (1921). Type: Zaire, Shaba [Katanga],
   Verdick (BR, holo.!)
   Fadogia viridescens De Wild. in F.R. 13: 138 (1914) & Notes Fl. Kat. 4: 91 (1914) & Contr. Fl.
   Katanga: 214 (1921). Type: Zaire, Shaba, Biano, *Homblé* 910 (BR, holo.!)
   F. giorgii De Wild. in Pl. Bequaert. 3: 203 (1925); Robyns in B.J.B.B. 11: 56 (1928)
   F. brachytricha (K. Schum.) Robyns in B.J.B.B. 11: 54 (1928)

   c. var. **gracilifolia** *Verdc.* in K.B. 36: 523 (1981). Type: Tanzania, Ufipa District, Sumbawanga,
Chapota, *Richards* 8502 (K, holo.!)

Plant glabrous save for inside of corolla-tube and stipule-sheath. Leaves mostly much smaller and
narrower, 1.7–5 cm. long, 0.8–1.8 cm. wide. Stems mostly with over 4 leafy nodes, up to 1 m. tall.
Inflorescences usually pedunculate, several-flowered cymes.

TANZANIA.   Kigoma District: Sabaga, 5 June 1975, *Kibuwa & Mungai* 2746!; Chunya District: 152 km.
   N. of Mbeya on Itigi road, Lupa N. Forest Reserve, 5 Mar. 1963, *Boaler* 861! & Igila Hill, Kipembawe
   [Kepembawa], 22 Mar. 1965, *Richards* 19803!
DISTR.   T 4, 7; Zambia
HAB.   *Brachystegia* woodland, scrub, on sandy soil; 1400–1650 m.

NOTE.   *McCallum Webster* T502 (Tanzania, between Kasanga and Sumbawanga, 30 Mar. 1959) has
   more distinctly woody stems.

   d. var. **pubicaulis** *Verdc.* in K.B. 36: 524 (1981). Type: Tanzania, Ufipa District, Chapota, *Bullock*
2000 (K, holo.!, BR, iso.!)

Stems and sometimes inflorescence parts sparsely to densely pubescent. Leaves glabrous, mostly
narrow as in var. *gracilifolia*.

TANZANIA.   Biharamulo Forest Reserve, Nyakanazi, Nov. 1954, *Carmichael* 489!; Buha District: 6.5
   km. from Kibondo, 15 Nov. 1962, *Verdcourt* 3320!; Mpanda District: Mahali Mts., Itemba, 1 Oct.
   1958, *Jefford & Newbould* 2812!
DISTR.   T 1, 4; Zambia
HAB.   *Brachystegia, Isoberlinia, Parinari* woodland, coarse grassland; 750–1800 m.

NOTE.   Numerous intermediates link this with var. *giorgii*, e.g. *Verdcourt* 3388 from Kigoma which has
   some leaves slightly ciliate.

   e. var. **peteri** *Verdc.* in K.B. 36: 524 (1981). Type: Tanzania, Mbeya District, Magangwe Rangers
Post, *Greenway & Kanuri* 14335 (K, holo.!, EA, iso.)

Plant glabrescent or with stems and foliage ± pubescent. Leaf-blades (1–)3–6.5 cm. long, (0.7–)1.2–
3.7 cm. wide. Stems mostly with only 2–4 leafy nodes, 10–40 cm. tall. Flowers solitary or inflorescences
2–3-flowered, the peduncles usually short or obsolete.

TANZANIA. Tabora District: Kombe, E. of Kaliua [Kaliuwa], 27 Jan. 1926, *Peter* 35674!; Dodoma District: Chaya, 7 Jan. 1926, *Peter* 34387!; Mbeya District: Ruaha National Park, Magangwe, Rangers' Post, 13 Apr. 1970, *Greenway & Kanuri* 14335! & same locality, 9 Mar. 1972, *Bjørnstad* 1463!
DISTR. **T** 4, 5, 7; not known elsewhere
HAB. *Brachystegia* woodland, *Julbernardia*, *Combretum*, *Terminalia* and similar open woodlands, bushland; 1060–1400 m.

SYN. ? *F. monantha* Robyns in B.J.B.B. 11: 53 (1928). Type: either "opposite Zanzibar I. or at Nyassa", *Kirk* or *Bishop E. Steere* (K, holo.!)

NOTE. Peter mentions a height of 80 cm. but his specimen (34387) is not above 30 cm.; his label to 35831 (Uvinsa, near Malagarasi) states 5 m. tree and must have the wrong label for the specimen which is undoubtedly the above plant. The type of *F. monantha* is a poor fragment and has 3 apparently genuinely solitary flowers at the only flowering node preserved. Since even its provenance is uncertain I have doubts about its identity with the above but it may well be the same — it is certainly some form of *F. triphylla*.

NOTE. (on species as a whole). I have taken a wide view of this species and although the variants may appear distinctive and, in some cases might be better considered local subspecies, the facies is due to such things as leaf-size, shape and pubescence which varies considerably and is too inconsistent to delimit species. A number of other names may have to be added to the synonymy.

10. **F. sp. A**; Verdc. in K.B. 36: 525 (1981)

Subshrub 30–90 cm. tall, with many erect densely shortly ± bristly pubescent distinctly angular stems from a woody stock; side-branches slender. Leaves in whorls of 3; leaf-blades ovate or broadly elliptic, 1.7–3.8 cm. long, 0.8–3.3 cm. wide, ± obtuse, acute or very shortly obtusely to acutely acuminate at the apex, broadly cuneate to rounded or almost subcordate at the base, densely shortly almost velvety pubescent; venation obscure but closely reticulate beneath; petioles 1–2 mm. long or almost obsolete; stipules with broad base ± 1 mm. long, with central subulate appendage ± 2 mm. long. Inflorescences shorter than to well exceeding the leaves, 1.5–?7 cm. long (see note), 3–5-flowered; peduncles 1–2.5(–3) cm. long; pedicels 0.3–2 cm. long; all parts densely shortly pubescent. Calyx-tube ± 2 mm. long including narrow truncate rim. Corolla greenish white; tube campanulate, 4 mm. long and wide, glabrous outside; lobes oblong-lanceolate, 5 mm. long, 2 mm. wide, with some scattered hairs outside. Young fruits globose, 8 mm. long.

TANZANIA. Ufipa District: Mmemya Mt., 14 Feb. 1951, *Bullock* 3692!
DISTR. **T** 4; known only from the above collection
HAB. Grassland; 1980 m.

NOTE. The only specimen seen is abnormal, leafy shoots having developed from abnormal flowers the 'petals' of which have formed a whorl of small leaves. Even in apparently normal flowers there are 3 separate styles and stigmatic clubs. Normal material is needed before the status of this taxon can be decided. It must be related to *F. triphylla*.

11. **F. vollesenii** *Verdc.* in K.B. 36: 525 (1981). Type: Tanzania, Kilwa District, Selous Game Reserve, Nakilala valley, *Vollesen* in M.R.C. 4786 (K, holo.!, C, EA, MRC, iso.)

Subshrubby herb ± 30 cm. tall; stems slender but distinctly woody, with spreading rather bristly hairs on younger parts. Leaves in whorls of 3; blades narrowly to broadly elliptic, 1–3.8 cm. long, 0.8–2 cm. wide, rounded, subacute or very shortly acuminate at the apex, cuneate at the base, densely pubescent in very young state but soon only sparsely so, ciliate; petiole up to 2.5 mm. long; stipules with broad base 1.5 mm. long and linear apex 2 mm. long. Inflorescence densely spreading hairy, 1–several-flowered; peduncles up to 6 mm. long; pedicels up to 3.5 mm. long. Calyx spreading hairy; tube ± 1 mm. long; limb-tube very short, 0.2 mm. long, pale, truncate or with small widely separated denticles up to 0.2 mm. long. Buds densely spreading hairy, ± acute. Corolla yellowish green; tube 1.5–2 mm. long, ± 2.5 mm. wide at the throat; lobes ± 3 mm. long, 1 mm. wide. Style exserted ± 1.5 mm., the stigmatic club cylindrical, 0.8 mm. long. Fruit unknown.

TANZANIA. Kilwa District: Selous Game Reserve, Nakilala Valley, 17 Nov. 1977, *Vollesen* in M.R.C. 4786!
DISTR. **T** 8; not known elsewhere
HAB. *Brachystegia* woodland; 400 m.

12. **F. verdcourtii** *Tennant* in K.B. 22: 441 (1968); Verdc. in K.B. 36: 526 (1981). Type: Tanzania, Buha District, Kasakela Chimpanzee Reserve, *Verdcourt* 3384 (K, holo.!, BR, EA, iso.!)

Subshrubby herb 60–90 cm. tall, with slender pale reddish brown to dark, densely pubescent unbranched or branched stems, probably from a thick woody stock. Leaves in whorls of 2–4; blades broadly to narrowly elliptic, 4–8.2 cm. long, 1.4–3.2 cm. wide, acute or slightly acuminate at the apex, cuneate at the base, glabrous or slightly ciliate, or densely pubescent on both surfaces, the venation often drying reticulate beneath; petioles 1–2 mm. long; stipules vaginate below, ± 1 mm. long, densely pilose within, caudate, the appendage 2–4 mm. long. Inflorescences 2–3-flowered; peduncles 1–3 cm. long, spreading pubescent; pedicels 1–2 cm. long, similarly pubescent. Buds clavate, distinctly acuminate-apiculate. Calyx-tube 1.5–2.5 mm. long, the limb 0.5 mm. long, truncate or with vestiges of narrow teeth. Corolla-tube green, cylindrical-funnel-shaped, very slightly curved, 0.9–1.4 cm. long, glabrous outside; lobes deep green, lanceolate-triangular, 5.5 mm. long, 2–2.5 mm. wide, venose in dry state. Ovary 5-locular; style 1.3–1.4 cm. long; stigmatic club coroniform, 2.3 mm. long. Immature fruits subglobose, ± 9 mm. diameter, with 3 pyrenes.

var. **verdcourtii**; Verdc. in K.B. 36: 526 (1981)

Leaf-blades glabrous or with a few submarginal ciliae.

TANZANIA.   Buha District: Kakombe valley, 7 Jan. 1964, *Pirozynski* P164! & Kasakela Chimpanzee Reserve, 20 Nov. 1962, *Verdcourt* 3384! & same locality, 6 Apr. 1961, *Siwezi* 125!
DISTR.   **T** 4; not known elsewhere
HAB.   Grassland with *Protea, Vitex, Diplorhynchus*, etc. on black soil; 975 m.

var. **pubescens** *Tennant* in K.B. 22: 442 (1968); Verdc. in K.B. 36: 526 (1981). Type: Tanzania, Buha District, between Kigoma and Kasulu, *Verdcourt* 3332 (K, holo.!, EA, iso.!)

Leaf-blades shortly softly pubescent on both surfaces.

TANZANIA.   Buha District: between Kigoma and Kasulu, near Kasulu, 16 Nov. 1962, *Verdcourt* 3332!
DISTR.   **T** 4; not known elsewhere
HAB.   Cleared ground in *Brachystegia, Piliostigma* woodland on stony red soil with black patches; 1230 m.

NOTE.   The typical variety of this species at any rate is subjected to a characteristic galling of the flowers; the galls are globose, pale green with green stripes, about 1.5 cm. diameter and are swollen ovaries, the corolla protruding from the apex.

13.   **F. ancylantha** *Hiern* in F.T.A. 3: 155 (1877); Verdc. in K.B. 36: 526 (1981); Fl. Pl. Lign. Rwanda: 558, fig. 208.2 (1982); Fl. Rwanda 3: 195 (1985); Verdc. in K.B. 42: 125, fig. 1M (1987). Type: Sudan, Niamniam, Nganye's village, *Schweinfurth* 3936 (K, lecto.!, BM, isolecto.!)

Subshrubby herb 0.4–1.8 m. tall, with 2–6 branched or unbranched glabrescent to shortly pubescent rounded or triangular often quite woody stems from the crowns of a woody stock up to 2 cm. in diameter. Leaves paired or in whorls of 3, elliptic to rounded-obovate or ± round, 2.3–11 cm. long, 1–7 cm. wide, obtusely often abruptly acuminate at the apex, cuneate to almost truncate at the base, rather thin, glabrous to shortly pubescent; petioles absent or short, under 2 mm. long; stipules with broad base, 1–2 mm. tall and filiform appendage 1.5–7 mm. long, the base densely pilose within. Inflorescences shortly stalked, mostly 2–3(–4)-flowered or flowers sometimes solitary; peduncles 0.2–1.6 cm. long, sometimes supra-axillary and decurrent; pedicels 0.6–2.2 cm. long, glabrous to shortly pubescent. Calyx 2.5–3 mm. long, the very short rim-like limb truncate. Buds obtuse to shortly acuminate at the apex. Corolla usually distinctly curved but sometimes straight (at least when dry), greenish yellow, the tube usually greener and the lobes green outside, whitish or pale yellow inside; tube ± cylindrical, 1.7–2.8 cm. long, 5.5–7 mm. wide at apex, 2.5–3.5 mm. wide at the base, usually very distinctly widened above, glabrous outside but with ring of hairs inside near base; lobes triangular, 5–6.5 mm. long, 2.5 mm. wide, with a short blunt appendage. Ovary 5-locular; style pale green, up to 3.4 cm. long; stigmatic club cylindrical, 2.5 mm. long, 10-angled, rounded 5-lobed at the apex. Fruit ± globose, 5-lobed when dry, ± 1.2 cm. diameter.

UGANDA.   Acholi District: Patiko [Paliko, Baker's Camp], 26 June 1940, *Maxwell Forbes* 166!
TANZANIA.   Bukoba District: Nyaishozi [Nyashozi], Dec. 1931, *Haarer* 2378!; Kondoa District: 8 km. S. of Bereko on Great North Road, 9 Jan. 1962, *Polhill & Paulo* 1109!; Mbeya District: 40 km. on Mbeya–Mbozi road, Jan. 1963, *Procter* 2224!; Iringa, 12 Feb. 1932, *Lynes* I.h.77!
DISTR.   **U** 1; **T** 1, 2, 4–8; Nigeria, Zaire, Burundi, Rwanda, Sudan, Malawi, Zambia and Zimbabwe

HAB.   Grassland with scattered trees, high grassland, *Acacia*, *Combretum* ecotone, *Brachystegia* and *Terminalia*, *Combretum*, *Julbernardia* woodland, also in old cultivations; 90–1890 m.

SYN.   *Rubiacea 481;* Thomson in Speke, Journ. Disc. Source Nile, App.: 636 (1863)
[*Fadogia fuchsioides* sensu Oliv. in Trans. Linn. Soc. 29: 85, t. 50* (1873); Hiern in F.T.A. 3: 155 (1877), pro parte quoad *Grant* 481 excl. lectotype]
F. obovata N.E.Br. in K.B. (1906): 105 (1906). Type: Zimbabwe, near Harare, Six-mile Spruit, *Cecil* 141 (K, lecto.!)
[*Temnocalyx fuchsioides* sensu Robyns in B.J.B.B. 11: 319 (1928), pro parte quoad *Grant* 481, *non* (Welw.) Robyns]
T. obovatus (N.E.Br.) Robyns in B.J.B.B. 11: 320, figs. 31, 32 (1928)
T. ancylanthus (Hiern) Robyns in B.J.B.B. 11: 323 (1928); F.P.S.: 463 (1952)
T. ancylanthus (Hiern) Robyns var. *puberulus* Robyns in B.J.B.B. 11: 323 (1928); F.P.S. 2: 464 (1952). Type: Sudan, Lado, Yei R., Yembi, *Sillitoe* 225 (K, holo.!)

NOTE.   *F. ancylantha* and *F. obovata* have always been retained as separate species and although several of the northern specimens have the leaves less obovate with more truncate bases and with more indumentum, the same combination of characters turns up again, but rarely, in Malawi. The variation is such that only one taxon can be recognised. Robyns' var. *puberula* is connected by many intermediates and seems not worth maintaining. Nevertheless the variation needs further study in the field. The Uganda and Sudan populations are slightly different from the rest with the exception of the one sheet from Malawi mentioned.

14. **F. fuchsioides** *Oliv.* in Trans. Linn. Soc. 29: 85, t. 50 (flowering part only) (1873), pro parte; Hiern in F.T.A. 3: 155 (1877) & Cat. Afr. Pl. Welw. 1: 482 (1898), pro parte excl. *Grant* 481; F.F.N.R., fig. 68A (1962); Verdc. in K.B. 36: 527 (1981). Type: Angola, Ambaca District, left bank of R. Lutete, forests of Cazella, *Welwitsch* 2568 (LISU, lecto., BM, isolecto.!)

Subshrub 0.3–1.5 m. tall, with several reddish or purplish angular or rounded, finely ridged (when dry) glabrous stems from a thick woody stock. Leaves in whorls of 3–4 (rarely 5–6); blades elliptic to obovate, 4.5–15 cm. long, 1.8–9 cm. wide, rounded, very obscurely rounded-acuminate or bluntly acute at the apex, cuneate at the base, rather thicker than in other species, glabrous, the venation sometimes reddish when dry and in life; petiole rather thick, 0.3–2 cm. long; stipules with a short broad base 1.5–2 mm. long, densely pilose within, and a subulate appendage 3.5 mm. long. Inflorescences glabrous, 1–3-flowered; peduncle red, 5–11 mm. long or up to 2.5 cm. long in solitary flower; pedicels 0.3–2.3 cm. long. Calyx red; tube 4–5 mm. long, the limb rim-like, truncate, ± 1.5 mm. long. Buds distinctly acuminate. Corolla-tube deep red, cylindrical and straight, 1.4–2.7 cm. long, (4–)6–8 mm. wide, glabrous inside and out or ± densely hairy inside; lobes 6–9, cream, buff or yellow within, oblong-lanceolate, 0.8–1.4 cm. long, 3–3.5 mm. wide, with a thick apical appendage 1–2 mm. long. Ovary 6–9-locular; style 2.8–3.4 cm. long; stigmatic club large, broadly obconic-coroniform, 4.5–6 mm. wide, 3–4 mm. long, ± 6–9-ridged or lobed. Fruits black, fleshy, ± globose or obovoid, 1.2–1.5 cm. diameter, containing 6–9 pyrenes.

TANZANIA.   Ufipa District: Kito Mt., 21 Nov. 1958, *Richards* 10221! & Kalambo Falls, 11 Jan. 1975, *Brummitt & Polhill* 13729!
DISTR.   **T** 4; Zaire, Malawi, Zambia and Angola
HAB.   *Brachystegia* woodland, grassland with scattered bushes subject to burning; 1500 m.

SYN.   *Temnocalyx fuchsioides* (Oliv.) Robyns in B.J.B.B. 11: 319 (1928), pro parte

NOTE.   Fruits are said to be edible.

99a. ANCYLANTHOS

Desf. in Mém. Mus. Paris 4: 5, t. 2 (1818); Robyns in B.J.B.B. 11: 324 (1928), as "*Ancylanthus*"

Shrubs or one species a subshrubby herb, slightly to much branched. Leaves usually opposite or in two species in whorls of 3, shortly petiolate; stipules connate into a sheath below, densely hairy within, distinctly subulate-caudate at the apex, at least the base persistent. Flowers fairly conspicuous, solitary or in few–several-flowered simple or

---

* Figure is based on *Welwitsch* and *Grant* specimens; only the fruiting part refers to the present species.

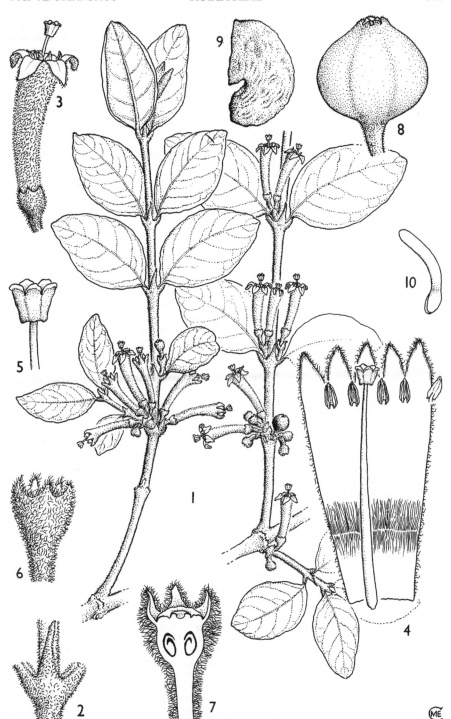

FIG. 143. *ANCYLANTHOS ROGERSII* — **1**, habit, × ⅔; **2**, stipule, × 2, **3**, flower, × 2; **4**, corolla opened out with style and stigmatic knob, × 3; **5**, stigmatic knob, × 6; **6**, calyx, × 6; **7**, longitudinal section through ovary, × 6; **8**, fruit, × 3; **9**, pyrene, × 4; **10**, embryo, × 4. 1, 2, from *Angus* 657; 3–7, from *Brenan* 7963; 8–10, from *Angus* 847. Drawn by Mrs M.E. Church.

branched peduncled pubescent to tomentose cymes; bracts small. Buds elongate, ferruginous or whitish tomentose or pubescent outside. Calyx-lobes 5, usually well developed and much longer than the tube, sometimes distinctly leafy, but in one species shortly triangular. Corolla-tube cylindric, elongate, curved, open and glabrous at the throat but with a ring of deflexed hairs inside at base and some scattered long hairs elsewhere; lobes lanceolate, reflexed, much shorter than the corolla-tube. Stamens inserted at the throat, the anthers slightly exserted. Ovary globose or semi-globose, 4–5-locular, each locule with a pendulous ovule; style slender, glabrous, shortly exserted; stigmatic club large, cylindric or less often subcoroniform, 5-lobed at the apex. Fruit yellow or red, globose, fleshy, crowned with calyx-limb, containing 5 pyrenes.

A small central African genus of 5 species. Fruits are mostly edible. *Ancylanthos rubiginosus* Desf. with very well-developed calyx-lobes and the embryo with exceptionally long cotyledons differs from the other 4 which belong to a separate section.

A. rogersii (*Wernham*) *Robyns* in B.J.B.B. 11: 326 (1928); F.F.N.R.: 400 (1962); Verdc. in K.B. 42, fig.2F (1987). Type: Zaire, Shaba, Lubumbashi [Elisabethville], *F.A. Rogers 10221* (BM, holo.!)

Shrub 0.3–4 m. tall; stems with dense white or grey cottony tomentum, at length greyish but persistently tomentose; older stems bearing the new year's shoots can be blackish and glabrous; bark on tall specimens smooth and brown. Leaves paired or in whorls of 3; blades ovate to broadly elliptic, 1.7–7 cm. long. 1–4 cm. wide, rounded to acute at the apex, rounded to cuneate at the base, very discolorous, densely covered with floccose white or grey cottony tomentum but becoming glabrescent above and green surface is scarcely obscured but persistently thickly white beneath; venation deeply impressed above; petiole 2–5(–7) mm. long; stipules triangular, 1–2 mm. long, with subulate appendage ± 6 mm. long. Cymes 3–6-flowered, but sometimes inflorescences at leafless nodes bearing 3 abbreviated new shoots appearing as dense verticillasters; peduncle 0–3(–15) mm. long; pedicels 3–4(–8) mm. long; bracts lanceolate, 3.5–5 mm. long. Buds densely cottony tomentose. Calyx-tube ± globose, 3–4 mm. long, appearing ± truncate but with thin teeth ± 1 mm. long, lacking tomentum. Corolla creamy white, pale green or distinctly yellow, 1.5–2 cm. long; lobes ovate-triangular, 3–6 mm. long, 2.2–2.5 mm wide, ± acute. Stigma green, exserted 1–2.5 mm. beyond the throat, the stigmatic club green, coroniform, deeply 5-lobed at apex. Fruit subglobose, ± 10 × 8 mm., distinctly 4-lobed in dry state, with scattered cottony tomentum. Figs. 132/10, p. 754 & 143, p. 801.

TANZANIA. Ufipa District: Mbizi Forest Reserve, 9 Nov. 1987, *Ruffo & Kisena* 2683!
DISTR. T 4; S. Zaire, N. and W. Zambia
HAB. Not stated but probably *Brachystegia* woodland; 2090 m.

SYN. *Fadogia rogersii* Wernham in S. Moore in J.B. 51: 208 (1913); De Wild., Etudes Fl. Katanga 2: 153 (1913) & Notes Fl. Kat. 7: 73 (1921) & Contr. Fl. Katanga: 213 (1921)

100. **RYTIGYNIA**

Blume, Mus. Bot. Lugd. Bat. 1: 178 (1850); Robyns in B.J.B.B. 11: 132 (1928); Verdc. in K.B. 42:145 (1987)

Shrubs or small trees, occasionally somewhat scrambling; stems spiny in a few species, usually distinctly lenticellate. Leaves opposite or occasionally in whorls of 3, particularly in some forms of *R. celastroides*, petiolate, often with domatia in axils of nerves beneath; stipules connate at the base, oblong or triangular, villous within, often ± persistent, ending in a mostly linear or subulate ± deciduous appendage. Flowers mostly 5-merous, small, usually white, yellowish or greenish, axillary, solitary or in 2–10(–15)-flowered sometimes umbel-like cymes; peduncles and pedicels mostly well developed; bracts and sometimes bracteoles present, small; sometimes inflorescence-axes scarcely developed and flowers appear fasciculate. Buds obtuse, acute or conspicuously long-apiculate, glabrous or pubescent outside. Calyx-tube ± subglobose, the limb short, truncate or denticulate only, distinctly lobed in a few species mostly in distinctive subgenera, mostly persistent on the young fruit. Corolla-tube cylindrical, glabrous or hairy within for part or most of length or with a ring of deflexed hairs in the middle; lobes shorter than to longer than the tube, acute to very distinctly long-apiculate or with a filiform appendage.

Stamens slightly to distinctly exserted; anthers often mucronate or slightly appendaged, often papillate; filaments very short. Ovary mostly 3–5(–6)-locular, but 2-locular in some species; style usually exserted, slightly swollen at base; stigmatic club coroniform, subglobose or occasionally cylindrical, sulcate beneath where in contact with anther-thecae, mostly distinctly 2–5-lobed at apex. Disk depressed, glabrous. Fruit mostly globose or asymmetrical and compressed if only 1–2 pyrenes developed, ± 1 cm. diameter, with 1–5 pyrenes; pyrenes narrowly ± reniform or boat-shaped, the notch about one-third from the apex, the dehiscence line on a ± marked keel, often pitted.

About 60–70 species in tropical Africa, South Africa and Madagascar.

The central core of the genus is distinctive enough and easily recognisable but other species connect to other genera, *Canthium* and allies, *Fadogia*, *Vangueria* and *Pachystigma*. Some with 2-locular ovaries approach the *Canthium* group very closely, especially in fruit. *Canthium* never has appendaged corolla-lobes and the inflorescence is mostly of a different structure and, moreover, the ovary is almost without exception 2-locular.

Three subgenera have been recognised — subgen. *Rytigynia* (species 1–42), subgen. *Fadogiopsis* Verdc. (species 43) and subgen. *Sali* Verdc. (species 44, 45). Inadequately known species are placed at the end (species 46–51).

1. Branches with distinct straight supra-nodal spines* 2
   Branches without spines (species in which spines may be
      present or absent are included in the key again) 6
2. Calyx-lobes linear-lanceolate (fig. 131/9, p. 752); corolla-
      lobes much shorter than corolla-tube . . . . . 41. *R. bugoyensis*
   Calyx-lobes not as above; corolla-lobes longer than
      corolla-tube 3
3. Calyx-limb with very distinct teeth, ± 1 mm.long . . . 27. *R. mrimaensis*
   Calyx-limb truncate or with less distinct teeth 4
4. Mature leaves with a ± bullate venation, sunk above and
      raised beneath, glabrous or with scattered hairs on
      venation beneath; peduncles 1–2.5 cm. long; corolla-
      lobe appendages ± 1.5 mm. long (U 1) . . . . . 26. *R. pauciflora*
   Mature leaves not bullate, the venation scarcely raised,
      glabrous to densely pubescent or velvety; corolla-lobe
      appendages very small (eastern Kenya and Tanzania) 5
5. Leaves small, 1–2.5 × 0.5–1.5 cm., rounded at apex, mostly
      glabrescent; branchlets with white or greyish white
      epidermis (north Kenya coast) . . . . 29. *R. parvifolia*
   Leaves larger, 2–6(–9) × 0.8–2.5(–4) cm., acuminate at the
      apex or at least narrowed to a rounded tip, glabrous to
      densely velvety; branchlets mostly brown. . . . . 28. *R. celastroides*
6. Leaves densely pubescent to quite velvety especially
      beneath and often discolorous** 7
   Leaves glabrous save for domatia on undersurface or with
      a few scattered hairs but not densely pubescent
      (rigorous choice not essential since doubtful species
      are included in key twice) 22
7. Corolla-lobes with long subulate appendages over 1 mm.
      long and often much longer 8
   Corolla-lobes without or with very short appendages 13
8. Calyx-lobes linear, 1.2 mm. long; corolla-tube 6 mm. long,
      the 4 mm. long appendages longer than the corolla-
      lobes (T 6, Nguru & Ukaguru Mts.) . . . . . . 37. *R. pseudolongi-*
                                                                caudata

---

* *Homewood* 62 & 64 (48, *R. sp. I*) from Kenya, Tana R., Ozi, has the habit of *Meyna*, with usually the leaves fastigiate on cushion-shoots below the spines, but much smaller fruits; on some branches the shoots are developed.
** *Semsei* 2034 (42, *R. sp. F*) with ferruginous pubescence, said to be a tree from Morogoro District, Chigurofumi Forest Reserve, only known in fruit (up to 5 pyrenes) will come here; the inflorescences are apparently 3–8-flowered with peduncle 1–3 mm. long; so also will *Gilchrist* in *F.H.* 1701 (49, *R. sp. J*), a sterile specimen from Iringa District, Kigogo Forest, Mgololo, 1980 m., with dense spreading ferruginous pubescence on the stems and rather velvety discolorous leaves — almost certainly a *Rytigynia*.

Calyx-lobes obsolete or at least not linear, if filiform-
subulate then corolla-tube shorter 9
9. Leaves ovate or elliptic, densely softly grey velvety beneath,
the leaf-surface practically obscured; corolla-tube 3.5
mm. long . . . . 40. *R. griseovelutina*
Leaves narrowly elliptic or oblong to lanceolate, pubescent
to ± velvety beneath but surface hardly obscured 10
10. Calyx-lobes ± 10, linear-triangular, 1–2.5 mm. long; corolla-
tube short, 2.5 mm. long, the lobes 4 mm. long with
additional subapical appendages 3 mm. long; flowering
pedicels 2.5 mm. long (T 7, Uzungwa Mts.) . . . 38. *R. hirsutiflora*
Calyx-lobes 5, obsolete to filiform-subulate and without
other character dimensions combined 11
11. Stems and leaves covered with ± dense rather long rusty
seta-like hairs but surfaces not obscured; corolla-lobe
appendages 4–5 mm. long, exceeding the 3 mm. long
corolla-lobes (T 3, E. Usambaras) . . . . 39. *R. xanthotricha*
Stems and leaves with shorter softer indumentum, the
undersides of the leaves almost velvety and partly
obscured or much less pubescent 12
12. Leaves shortly densely pubescent to almost glabrous;
peduncles scarcely developed but pedicels up to 1.8 cm.
long in fruit; corolla-lobe appendages 2.5 mm. long (T 7,
inadequately known) . . . . 31. *R. flavida*
Leaves pubescent to almost velvety beneath with the
surface partly obscured; pedicels under 1 cm. long;
corolla-lobe appendages 3.5–5 mm. long (U 2, T 3) 32. *R. bagshawei*
13. Corolla-tube 6–7 mm. long, the lobes much shorter 2–3
mm. long; glabrous ovate calyx-lobes contrasting with
the densely pubescent tube; stems densely hairy; leaves
very discolorous, velvety beneath; cymes subsessile (T
6, Uluguru Mts.) . . . . 22. *R. lichenoxenos*
Corolla-tube usually shorter or if not then without other
characters combined 14
14. Calyx-limb with distinct lobes, 1.5–5 mm. long; ovary 2–3-
locular 15
Calyx-limb truncate or with much less distinct lobes; ovary
2- or more locular 16
15. Leaves very discolorous, velvety beneath; calyx-lobes
filiform; ovary 2-locular . . . . 43. *R. decussata*
Leaves not discolorous but stems and foliage densely
ferruginous bristly-pubescent; calyx-lobes linear to
linear-lanceolate; ovary 3-locular . . . . 45. *R. fuscosetulosa*
16. Stipule-appendages laterally flattened and somewhat
scimitar-shaped rather than subulate; cymes not sessile
or subsessile, the peduncles and pedicels often well
developed, over 1 cm. long; calyx-lobes short, ±
triangular, 0.5 mm. long or limb truncate; corolla
usually hairy on lobes outside with base of tube
glabrous, but some forms, especially outside the Flora
area, have corolla glabrous; ovary 4-locular . . . 30. *R. monantha*
Stipule-appendages ± subulate; cymes ± sessile or peduncles
only 2–3 mm. long; pedicels short or long; ovary 2–4-
locular 17
17. Corolla-tube 5–6 mm. long; leaves oblong to ovate,
pubescent to velvety 18
Corolla-tube 1.5–3 mm. long and lobes ± 2 mm. long; leaves
oblong-elliptic to oblong-lanceolate 19
18. Leaves oblong or ovate, 1.5–3.2 × 0.8–2.2 cm.; ovary 3-
locular; flowers solitary or in 2–3-flowered cymes; fruit
unknown (T 7, Sao Hill & Umalila) . . . . . . 21. *R. pubescens*

Leaves ovate-oblong, 3–7.5 × 1.3–5 cm.; ovary 2-locular;
flowers in 3–4-flowered cymes; fruit large 1.5–2 cm.
long with 2 pyrenes ± 1.5 cm. long (**K** 4,6; **T** 2)　　　24. *R. induta*

19. Inflorescence several-flowered, usually with secondary
branches produced, usually not a simple umbel or
fascicle; branchlets very often in whorls of 3; buds
pubescent or less often glabrous; corolla-lobes
exceeding the short 2–3 mm. long tube; pyrenes and
embryo characteristic (eastern lowlands of Kenya and
Tanzania to ± 1000 m.) . . . . 　28. *R. celastroides*
Inflorescence several-flowered but secondary branches
not developed; branchlets paired or in 3's in species 1
and 11 20

20. Leaves mostly distinctly discolorous and velvety beneath;
ovary 2-locular; fruits 8–10 mm. long and wide with 1–2
pyrenes (**U** 2) . . . . 　11. *R. kiwuensis*
Leaves sparsely to densely pubescent, not very discolorous;
ovary 3–5-locular; fruits often smaller 6–7(–9) mm. long
and wide with 2–5-pyrenes 21

21. Leaves more abruptly acuminate, the marginal curvature
from the main blade into the acumen distinctly greater;
indumentum of longer seta-like hairs (**U** 4; **T** 4)　　2. *R. umbellulata*
Leaves more gradually acuminate, the margin distinctly
less curved; indumentum of shorter hairs (**U** 2; **T** 4)　1. *R. beniensis*

22. At least some indumentum on the stems and/or leaves 23
Stems and leaves quite glabrous save for domatia beneath
and densely pilose insides of stipule-bases* 30

23. Leaves oblong to oblong-elliptic, 2.3–7 × 0.8–3.6 cm., the
margins with characteristic ciliation and also some
scattered hairs above and on venation beneath (known
only from one sterile specimen from **T** 3, Mt. Mlinga)　18. *R. sp. D*
Leaves without characteristic ciliation 24

24. Corolla-lobes with appendages up to 2.5 mm. long; fruiting
pedicels up to 1.8 cm long 25
Corolla-lobes without or with minute appendages (perhaps
longer in 20, *R. adenodonta*) 27

25. Slender-stemmed plant; leaves thin, 1.2–6.5 × 0.8–2.5 cm.,
with short seta-like hairs on midrib beneath, margins
and sometimes upper surface; petiole with similar
hairs; corolla-lobes 2 mm. long with appendage 2 mm.
long, spreading, setulose outside (**T** 8) . . . . . 　10. *R. pergracilis*
More robust plant with different leaf-indumentum and
larger corolla without seta-like hairs 26

26. Corolla glabrous or with very few scattered hairs; leaves
lanceolate to oblong-lanceolate, glabrescent save for
pubescence along the main nerves . . . . . . 　30. *R. monantha*
　　　　　　　　　　　　　　　　　　　　　　　var. *lusakati*

Corolla pubescent; leaves oblong, densely shortly-
adpressed pubescent on both surfaces . . . . . 　31. *R. flavida*

27. Inflorescences usually with secondary branches shortly
developed; corolla-tube 2–3 mm. long, with slightly
longer lobes; pyrenes characteristic . . . . See couplet 5
Inflorescences shortly pedunculate umbel-like cymes,
fascicles, 2-flowered cymes or flowers solitary 28

---

* A very poor specimen from **T** 8 (Rondo Plateau), with fruiting pedicels 2.5–3 cm. long and
peduncles 2 mm. long, comes here (6, *R. sp. C*(see addenda, p. 927)). Species 46 (*R. sp. G*) will also key
here; the purplish brown epidermis on the young stems flaking to reveal an almost powdery surface
and very long-acuminate leaves will identify it but the only specimen, *Perdue & Kibuwa* 11248 from
Iringa District, Mufindi, Nyalawa R., is quite sterile. Species 47 (*R. sp. H*) will also come in this section
of the key and has narrowly acuminate leaves, it is a miombo woodland species from Tabora.

28. Young stems densely shortly setulose pubescent with
spreading ferruginous hairs; inflorescences almost
sessile; pedicels short, ± 1 mm. long; corolla-tube 6 mm.
long; leaves with costa very shortly setulose pubescent
above and beneath (**T** 6, Uluguru Mts.)   .   .   .   .   .   23. *R. nodulosa*
Young stems adpressed pubescent or with only scattered
hairs or ± glabrous 29
29. Leaves glabrous, usually small, 1.3–4(–6.2) × 0.4–2(–2.8) cm.
but mostly 2 × 1 cm. in typical variety, elliptic or broadly
ovate-elliptic, shortly obtusely acuminate at the apex;
stems adpressed pubescent (in typical variety) to
glabrous; flowers usually solitary (**T** 4,7)   .   .   .   20. *R. adenodonta*
Leaves sparsely pilose or shortly pubescent, 1.8–10 × 0.7–5
cm., thin, more distinctly acuminate at the apex; stems
glabrous to pubescent; inflorescences 1–8-flowered;
habit mostly graceful with slender branchlets   .   .   .   .   return to couplet 22
30. Corolla-lobes with well-developed subulate appendages,
the buds acuminate or with 5 distinct tails* 31
Corolla-lobes without or with very indistinct appendages 37
31. Corolla-tube ± 9 mm. long; stipule-appendage ligulate, ±
thickened and held in a radial plane (**T** 6, Ngurus)   34. *R. longituba*
Corolla-tube shorter and stipules not as above 32
32. Appendages very well developed ± equalling or longer
than the lobes 33
Appendages shorter 34
33. Appendages almost or quite equalling the corolla-lobes;
tube sparsely pubescent outside and at throat, otherwise
glabrous within; leaves cuneate to rounded at the base
(**T** 3, W. Usambaras)   .   .   .   .   33. *R. longicaudata*
Appendages 2–3 times as long as the corolla-lobes; tube
glabrous outside but with a ring of deflexed hairs at
throat; leaves rounded to usually subcordate at the base
(**T** 7, Uzungwa Mts.)   .   .   .   .   36. *R. caudatissima*
34. Venation very evident beneath, reticulate (but not
distinctly raised) beneath when dry; buds acuminate to
distinctly 5-tailed, the corolla-lobe appendages 1.5–2.5
mm. long; peduncles 0–4 mm. long (**U** 2)   .   .   .   .   15. *R. kigeziensis*
Venation not so reticulate beneath, usually rather obscure
but if not then peduncles longer; corolla-lobe
appendages short, 1–1.5 mm. long 35
35. Inflorescences lax, 2(–3)-flowered, the peduncles 1.3–2.5
cm. long and pedicels 0.5–2 cm. long; leaves narrowly
acuminate, drying a pale olive-green, the venation
usually finely raised beneath (**T** 6,8)   .   .   .   .   .   .   25. *R. binata*
Inflorescences shorter, either 2-flowered with peduncles
4–8 mm. long and pedicels 1.5–3.5 mm. long or flowers
solitary with pedicels 5–12 mm. long 36
36. Inflorescences 2-flowered; leaves acutely acuminate, the
acumen usually well developed; ovary 2-locular, the
fruit compressed, ± obcordate (**U** 2, Kigezi)   .   .   .   14. *R. bridsoniae*
Flowers solitary; leaves ± obtusely acuminate, the acumen
short; ovary 3–4-locular, the fruit subglobose (**T** 7,
Kyimbila; almost unknown species)   .   .   .   .   .   .   8. *R. obscura*
37. Calyx-lobes well developed, lanceolate, 1.5–3 mm. long;
inflorescences ± 5-flowered   .   .   .   .   44. *R. saliensis*
Calyx-lobes less developed, mostly under 1 mm., or limb ±
truncate 38
38. Inflorescence a (1–)2-flowered cyme with usually distinct
pedicels and peduncle** 39

* Certain glabrous variants of the protean *R. celastroides* could key here or to both couplets of 36. If
no obvious fit is found for the plant under investigation it may be that species.
** See also *R. longipedicellata*, p. 927

Inflorescence with several flowers in sessile fascicles or
   shortly pedunculate umbel-like cymes but some
   inflorescences may be reduced to few or even 1 flower 42
39. Ovary 2-locular; corolla-tube about same length or shorter
   than the lobes; fruit where known larger, the pyrenes
   up to 10 × 8 mm. 40
   Ovary 3–5-locular; corolla-lobes usually much shorter than
   the tube; fruit smaller, the pyrenes ± 7 × 4 mm. 41
40. Petioles ± 1 mm. long; leaf-blades 4–5 × 1–3 cm., without
   domatia in single specimen seen; branchlets dark
   purple-brown (**T** 5, Kiboriani Mts.).   . . . .   16. *R. ignobilis*
   Petioles 4–8 mm. long; leaf-blades 4–14.5 × 1.7–6.8 cm.,
   with small pubescent domatia; branchlets mostly grey-
   brown (**T** 3, Usambaras)   . . . .   17. *R. eickii*
41. Leaves thicker and more discolorous; corolla more robust,
   mostly slightly larger and drying darker with longer
   lobes, 3–3.5 mm. long; calyx-lobes sometimes more
   developed (**T** 7)   . . . .   20. *R. adenodonta* var.
                                                      reticulata
   Leaves thinner and less discolorous; corolla thinner and
   slightly smaller, drying pale, with shorter lobes 2 mm.
   long; calyx-limb truncate, poorly developed (Kenya &
   Tanzania, widespread in highlands)   . . . . .   19. *R. uhligii*
42. Pedicels in one inflorescence of unequal length, one much
   longer than the others, sparsely pubescent; corolla-
   tube 2mm. long, the lobes ± 4 mm. long; slender shrub
   to 2.5 m. with thin, oblong-elliptic narrowly acuminate
   leaves, 5–9 × 1.5–4.5 cm. (**U** 2, Ankole, Ruampara —
   only known from a single specimen)   . . . . .   3. *R. sp.* A
   Pedicels more equal and corolla-lobes not usually twice
   length of the tube 43
43. Ovary 2-locular (rarely 3 in 13, *R. ruwenzoriensis*) 44
   Ovary mostly 3–5-locular 46
44. Inflorescence 9–14-flowered; virtually unknown species
   from **T** 7   . . . .   9. *R. stolzii*
   Inflorescence 1–5(–7)-flowered (**U** 1,2; **K** 1) 45
45. Leaves oblong-elliptic; stems grey (northern species from
   U 1 & **K** 1)   . . . .   12. *R. neglecta*
   Leaves distinctly more lanceolate; stems dark purple (**U** 2,
   Ruwenzori)   . . . .   13. *R. ruwenzoriensis*
46. Corolla-lobes distinctly much shorter than corolla-tube
   (upland forest)   . . . . return to 41
   Corolla-lobes not or little shorter than corolla-tube
   (lowland or upland forest) 47
47. Leaves elliptic-lanceolate, 2.5–4.5 × 0.5–1.5 cm.;
   inflorescences sessile and pedicels 1–3 mm. long; bark
   dark purple-red (**U** 3, Lolui I., known only from one
   specimen)   . . . .   4. *R. sp.* B
   Leaves distinctly less lanceolate 48
48. Graceful species with slender stems; fruits small, ± 7 mm.
   diameter, nearly always with 5 pyrenes; stems and
   leaves nearly always with some scattered hairs but
   occasionally glabrous; bark usually pale brown not
   breaking down into a red powder   . . . . . .   2. *R. umbellulata*
   Coarser species with stouter stems; fruits often larger, 8–9
   mm. diameter, often with 1–3 pyrenes; stems and leaves
   glabrous; bark usually pale brown or dark purplish and
   powdery 49
49. Bark dark purplish, often breaking down into a red powder
   on the older branchlets; mostly a plant of upland forest
   (save in Bukoba area), 1110–2400 m.   . . . . . .   7. *R. acuminatissima*

Bark pale brown or grey-brown, not breaking down to red
powder but sometimes flaky; plant of lowland forest,
often in seasonally wet places 1110–1500 m.   . . .   5. *R. dubiosa*

1. **R. beniensis** (*De Wild.*) *Robyns* in B.J.B.B. 11: 170 (1928) & F.P.N.A. 2: 349 (1947); Fl. Pl.
Lign. Rwanda: 597, fig. 203.2 (1982); Fl. Rwanda 3: 216, fig. 68.2 (1985); Verdc. in K.B. 42:
151 (1987). Type: Zaire, Mayolu, *Bequaert* 3982 (BR, lecto.!)

Small symmetrical divaricately branched shrub 0.9–4.2 m. tall, with branchlets either
opposite or ternate or occasionally one suppressed and apparently alternate, slender,
pubescent when young, later ± glabrous and with ridged dark red-brown bark densely
dotted with pale lenticels. Leaves paired or ? sometimes in whorls of 3; blades elliptic,
narrowly oblong-elliptic or oblong-lanceolate, 1.2–5.5 cm. long, 0.7–2.2 cm. wide, mostly
rather long-acuminate at the apex, cuneate to rounded at the base, the actual apex
narrowly rounded, with scattered short ± stiff hairs mainly on the venation; margins
sometimes undulate in dried material; petiole 1–2 mm. long; stipules joined to form a
pubescent sheath 2–2.5 mm. long; appendages subulate 1.5–2.5 mm. long; later the
sheath splits and appears lacerated, densely pilose within, the bases ± persistent. Flowers
2–6 in stalked fascicles supported by paired scarious ovate bracts 2 mm. long and wide;
peduncle 1–3 mm. long; pedicels 3–5(–7) mm. long, usually short, pubescent. Buds
acuminate. Calyx-tube subglobose, 1–1.3 mm. long, the limb ± scarious, 0.5 mm. long,
subtruncate or 5-denticulate, the teeth triangular ± 0.5 mm. long. Corolla cream, white or
green; tube 1.5–2 mm. long, glabrous outside, with a ring of deflexed hairs inside; lobes
triangular, 2 mm. long, 1.3 mm. wide, subacute or very shortly acuminate. Ovary 3(–4)-
locular; style exserted 1–1.5 mm., the stigmatic club ± 1 mm. long, usually 3-lobed. Fruit
subglobose, 6 mm. long, 7 mm. wide, 4–7 mm. thick, crowned with calyx-limb, with (1–)2–3
pyrenes, each ellipsoid-reniform, 6 × 4 × 3.5 mm., notched one-third from the apex on
inner angle.

UGANDA. Ankole District: Mbarara, banks of Ruizi R., Oct. 1925, *Maitland* 860!; Kigezi District:
    Ruzhumbura, Nyakagyeme, Nov. 1946, *Purseglove* 2233!; Mengo District: Kampala, Apr. 1923,
    Maitland 670!
DISTR. **U** 2, 4; Zaire and Rwanda
HAB. Forest edges, thicket woodland or grassland, often in swampy places or on termite mounds;
    also reported from rocky areas and eroded land; (?990–)1240– 1500 m.
SYN. [*Vangueria acuminatissima* sensu K. Krause in Z.A.E. 1907–8, 2: 326 (1911), *non* K. Schum. (1895),
        fide Robyns]
    V. beniensis De Wild. in B.J.B.B. 8: 43 (1922) & Pl. Bequaert. 2: 267 (1923)
NOTE. Very close to *V. umbellulata* but the subtle differences in leaf-shape and indumentum
    probably indicate they are distinct.

2. **R. umbellulata** (*Hiern*) *Robyns* in B.J.B.B. 11: 184 (1928); F.F.N.R.: 420 (1962); Hepper
in F.W.T.A., ed. 2, 2: 186 (1963); Verdc. in K.B. 42: 152, fig. 1E, 8C (1987). Type: Ghana,
Cape Coast, *W. Brass* (BM, holo.!)

Scrambling or erect shrub or small tree 1–5(–9) m. tall, often deciduous, the branches
usually slender; bark grey-black; branchlets grey to purple-brown, with or without
lenticels, glabrous or pubescent or setulose on young parts. Leaf-blades oblong-elliptic to
ovate or lanceolate, 1.8–11 cm. long, 0.7–5 cm. wide, acuminate at the apex, the tip ±
rounded, broadly cuneate to rounded at the base, glabrous save for barbellate domatia or
sparsely to fairly densely adpressed pilose or setulose particularly on the midrib, usually
distinctly thin; petiole 2–7 mm. long; stipules with triangular-truncate bases 1.5–3 mm.
long, ± connate at the base, scarious but sometimes becoming woody, ± villous inside,
with a subulate appendage 2–8 mm. long or rarely ± lacking. Cymes umbel-like, 2–8-
flowered, often produced with young leaves; peduncle 0.5–4 mm. long or ± obsolete;
pedicels 2–10 mm. long; bracteoles 1–2 mm. long, connate to form a scarious involucre.
Buds rounded to subapiculate at the apex. Calyx-tube 1–1.5 mm. long, the limb 0.3 mm.
long, truncate or obscurely toothed, scarious. Corolla creamy, yellow or greenish white;
tube 2.5–4 mm. long, with a ring of deflexed hairs within; lobes triangular or oblong, 1.8–3
mm. long, subapiculate. Ovary (4–)5-locular; style 4–4.5 mm. long; stigma described as
yellow below, green in middle and blue on top, coroniform, 5-lobed. Fruit black,
subglobose, 6–9 mm. diameter, with (2–)4–5 pyrenes; pedicels 7–12 mm. long; pyrenes
mostly strongly pitted.

UGANDA. Masaka District: Sese [Sesse] Is., Bugala I., Daje, 28 Feb. 1933, *A.S. Thomas* 907! & Sese Is., Nkose I., 21 Jan. 1956, *Dawkins* 843!; Mengo District: Entebbe, Oct. 1931, *Eggeling* 17!

TANZANIA. Buha District: Gombe Stream Reserve, 17 Dec. 1971, *Harris* 6038! & Kakombe valley, 7 June 1964, *Pirozynski* 193!; Mpanda District: Mahali Mts., Kasiha [Kasieha], 26 Sept. 1958, *Jefford et al.* 2700!

DISTR. U ?2, 4; T ?1, 4; Guinea Bissau to Cameroon and Angola, Zaire, Botswana, Zambia and Mozambique

HAB. Evergreen forest, woodland and thicket, frequently by lakesides, often in rocky places, ravines, etc.; 780–1170(–1290) m.

SYN. *Vangueria concolor* Hiern in F.T.A. 3: 150 (1877); De Wild. in B.J.B.B. 8: 330 (1922). Type: Principe, *Mann* 1128 (K, lecto.!)
     V. umbellulata Hiern in F.T.A. 3: 150 (1877) & Cat. Afr. Pl. Welw. 1: 480 (1898)
     V. sparsifolia S. Moore in J.L.S. 40: 92 (1911); De Wild. in B.J.B.B. 8: 64 (1922). Type: Mozambique, Gazaland, Madanda Forest, *Swynnerton* 551 (K, holo.!)
     V. ituriensis De Wild. in B.J.B.B. 8: 56 (1922) & in Pl. Bequaert. 2: 275 (1923). Type: Zaire, Ituri R., Penge, *Bequaert* 2479 (BR, holo.!)
     Rytigynia perlucidula Robyns in B.J.B.B. 11: 175 (1928). Type: Zaire, R. Ituri, Lesse, *Bequaert* 3181 (BR, holo.!)
     R. sparsifolia (S. Moore) Robyns in B.J.B.B. 11: 180 (1928)
     R. concolor (Hiern) Robyns in B.J.B.B. 11: 190 (1928)
     R. welwitschii Robyns in B.J.B.B. 11: 193 (1928). Type: Angola, Pungo Andongo, Calemba I., *Welwitsch* 5349 bis (K, holo.! & iso.!, BM, iso.)

NOTE. As accepted here *R. umbellulata* is a very variable species widespread in tropical Africa. The type is glabrous but variation in indumentum is such that it is not even of varietal significance. *R. beniensis* maintained as a separate species in this account is closely allied. I have seen no authentic material of *R. acuminatissima* but in this account have associated the name with another species (no. 7) although from the description it might equally belong here. Robyns had first intended keeping De Wildeman's *Vangueria ituriensis* as a distinct species of *Rytigynia* but in his revision finally sank it into *R. neglecta*, but that is definitely incorrect. *Rytigynia gracilipetiolata* (De Wild.) Robyns is possibly the same but the leaves have a different facies. *Clutton Brock* 9 from Gombe Stream Reserve is described as having lateral spines on the trunk but no other reference to this has been found and it needs confirmation. *Bidgood et al.* 2032 (Masasi, 17 Mar. 1991, 500 m.) belongs near here and has leaves up to 14.5 × 8 cm.

3. R. sp. A

Slender shrub 2.4 m. tall with long graceful branches covered with purplish bark, shiny and lenticellate. Leaf-blades oblong-elliptic, 5–9 cm. long, 1.5–4.6 cm. wide, narrowly acuminate at the apex, ± rounded at the base, thin, scarcely discolorous, glabrous save for sparsely pubescent domatia; petiole slender, 5 mm. long; stipule-bases densely pilose inside, ± 2 mm. long, joined to form a tubular sheath; appendages lanceolate, (0.5–)5 mm. long. Inflorescence 3–4-flowered; peduncle 3 mm. long; pedicels slender, 3–8 mm. long, usually 1 long and 2 short; bracts ovate, 1.5–2 mm. long. Calyx-tube obconic, 1 mm. long, the limb subtruncate or slightly toothed. Corolla greenish; tube 2 mm. long; lobes 4 mm. long, not appendaged.

UGANDA. Ankole District: Rwampara [Ruampara], Kigarama Hill, Oct. 1932, *Eggeling* 646!
DISTR. U 2; not known elsewhere
HAB. Stream washout; 1650 m.

SYN. *R. sp. B* sensu Verdc. in K.B. 42: 153 (1987)

NOTE. Only one specimen has been seen of this and it does not appear to be referable to any known species. Unfortunately the material is not adequate to describe and the duplicates in Uganda may have been destroyed.

4. R. sp. B

Small much branched tree with distinctive dark purple-red bark; branchlets glabrous, ridged, lenticellate, with ± peeling bark. Leaf-blades narrowly oblong-elliptic to elliptic-lanceolate, 2.5–4.5 cm. long, 0.5–1.5 cm. wide, tapering acuminate to the apex, cuneate at the base, slightly discolorous, glabrous, without domatia; petiole ± winged, 2–4 mm. long; stipules with persistent cup-like bases 2.5 mm. long, pilose within, the appendages subulate, 2–4 mm. long. Inflorescence very short, 2–5-flowered, peduncle suppressed or not projecting beyond the stipule-sheath; pedicels short, 1–3 mm. long; bracts ± 1 mm. long. Buds obtuse. Calyx-tube 1 mm. long; limb 0.3 mm. long, ± truncate. Corolla greenish; tube 1.8 mm. long; limb 2 mm. long. Ovary 3-locular. Fruit not seen.

UGANDA. Busoga District: Lake Victoria, Lolui I., 16 Dec. 1964, *G. Jackson* 12641!
DISTR.   U 3; not known elsewhere
HAB.   Rocky shore; 1130 m.

SYN. *R. sp. C* sensu Verdc. in K.B. 42: 153 (1987)

NOTE.   Only one gathering of this has been seen. It may be an extreme variant of *R. umbellulata* and
also shows some resemblance to *R. junodii* (Schinz) Robyns, which itself is probably only a form of
R. umbellulata from Delagoa Bay, but in the absence of more material, particularly fruit I have
preferred to leave it as an unidentified taxon.

5. **R. dubiosa** (*De Wild.*) *Robyns* in B.J.B.B. 11: 216 (1928); Verdc. in K.B. 42: 153 (1987).
Type: Zaire, Penge, *Bequaert* 2231 (BR, holo.!)

Shrub or small tree to 2.4 m. tall; branches glabrous, the branchlets lenticellate, with
mostly pale brown or chestnut bark which peels to reveal a grey or grey-brown surface,
probably sometimes ± powdery. Leaf-blades oblong?-elliptic to elliptic, 6.5–18 cm. long,
2.5–9 cm. wide, acuminate at the apex, cuneate to sometimes rounded at the base,
glabrous save for barbellate domatia in nerve-axils beneath but these may be reduced or
almost absent; petioles 4–7 mm. long; stipules joined to form basal sheath 2–3.5 mm. long,
becoming accrescent and ± triangular, densely pilose inside, with compressed ultimately
deciduous subulate appendage 5–6 mm. long. Inflorescences subumbellate or with
secondary branches slightly developed, 7–16-flowered; peduncle, 4–6 mm. long, with
paired ovate bracts 1.5 mm. long; pedicels 2.5–6 mm. long. Buds ovoid, rounded or very
slightly acuminate. Calyx-tube 1 mm. long, the limb 0.3 mm. long, scarious, minutely
denticulate. Corolla green, glabrous; tube campanulate, 2.5–3 mm. long, with a row of
deflexed hairs inside below the throat; lobes narrowly triangular, 2–3 mm. long with
vestigial appendage. Ovary 3–4-locular; style 4.5 mm. long, slightly swollen at base,
exserted; stigmatic club coroniform, sulcate, 4-lobed at the apex, 1.25 mm. wide. Fruits
subglobose, 7–8 mm. long, 9–11 mm. wide, often with only 2 pyrenes developed; pedicels
up to 1.2 cm. long.

UGANDA. Masaka District: Sese Is., Bunyama I., June 1925, *Maitland* 887! & Bukasa I., Dec. 1922,
   Maitland 579!; Mengo District: 12 km. on Kampala–Masaka road, Oct. & Dec. 1937, *Chandler* 1943!
DISTR.   U ?2, 4; Zaire and Cameroon (*fide* Robyns)
HAB.   Evergreen forest edges; 1110–1170(?–1500) m.

SYN. *Plectronia dubiosa* De Wild., Pl. Bequaert. 3: 182 (1925)
   [*Rytigynia neglecta* sensu Robyns in B.J.B.B. 11: 183 (1928) quoad *Maitland* 887! & ?264, *non*
   (Hiern) Robyns sensu stricto]

NOTE.   I am not certain of the identity of this plant although I believe it is correctly associated with
De Wildeman's taxon. It is possibly no more than a variant of *R. umbellulata* in which case many
other names for plants described from Zaire will also have to disappear into synonymy. What I
have taken to be *R. acuminatissima* (q.v.) has different bark and occurs in general at higher
altitudes. This whole group of typical *Rytigynia* is difficult to study in the herbarium and field work
may solve the problems. *Osmaston* 2847 (Ankole District, Kalinzu Forest Reserve, Feb. 1953, 1500
m.) may belong here rather than to *R. acuminatissima*.

6. **R. sp. C**

Small tree to 4.5 m.; stems with dull purplish brown flaky bark with some reddish
powder beneath, glabrous. Leaf-blades oblong-elliptic, sometimes asymmetric, 3–11 cm.
long, 2–4.7 cm. wide, shortly acuminate at the apex, cuneate at the base, glabrous, very
thin; petiole 4 mm. long; stipule-sheath brown and hyaline, 2 mm. long, densely setose-
pilose inside, with diverging subulate appendage 1.5 mm. long. Inflorescences 1–2-
flowered but no flowers seen; peduncle 2 mm. long; pedicels 2.5–3 cm. long. Immature
fruits globose with 1–2 pyrenes.

TANZANIA. Lindi District: Rondo Plateau, Mchinjiri, Mar. 1952, *Semsei* 700!
DISTR.   T 8; known only from the above gathering
HAB.   Not recorded

SYN. *R. sp. D* sensu Verdc. in K.B. 42: 153 (1987)

NOTE.   Material too poor for identification but not matched with any known species. The very long
pedicels are distinctive. See addendum p. 927.

7. **R. acuminatissima** (*K. Schum.*) *Robyns* in B.J.B.B. 11: 169 (1928); T.T.C.L.: 530 (1949); Fl. Rwanda 3: 216, fig. 68.3 (1985); Verdc. in K.B. 42: 153 (1987). Type: Tanzania, Bukoba, *Stuhlmann* 1014 (B, holo. †)

Shrub or small tree 1.8–9(–12, ? rarely–20) m. tall, ? sometimes ± scandent; branches mostly with dark purplish red bark often breaking down in small pieces to reveal a dark red-brown powdery surface; branchlets lenticellate, glabrous, ridged. Leaf-blades oblong-elliptic, elliptic or obovate, 3–14.5(–15) cm. long, 0.6–6(–8) cm. wide, narrowly and distinctly acuminate at the apex, the actual tip usually narrowly rounded, rarely acute, cuneate at the base, glabrous save for ± sparsely pubescent to hairy domatia, slightly discolorous, costa and main nerves sometimes drying yellowish, margins sometimes crinkly on drying; petiole 3–6 mm. long; stipule-bases ovate, connate, 2–4 mm. long, densely pilose inside and ± persistent, with a long subulate ± laterally compressed deciduous appendage 3.5–7 mm. long. Inflorescences (1–)2–6(–13)-flowered, often borne at leafless nodes; common peduncle usually short, 1–2(–10) mm. long; pedicels 1.5–8(–12 in fruit) mm. long; bracts ovate, 1.5–2 mm. long, forming an involucre ± 4 mm. wide. Buds clavate, obtuse. Calyx-tube 1–1.8 mm. long; limb 0.25 mm. long, hyaline, truncate or with traces of teeth. Corolla whitish or pale green; tube sometimes urceolate, 2–3.2 mm. long, with a ring of deflexed hairs inside; lobes ovate-triangular, 2–3 mm. long, 1.2–1.5 mm. wide. Ovary 3–4-locular; style 3 mm. long. Fruit subglobose, 0.7–1.3 cm. long, 6–10 mm. wide according to number of pyrenes; pyrenes 1–3(–4), ± 7.5 × 4.5 × 3.5 mm., ± reniform, angled above the central internal notch.

subsp. **acuminatissima**; Verdc. in K.B. 42: 154 (1987)

Common peduncle of inflorescence short, 2–4 mm. long.

UGANDA. Elgon, Bugisu [Bugishu], Bubungi, July 1926, *Maitland* 1231! & 1245! & Butandiga, 14 July 1924, *Snowden* 907b!
KENYA. Trans-Nzoia District: N. Elgon Forest Station, May 1933, *Dale* in *F.D.* 3119!; N. Kavirondo District: Kakamega Forest, Apr. 1934, *Dale* in *F.D.* 3248!; Kisumu-Londiani District: Tinderet Forest Reserve, 15 June 1949, *Maas Geesteranus* 4981!
TANZANIA. Bukoba District: Rubare Forest Reserve, Nov. 1958, *Procter* 1058! & Bukoba, Aug. 1931, *Haarer* 2128! & Minziro Forest, June 1952, *Procter* 60! & Kaagya, *Gillman* 358!
DISTR. U 3; K3,5; T 1; Rwanda
HAB. Evergreen forest understorey and edges, in Kenya especially in *Prunus, Podocarpus, Acacia lahai* associations; (1110–)1650–2400 m.

SYN. *Vangueria acuminatissima* K. Schum. in P.O.A. C: 385 (1895); De Wild. in B.J.B.B. 8: 42 (1922)
    [*Rytigynia neglecta* sensu Robyns in B.J.B.B. 11: 183 (1928), pro parte quoad *Maitland* 938, 1231, 1245 & *Snowden* 970b; K.T.S.: 473 (1961), pro parte, *non* (Hiern) Robyns sensu stricto]
    R. sp. C sensu Fl. Pl. Lign. Rwanda: 602, fig. 203.2 (1982)

NOTE. Unfortunately the type of *V. acuminatissima* has been destroyed and it is possible I have misidentified it. There is no doubt in my mind that the material from Elgon and W. Kenya is the same taxon as that occurring in Rwanda; the populations are very uniform with dark bark usually flaking to reveal a powdery undersurface; some sheets bear a manuscript name of A. Bullock's derived from Mt. Elgon although curiously there is no mention of the plant under any name in his account of the flora of Elgon (K.B. 1933: 49–106 (1933)). Robyn's description based on the original material describes the leaves as lanceolate, up to 1.5 cm. wide, which agrees, but Stuhlmann probably collected his specimens near the Lake and all the Bukoba material I have seen (which surely must be the same taxon as Stuhlmann's specimen) comes from 1100–1200 m. Ecologically it is more likely to be conspecific with the Sese Is. and other lowland Uganda material here treated as *R. dubiosa* and *R. umbellulata*; on the other hand the upland one does descend to 1300 m. in Rwanda. A study of *Rytigynia* populations in Bukoba, Masaka, Mengo and Busoga districts and their relation to more upland populations is needed and also a study of *R. umbellulata* throughout tropical Africa. It may well be that there is one variable lowland species *R. umbellulata* with an upland subspecies. Oversimplification, although frankly this is what the herbarium material suggests, would obscure the problems to be solved but may well ultimately be the correct solution. *Osmaston* 2847 (Uganda, Ankole District, Kalinzu Forest Reserve, Feb. 1953, 1500 m.) is difficult to place but has been referred to *R. dubiosa*. A sterile specimen *Uchara* 187! from T 4 (Mahali Mts.) may belong here. Subsp. *pedunculata* Verdc. with peduncles 5–10 mm. long occurs in Burundi.

8. **R. obscura** *Robyns* in B.J.B.B. 11: 149 (1928); T.T.C.L.: 529 (1949); Verdc. in K.B. 42: 157 (1987). Type: Tanzania, Rungwe District, Mwakalila, *Stolz* 2295 (B, holo. †, W, iso.!)

Glabrous shrub 2–3 m. tall, much branched; stems with wrinkled finely closely fissured grey bark. Leaves paired; blades elliptic, 2.3–6 cm. long, 1.2–2 cm. wide, acuminate at the apex, the acumen narrowly rounded or subacute, cuneate at the base into a margined petiole 2–4 mm. long, glabrous save for barbellate domatia in the axils of some nerves of some leaves beneath; margins wrinkled at least when dry; stipules with connate ± coriaceous ovate-triangular base, 2.5 mm. long, villous inside, with subapical subulate appendage 2–2.5 mm. long, ultimately deciduous. Buds acuminate. Flowers solitary in the axils of apices of younger branchlets; pedicels 5–6 mm. long, attaining 1.2 cm. in fruit. Calyx-tube hemispherical, 1.5 mm. long; limb denticulate, the teeth triangular, 0.5 mm. long, acute. Corolla greenish white; tube narrowly cylindrical, 6 mm. long, with a ring of deflexed hairs below the throat; lobes oblong-linear, 3 mm. long, the apical appendages 1 mm. long. Ovary 3–4-locular; style 6–7 mm. long; stigmatic club coroniform, 1–1.3 mm. long, 3–4-lobed. Fruit subglobose, 8 mm. long, 12 mm. wide, with 2–4 pyrenes.

TANZANIA.   Rungwe District: Mwakalila, Nov. 1913, *Stolz* 2295!; Njombe District: Madehani, Aug. 1913, *Stolz* 2104!
DISTR.   T 7; not known elsewhere
HAB.   Bamboo forest; 2000 m.
NOTE.   A virtually unknown species apparently never recollected.

9. **R. stolzii** *Robyns* in B.J.B.B. 11: 224 (1928); T.T.C.L.: 531 (1949); Verdc. in K.B. 42: 157 (1987). Type: Tanzania, Rungwe District, Masoko, *Stolz* 446 (B, holo. †, BM, K, PRE, iso.!)

Shrub 2–3.6 m. tall, usually much branched*, sometimes ± scandent; stems with very dark purplish brown ± minutely flaking bark, glabrous. Leaf-blades oblong to oblong-elliptic or lanceolate, 3–11 cm. long, 1.2–6 cm. wide, acuminate, rounded at the base, thin, ± glabrous or with a few very scattered seta-like hairs and pubescent domatia beneath; petiole 3.5–5 mm. long; stipules triangular, 3 mm. long, connate at the base, persistent, densely villous inside; appendage broadly subulate, ± fleshy, 4–8 mm. long, soon deciduous or sometimes completely lacking. Inflorescence 9–14-flowered; peduncle 2–4 mm. long; pedicels 2–4.5 mm. long; bracts 1.5 mm. long, connate into an involucre. Buds shortly acuminate, glabrous. Calyx-tube 1.5–2 mm. long; limb ± 0.2 mm. long, truncate. Corolla greenish yellow; tube 2.5–3 mm. long, with a ring of deflexed hairs inside; lobes 3–3.5 mm. long, 1.5–1.8 mm. wide, not or very shortly appendaged. Ovary 2-locular. Fruit black, bilobed in dry state, 8 mm. long, 7.5 mm. wide, with 2 pyrenes.

TANZANIA.   Chunya District: Rungwa Game Reserve about 1 km. W. of Itigi–Mbeya road, 28 Jan. 1969, *Yohannes* in *C.A.W.M.* 5061!; ?Mbeya District: N. slope of Poroto Mts., below sawmill near Iringa–Mbeya road, 29 Feb. 1932, *St. Clair-Thompson* 556! & 8 Mar. 1932, *St. Clair-Thompson* 739!; Rungwe District: Masoko, 30 Nov. 1910, *Stolz* 446!
DISTR.   T 7; not known elsewhere
HAB.   Evergreen thicket on termite mounds, *Brachystegia* woodland, edge of *Allophylus, Osyris, Carissa, Combretum* thicket, edge riverine or dry evergreen forest; 1380–1950 m.
NOTE.   There is considerable variation in leaf-shape but the material cited seems conspecific but the type is in flower whereas the other three sheets are in fruit and two are rather different in foliage from the type. Correlated material is needed to confirm my identification.

10. **R. pergracilis** *Verdc.* in K.B. 42: 157 (1987). Type: Tanzania, Lindi District, Mlinguru, *Schlieben* 5829 (BR, holo.!, K, iso.!)

Shrub to 1 m.; stems slender, glabrous, the older parts with thin peeling purplish brown epidermis, the surface straw-coloured beneath. Leaves borne mainly on short lateral branches in 4–6 pairs; blades elliptic-lanceolate or elliptic, 1.2–6.5 cm. long, 0.8–2.5 cm. wide, rounded to acuminate at the apex, the tip rounded, cuneate at the base, thin, glabrous above or with short seta-like hairs on midrib and with similar hairs on the midrib beneath and on petiole, also the margins ciliate; petiole 0.5–1 mm. long; stipule-bases 1–1.5 mm. long, with divergent subapical subulate appendage 1–4 mm. long, the longer ones rather curved. Cymes 2-flowered; peduncle ± 1–4 mm. long; pedicels 2.5–6 mm. long, elongating to 2.3 cm. in young fruit, sparsely bristly pubescent. Calyx-tube narrow, 0.5 mm. long, with spreading ± lobed limb 0.5 mm. tall. Buds yellow, subglobose, shortly acutely acuminate to prominently caudate, the appendages ± 1–2 mm. long, very sparsely pilose outside. Corolla-tube infundibular, 2 mm. long, with ring of deflexed hairs inside; lobes 2 mm. long, 1.5 mm. wide, spreading setulose outside, minutely puberulous inside,

---

* Robyns 'parce ramosa' was based only on the herbarium specimen and not on field data.

with appendage to 2 mm. long. Ovary 2–4-locular. Fruit globose, 7 mm. long and wide, the pedicels 8–10 mm. long; pyrenes 1–4.

TANZANIA. Kilwa District: Matumbi Hills, Nov. 1989, *Kingdon* 43!; Lindi District: Lake Lutamba and/or Mlinguru, 4 Jan. 1935, *Schlieben* 5829! & Rondo Plateau, 11 Feb. 1991, *Bidgood et al.* 1471!
DISTR. T 6, 8; not known elsewhere
HAB. Deciduous woodland; 240–800 m.

NOTE. The holotype bears the locality Mlinguru, but the Kew isotype label states Lake Lutamba. When I altered my original choice of the Kew material as holotype after seeing the fuller BR material I did not notice this label difference and gave the type locality as Lake Lutamba. Whether there is an error or the gathering was made up of material from different localities is not known.
   Bidgood et al. 1663 (Rondo Plateau, 750 m.) is a distinctive variant with abbreviated short lateral shoots and glabrous buds and pedicels.

11. **R. kiwuensis** (*K. Krause*) *Robyns* in B.J.B.B. 11: 156 (1928); Fl. Pl. Lign. Rwanda: 598, fig. 203.4 (1982); Fl. Rwanda 3:217, fig 68.4 (1985); Verdc. in K.B. 42: 158 (1987). Type: Rwanda, near Lake Kalago, *Mildbraed* 1519 (B, holo. †)

Shrub 1–2.4 m. tall, much branched; stems glabrous to sparsely to ± densely pubescent above with spreading ferruginous hairs, the older stems dark purplish grey, ridged and lenticellate, glabrescent; branches opposite or ternate. Leaf-blades oblong-elliptic, 2–10.5 cm. long, 0.8–4 cm. wide, acuminate at the apex, the actual acumen mostly narrowly rounded at the tip, broadly cuneate to rounded at the base, discolorous, shortly pubescent above, more densely hairy beneath, particularly on the venation with longer, softer, grey or ferruginous hairs, usually ± velvety; petiole 2–7 mm. long, densely pubesent; stipules ovate, 3–5 mm. long, ± connate, densely ferruginous pubescent outside, long pilose inside, with slightly subapical subulate appendage 3–6 mm. long, ± persistent. Flowers in several-flowered dense fascicles ± forming verticils at the nodes; peduncles obsolete; pedicels 1–2(–3) mm. long. Calyx pubescent or ± glabrous; tube campanulate, 1.2 mm. long; limb brown, glabrous, ± scarious, ± 1 mm. long, divided into triangular acuminate teeth 0.5 mm. long. Buds obovoid, obtuse or ± acuminate, glabrous. Corolla white, cream, yellow or greenish; tube 2.5–3.5 mm. long, glabrous inside; lobes ovate, 2–3.5 mm. long, 1–1.2 mm. wide, acute but without appendages. Ovary 2-locular; style ± 3.5 mm. long, thickened below; stigmatic club ellipsoid, 1.2 mm. long, sulcate beneath, apically bilobed. Fruits compressed-subglobose, 8 mm. long, 9–10 mm. wide (10 × 12 in life), 4 mm. thick, with (1–)2 pyrenes; pyrenes ± reniform, 7 × 4.5 × 4 mm., rugulose.

UGANDA. Kigezi District: Rubaya, 4 July 1945, *A.S. Thomas* 4257!
DISTR. U 2; Rwanda, Burundi, and Zaire
HAB. Evergreen forest; 1950 m.

SYN. *Vangueria kiwuensis* K. Krause in E.J. 57: 35 (1920)

12. **R. neglecta** (*Hiern*) *Robyns* in B.J.B.B. 11: 183 (1928), pro parte; E.P.A.: 1008 (1965); Verdc. in K.B. 42: 158 (1987). Ethiopia, "7000–8000 ft." [Begemder], *Schimper* 1106 (K, holo.!, BM, E, iso.!)

Shrub or small tree 1–9 m. tall, with spreading branches often flowering when leaves are ± unexpanded; young shoots brown, ridged and lenticellate, the older with pale grey corky bark. Leaf-blades elliptic, 2–14.5 cm. long, 1.5–6.5 cm. wide, tapering acuminate at the apex, the tip acute, cuneate to ± rounded at the base, mostly drying pale green with costa yellowish, glabrous save for sparsely hairy domatia or with sparse pubescence; petiole 5–8 mm. long, yellowish; stipule-bases joined to form a sheath 3(–4) mm. long, pilose within and usually a decurrent appendage projecting 2–3(–7) mm. long. Cymes umbellike or fasciculate, up to 10-flowered but usually (1–)3–6: peduncle 1–2 mm. long or ± suppressed; pedicels 1–3 mm. long, up to 4–7 mm. long in fruit; bracts forming an involucre which when peduncle is poorly developed appear sessile at the node. Buds completely rounded. Calyx-tube 1.5 mm. long; limb 0.3–0.5 mm. long, minutely denticulate. Corolla green or yellow; tube 2 mm. long, with a ring of deflexed hairs within. Ovary 2-locular; style 3.5 mm. long; stigmatic club large and bilobed, 1.2 mm. wide. Fruit black, didymous and closely resembling many species of *Canthium*, 8–9 mm. long and wide, with 1–2 pyrenes.

var. **neglecta**; Verdc. in K.B. 42: 158 (1987)

Leaves glabrescent.

UGANDA. Karamoja District: Mt. Morongole, Feb. 1959, *J. Wilson* 720! & Imatong Mts., Apr. 1938, *Eggeling* 3609! & Mt. Debasien, 31 May 1939, *A.S. Thomas* 2954!

KENYA. Northern Frontier Province: S. Mt. Kulal, 23 Feb. 1977, *Herlocker* H.408! & 6 Sept. 1944, *J. Adamson* 122! & 3 km. N. of Gatab, 18 Nov. 1978, *Hepper & Jaeger* 6911!

DISTR. **U** 1, ?3 (see note); **K** 1–3; Ethiopia, and Sudan

HAB. Montane forest of *Podocarpus, Juniperus, Teclea, Olea*, etc., also forest edges, often in rocky places; (1350–)1890–2400 m.

SYN. *Canthium neglectum* Hiern in F.T.A. 3: 135 (1877)

NOTE. Although usually glabrous, forms with some pubescence occur in Ethiopia and Kenya; the stems have a few hairs and the young leaves have scattered pubescence. The following Kenya material has been seen — Marsabit, July 1958, *T. Adamson* 2!; W. Suk, Kapenguria, 6 May 1954, *Padwa* 54! & Cherangani, between Kapenguria and Lelan Forest, 13 May 1979, *Bridson* 106. This form accounts for the 1350 m. recorded above and the records from **K** 2 and 3. Robyns was in error sinking *Vangueria ituriensis* into *R. neglecta*. Var. *vatkeana* (Hiern) Verdc., type: Ethiopia, Lake Tana, R. Reb [R. Repp], *Schimper* 1130 (K, holo.!), is a variety of *R. neglecta* which is much more densely pubescent. It is possible that *Webster* 1060! (Uganda, Mbale District, Bugisu, Sebei, Sipi, Oct. 1955) is this species in which case it is the southernmost record. There is every phytogeographical reason for it to occur on N. Elgon; the specimen is in fruit only.

13. **R. ruwenzoriensis** (*De Wild.*) *Robyns* in B.J.B.B. 11: 166 (1928) & F.P.N.A. 2: 348 (1947); Fl. Pl. Lign. Rwanda: 600, fig. 204.2 (1982); Fl. Rwanda 3: 217, fig. 69.2 (1985); Verdc. in K.B. 42: 158 (1987). Type: Zaire, Ruwenzori, Lamia valley, *Bequaert* 4299 (BR, lecto.!)

Shrub or small tree 2–6(–15) m. tall; older stems dark reddish purple with fine longitudinal fissures and ± peeling epidermis but not becoming powdery, lenticellate, glabrous; young shoots olive, closely longitudinally ribbed in the dry state, the internodes often short; slash red or chocolate-coloured. Leaf-blades narrowly elliptic to oblong-lanceolate or lanceolate, (2–)5–13 cm. long, 1–3.5 cm. wide, with a long slender acumen, cuneate at the base, somewhat discolorous, glabrous save for the distinct hairy acarodomatia; petiole 4–10 mm. long; stipules with sheath part 1.5–3 mm. long, pilose inside, usually with rather divergent subulate appendage 1–2 mm. long but sometimes ± absent. Cymes (2–)4–7(–12)-flowered; peduncle 1–2(–5) mm. long; pedicels 2.5–3 mm. long, attaining 1.2–2.7 cm. in fruit; bracts 1.5 mm. long, forming a small involucre. Buds obtuse or very slightly acuminate. Calyx-tube campanulate, 1–1.5 mm. long; limb 0.5–0.7 mm. long, ± truncate or with short lobes ± 0.3 mm. long. Corolla green, greenish white or cream; tube 2–3.5 mm. long, with a ring of short deflexed hairs inside; lobes oblong, 2–3.3 mm. long, 1.5 mm. wide, not or only with minute appendage. Ovary 2(rarely 3)-locular; stigmatic club bilobed. Fruit didymous, 1.1–1.2 cm. long, 1–1.2 cm. wide, compressed, with 2 pyrenes.

UGANDA. Toro District: Ruwenzori, Mahoma R., opposite Nyabitaba, 9 Sept. 1951, *Osmaston* 1168! & same locality, Jan. 1951, *Osmaston* 3675! & Mobuku valley, Bikoni, 30 Dec. 1934, *G. Taylor* 2726!; Kigezi District: S. Impenetrable Forest, Luhizha [Luhiza], June 1951, *Purseglove* 3663, pro parte!

DISTR. **U** 2; Zaire

HAB. Clearings in ridge-top forest and forest edges; (1500–)2000–2700 m.

SYN. *Vangueria ruwenzoriensis* De Wild. in B.J.B.B. 8: 61 (1922) & Pl. Bequaert. 2: 277 (1923)
    V. ruwenzoriensis De Wild. var. *breviflora* De Wild. in B.J.B.B. 8: 62 (1922) & Pl. Bequaert. 2: 278 (1923). Type: Zaire, Ruwenzori, *Bequaert* (BR, holo.!)
    Rytigynia ruwenzoriensis (De Wild.) Robyns var. *breviflora* (De Wild.) Robyns in B.J.B.B. 11: 167 (1928) & F.P.N.A. 2: 349 (1947)

NOTE. *Maitland* 938 (Ruwenzori, Ibanda, 1500–1800 m.) is annotated by Robyns as '*R. ituriensis*' but cited as *R. neglecta* into which he finally sank *Vangueria ituriensis* De Wild.; the label also states 'ovario 3–4-mero'. The specimen, however, has a 2-locular ovary and is *R. ruwenzoriensis* so possibly some confusion of specimens has occurred. *Mbaguta* 4 reports the tree reaching 15 m. *Katende* 576 (Kigezi, Impenetrable Forest, 30 Sept. 1970, 2432 m.) has peduncles to 6 mm. but probably belongs here. The Uganda material would all come under var. *breviflora* if that were retained. All I have seen but one have been 2-locular but I have of course been unable to check the ovaries of the original type thoroughly. The matter needs checking in the field. There may have been some confusion on De Wildeman's part. A drawing on the type sheet depicts three ovary sections, two with three locules and one with two yet Robyns says 2-locular in his revision and notes and does not mention this. De Wildeman also states 2–3-locular in his original description.

14. **R. bridsoniae** *Verdc.* in K.B. 40: 656 (1985), as '*bridsonii*'; Fl. Rwanda 3:317, fig. 69.3 (1985); Verdc. in K.B. 42: 159, fig. 9 (1987). Type: Rwanda, Wisumo, Gisovu, Swiss Forestry Centre, *Bridson* 407 (K, holo.!, BR, HNR, iso.)

Shrub or small tree (2–)3–6 m. tall; branches slender, glabrous or with sparse bifarious pubescence, brown and ridged and somewhat glossy, lenticellate. Leaf-blades elliptic or oblong to narrowly oblong, 2.5–10 cm. long, 0.8–4.2 cm. wide, acuminate at the apex, cuneate at the base, glabrous save for sparsely pubescent domatia beneath, scarcely discolorous; petiole 2–7 mm. long, slightly winged; stipule-bases 2–3 mm. long, pilose inside, the appendages short, subulate, ± 1.5 mm. long. Inflorescences 2-flowered; peduncles 4–8 mm. long; pedicels 1.5–3.5 mm. long, glabrous or sometimes sparsely pubescent; bracts ovate, 1.5–2 mm. long, acuminate. Calyx-tube 1 mm. long, the limb ± 1 mm. long, 5-toothed. Buds distinctly acuminate. Corolla greenish cream to yellow; tube 2–4.5 mm. long; lobes 2.5–3 mm. long, 1.2–1.5 mm. wide, with appendage 0.7–2 mm. long. Ovary 2-locular; stigmatic club distinctly 2-lobed above. Fruit violet to black, obcordate, 0.8–1.1 cm. long and wide, with 2 pyrenes; fruiting peduncle to 1.5 cm.

subsp. **bridsoniae**; Verdc. in K.B. 42: 160, fig. 9 (1987).

Leaves larger, thinner and more acuminate.

UGANDA. Kigezi District: Luhizha [Luhiza], June 1951, *Purseglove* 3664! & same locality and date, *Purseglove* 3663 in part! & Impenetrable Forest, Mar. 1947, *Purseglove* 2369!
DISTR.   U 2; Rwanda and Burundi
HAB.   Bamboo forest; 2400 m.

SYN. *R. sp. A* sensu Fl. Pl. Lign. Rwanda: 600, fig. 204.3 (1982)

NOTE.   Only sheet 2 of *Purseglove* 3663 at Kew belongs here, the rest being *R. ruwenzoriensis*. The species can easily be confused with *Canthium oligocarpum* Hiern. *Purseglove* 3664 has much bigger flowers than the type with wider calyx-lobes, longer corolla-tube, also pubescent inflorescence-axes; it may prove to be a distinct variant but more Uganda material is needed. Subsp. *kahuzica* Verdc. with smaller thicker leaves occurs in Zaire (S. Kivu).

15. **R. kigeziensis** *Verdc.* in K.B. 40: 656 (1985); Fl. Rwanda 3:217, fig 69.5 (1985); Verdc. in K.B. 42: 163, fig. 10 (1987). Type: Uganda, Kigezi, Luhizha [Luhiza], *Purseglove* 3665 (K, holo.!, EA, iso.!)

Shrub or small tree 1–8 (4.5–6 in Uganda) m. tall, with grey smooth persistent bark on the trunk; stems glabrous, lenticellate with dark red or greenish bark cracking into small flakes, very often peeling to reveal a red powdery layer. Leaf-blades narrowly ovate to elliptic-lanceolate, 1.5–9(–12) cm. long, 0.7–3.8(–5) cm. wide, long-acuminate at the apex, the actual tip narrowly rounded, rounded to cuneate at the base, paler beneath, often shiny above, glabrous and domatia often ± poorly developed with few hairs, the venation often characteristically reticulate beneath; petiole 1.5–6 mm. long; stipules with bases joined 1.5–3 mm., hairy inside, with appendage 0.5–4 mm. long. Inflorescence said to be sweet-scented, 2-flowered; peduncle 0.4 mm. long; pedicels 5–8.5(–10 in fruit) mm. long; bracts paired, ovate, 2 mm. long. Buds acuminate to distinctly 5-tailed. Calyx-tube 1.2 mm. long; limb ± 1 mm. long, hyaline, subtruncate to minutely denticulate. Corolla white or tinged green; tube 4–5 mm. long, 1.5–1.7 mm. wide, glabrous outside with deflexed hairs inside; lobes 3–4 mm. long excluding the finely tapering-acute appendage (0.5–)1.5–2.5(–3) mm. long. Ovary 2–3-locular; style ± 6 mm. long, with stigmatic club coroniform 1.2 mm. long. 1.5 mm. wide. Fruit 1–1.2 cm. long, 6–10 mm. wide, with 1–2 pyrenes, often drying ± curved when only 1 pyrene. Fig. 144, p. 816.

UGANDA. Kigezi District: Luhizha [Luhiza], June 1951, *Purseglove* 3665! & Mafuga Forest Reserve, Apr. 1969, *Hamilton* 1045!
DISTR.   U 2; Zaire, Rwanda and Burundi
HAB.   Evergreen forest of *Prunus, Polyscias, Albizia, Maesa, Xymalos*, etc.; 1800–2340 m.

NOTE.   One or two Zaire specimens have some bristly pubescence on the venation beneath.

SYN. *R. sp. B* sensu Fl. Pl. Lign. Rwanda: 602, fig. 204.5 (1982)

16. **R. ignobilis** *Verdc.* in K.B. 42: 164 (1987). Type: Tanzania, Mpwapwa District, Kiboriani Range, Kongwa, *Anderson* 583 (EA, holo.!)

FIG. 144. *RYTIGYNIA KIGEZIENSIS* — **A**, flowering branch, × 1; **B**, stipule, × 4; **C**, bud, × 4; **D**, flower, × 4; **E**, longitudinal section through flower, × 4; **F**, inside upper part of corolla-tube, × 8; **G**, anther, × 10; **H**, style and stigmatic knob, × 6; **J**, longitudinal section through ovary, × 10; **K**, transverse section through ovary, × 10; **L**, part of fruiting branch, × 1; **M**, fruit, × 2; **N**, pyrene, × 2. A–K, from *Bridson* 409; L–N, from *Lewalle* 3622. Drawn by Mrs M.E. Church.

Shrub ± 2 m. tall; stems dark purple-brown, minutely fissured and peeling in small pieces, glabrous. Uppermost young leaves often lanceolate or oblong-lanceolate, ± 4 × 1–1.5 cm., tapering at apex, cuneate at base, later leaves oblong up to 5 cm. long, 3 cm. wide, mostly shortly obtusely acuminate at the apex, ± rounded at the base, ± discolorous, glabrous, without domatia, venation evident and reticulate beneath in dry state but not raised; petioles short, scarcely 1 mm. long; stipule-bases joined to form a sheath 1–1.5 mm. long, white-villous inside, the appendage subulate, 3–5 mm. long, at length deciduous. Inflorescences 2-flowered; peduncle 3–11 mm. long; pedicels 3–7 mm. long; bracts small, 0.5–1.5 mm. long, not opposite. Buds subobtuse or minutely apiculate, glabrous. Calyx-tube campanulate, 1.5–2 mm. long, the limb narrow, hyaline, ± 0.2 mm. tall, minutely denticulate. Corolla cream; tube 1.8 mm. long, with ring of matted hairs at top inside; lobes oblong, 2.2 mm. long, 1 mm. wide, acute but not appendaged. Anthers ± 1 mm. long, the connective with a small muricate appendage. Ovary 2-locular; style 3.2 mm. long, slightly swollen at the base, the stigmatic club truncate-obconic, 0.8 mm. long, sulcate. Fruits not seen.

TANZANIA.   Mpwapwa District: Kiboriani range, Kongwa, 22 Jan. 1950, *Anderson* 583!
DISTR.   **T** 5; not known elsewhere
HAB.   *Brachystegia* woodland on hill summit in shallow soil over quartzite; 1530 m.

17. **R. eickii** (*K. Schum. & K. Krause*) *Bullock* in K.B. 1932: 389 (1932); T.T.C.L.: 529 (1949); Verdc. in K.B. 42: 165 (1987). Types: Tanzania, Lushoto District, near Kwai, *Eick* 86 (B, syn. †, K, fragment!) & near Muafa, *Buchwald* 611 (B, syn. †, BM, K, isosyn.!)

Shrub or small tree, 1.5–5 m. tall, the stems with pale shiny epidermis peeling to reddish grey beneath, glabrous. Leaf-blades oblong to ovate-oblong, 4–14.5 cm. long, 1.7–6.8 cm. wide, shortly obtusely acuminate at the apex, cuneate to ± rounded at the base, drying olivaceous, ± discolorous, glabrous save for small pubescent domatia beneath; petiole 4–8 mm. long; stipules joined to form 1.5–3 mm. long sheath, pilose inside, the appendages subulate 2–5 mm. long, blackish on drying, deciduous. Cymes 2-flowered; peduncle (1.5–)4–5 mm. long; pedicels 2–4 mm. long; bracts obsolete or ± 1 mm. long. Buds shortly acuminate. Calyx-tube 1–1.2 mm. long, the limb ± 0.5 mm. long, truncate or shortly toothed. Corolla pale greenish yellow; tube funnel-shaped, 3 mm. long, with a ring of scale-like deflexed hairs inside; lobes pale yellow inside, 2.5–3.5 mm. long, 2 mm. wide, shortly apiculate but not appendaged. Ovary 2-locular. Fruits usually didymous, 1–1.1 cm. long, 1.3–1.5 cm. wide, 6.5 mm. thick, the pedicels up to 1.5 cm. long and peduncle up to 7 mm.; pyrenes 1–2, 1 cm. long, 7–8 mm. wide, 5–6.5 mm. thick.

KENYA.   Teita Hills, Ngangao Forest, 11 Feb. 1977, *R.B. & A.J. Faden* 77/335! & 9 May 1985, *Faden et al.* 334! & 15 May 1985, *Faden et al.* 473!
TANZANIA.   Lushoto District: E. Usambara Mts., Monga Hill, 23 Mar. 1912, *Grote* in Herb. *Amani* 3589! & W. Usambara Mts., Jaegertal, 11 Jan. 1967, *Archbold* 901A! & Soni, 25 Feb. 1972, *Faulkner* 4706!; Morogoro District: W. Nguru Mts. above Maskati, 16 Mar. 1988, *Bidgood et al.* 508!
DISTR.   **K** 7; **T** 3,6; not known elsewhere
HAB.   Open bushland in granite areas, submontane forest, etc.; 950–1830 m.

SYN.   *Plectronia eickii* K. Schum. & K. Krause in E.J. 39: 538 (1907)
        Rytigynia biflora Robyns in B.J.B.B. 11: 178 (1928); T.T.C.L.: 530 (1949). Type: Tanzania, Usambara Mts., *Buchwald* 635 (B, holo. †, BM, K, iso.!)

NOTE.   *Peter* 52182 from the S. Pare Hills, Suji to Taë, 4 Mar. 1915, may be this species.

18. **R. sp. D**

Presumably a shrub; stems slender, quite glabrous, even on the youngest parts. Leaves elliptic to oblong-elliptic, 2.3–7 cm. long, 0.8–3.6 cm. wide, abruptly acuminate, the actual apex narrowly rounded, rounded at the base, thin, with characteristic ciliation around the margins of the lamina, also some scattered hairs above and on venation beneath; petiole 1.5–2 mm. long, margined; stipules joined at base into sheath 2.5–3 mm. long, densely pilose within, deciduous; appendage compressed, broadly subulate, divergent, 2.5–3 mm. long. Flowers and fruits unknown.

TANZANIA.   Tanga District: Mt. Mlinga, on summit, 2 Feb. 1917, *Peter* 19385!
DISTR.   **T** 3; not known elsewhere
HAB.   Probably rain-forest; 1080 m.

SYN.   *R. sp. E* sensu Verdc. in K.B. 42: 165 (1987)

NOTE. No other material has been seen from this well-known but rarely collected locality. An endemic plant there would be surprising but not impossible. Apparently now the natural vegetation has been entirely destroyed.

19. **R. uhligii** (*K. Schum. & K. Krause*) *Verdc.* in K.B. 42: 165, fig. 1F, G, 5F & 8D (1987). Type: Tanzania, Kilimanjaro, *Uhlig* 521 (B, holo. †, EA, iso.!)

Shrub or tree 0.9–9 m. tall, ? rarely scandent, laxly branched; bark smooth, pale brown to blackish grey; stems pale chestnut-brown or purplish, longitudinally ridged, lenticellate, glabrous, the epidermis or bark wearing off to reveal pale undersurface. Leaves mostly not fully developed at flowering time; blades elliptic to ovate, 1.5–11 cm. long, 0.5–4.8(–5.2) cm. wide, narrowly or abruptly acuminate at the apex, the tip usually rounded, cuneate, rounded or ± truncate at the base, glabrous save for obvious tufted domatia or rarely sparsely pubescent, ± thin, not or only slightly discolorous; petiole 2–6 mm. long; stipule-bases triangular or ovate, joined, 3–5 mm. long, densely pilose within, with ± spreading often compressed subulate appendage 1–4.5 mm. long, decurrent or sometimes completely lacking. Inflorescences 1–2-flowered, rarely more; peduncle either ± suppressed or up to 6(–10) mm. long particularly in fruit; pedicels (0.2–)0.6–1.2 cm. long, often 1.5–2.3(–3) cm. long in fruit; bracts up to 1.5 mm. long. Buds oblong, very obtuse, the limb part characteristically much shorter than the tube. Calyx-tube 1.2–1.5 mm. long, the limb a rim ± 0.2 mm. long, with obsolete or very small teeth. Corolla white to greenish yellow; tube mostly subglobose or urceolate-campanulate or sometimes ± cylindrical, 4.5–5.5 mm. long, with a ring of deflexed hairs inside; the limb usually ± closed; lobes mostly green, ovate, 2 mm. long, 1.5 mm. wide, acute but not appendaged. Ovary (2–)3–5-locular; style swollen at the base; stigmatic club just exserted or exserted up to 2.5 mm., 2–5-lobed. Fruit blue-black, globose, (7–)9–10(–? 13 in life) mm. diameter, containing 1–5 pyrenes 6–7 × 3–4 mm. Fig. 132/11, p. 754.

KENYA. Machakos District: Kilungu Forest, N. of Nunguni, 9 Jan. 1972, *Mwangangi* 1942!; Masai District: Chyulu Hills, 19 Oct. 1969, *Gillett & Kariuki* 18845!; Teita Hills, footpath between Wusi [Wuzi]–Ngerenyi road and Bura Bluff, 17 Sept. 1953, *Drummond & Hemsley* 4399!
TANZANIA. Arusha District: S. slope of Mt. Meru, 1 Nov. 1959, *Greenway* 9607!; Lushoto District: Monga, 24 Nov. 1916, *Zimmermann* in *Herb. Amani* 7916!; Iringa District: Mufindi, Luisenga [Lusenga] R., 17 Jan. 1971, *Paget-Wilkes* 858!
DISTR. **K** 1, 4, 6, 7 (Teita); **T** 2, 3, 5–8; ?Malawi, and Zimbabwe
HAB. Montane evergreen forest, particularly with *Zanthoxylum, Olea, Casearia, Diospyros abyssinica, Maesa, Neoboutonia, Allophylus, Celtis*, and derived grassland with scattered trees or thickets, sometimes riverine; (950–)1000–2250 m.

SYN. [*Vangueria neglecta* sensu K. Schum. in P.O.A. C: 384 (1895), *non* Hiern]
   V. uhligii K. Schum. & K. Krause in E.J. 39: 534 (1907); De Wild. in B.J.B.B. 8: 65 (1922)
   Plectronia kidaria K. Schum. & K. Krause in E.J. 39: 539 (1907). Type: Tanzania, W. Usambara Mts., Kwai, *Albers* 121 (B, holo. †)
   Rytigynia euclioides Robyns in B.J.B.B. 11: 146 (1928); T.T.C.L.: 529 (1949); K.T.S.: 472 (1961). Type: Tanzania, Mbulu [Umbulu], *Holtz* 3341 (B, holo. †)
   R. schumannii Robyns in B.J.B.B. 11: 158 (1928); T.T.C.L.: 530 (1949); K.T.S: 473 (1961). Type: Tanzania, Kilimanjaro, Marangu, *Volkens* 1450 (B, holo. †)
   R. schumannii Robyns var. *uhligii* (K. Schum. & K. Krause) Robyns in B.J.B.B. 11: 158 (1928); T.T.C.L.: 530 (1949)
   R. undulata Robyns in B.J.B.B. 11: 159 (1928); T.T.C.L.: 530 (1949). Type: Tanzania, E. Usambara Mts., Monga, *Grote* in *Herb. Amani* 3898 (B, holo. †, EA, iso.)
   R. lenticellata Robyns in B.J.B.B. 11: 181 (1928); T.T.C.L.: 530 (1949). Type: Tanzania, Uluguru Mts., Luhangala [Lussangeli], *Stuhlmann* 8739 (B, holo. †)
   R. kidaria (K. Schum. & K. Krause) Bullock in K.B. 1932: 389 (1932); T.T.C.L.: 529 (1949)
   Canthium sarogliae Chiov., Racc. Bot. Miss. Consol. Kenya: 55 (1935). Types: Kenya, Meru, *Balbo* 17 (TOM, syn.!) & Nyeri, *Balbo* 826 (TOM, syn.!)
   Rytigynia murifolia Gilli in Ann. Naturhist. Mus. Wien 77: 26, fig. 4 (1973). Type: Tanzania, Songea District, Matengo Hills, *Zerny* 347 (W, holo.!, K, photo.!)

NOTE. The variation in this widespread and common species is considerable and in some areas rather diffuse races might be recognized, but the characters do not hold up well enough for this to be practical. In the W. Usambaras forms with short pedicels scarcely 2 mm. long have been called *R. kidaria* (e.g. Lushoto District, Shume, World's View, 1 June 1957, *Verdcourt* 1758!); these occasionally have up to 4-flowered inflorescences. Specimens with several-flowered inflorescences and often lenticellate shoots have also been collected in the Uluguru Mts. (e.g. *E.M. Bruce* 197, Kitundu, 22 Nov. 1935, and *Harris & Pócs* 3275, Bondwa, 7 Sept. 1969). A few specimens have been seen (e.g. *B.D. Burtt* 1281 from Kondoa District, Irangi, *B.D. Burtt* 974 from Kondoa, Ghost Mt., *Wilson* 34 from North Pare and *Ritchie* from Moshi, Uru) in which the stipules have no appendages, but I doubt if this is a constant character; one of the syntypes of *Canthium sarogliae* is a similar form.

Iversen et al. 85703 (W. Usambaras, summit of Ndamanyilu, 2200–2270m. in rocky dry evergreen forest) has very narrow leaves but is sterile; it probably belongs here. *R. murifolia* Gilli is a form with the leaf-acumen more acute than usual and *V.G. van Someren* 133 (Masai District, Emali Hill, 15 Mar. 1940) has ± ovate leaves with truncate or even subcordate bases but is probably only a form of *R. uhligii.* I am not certain of the identity of *Vanguera neglecta* Hiern var. *puberula* K. Schum. (P.O.A. C: 384 (1895); type: Tanzania, Usambaras, *Holst* 247 (B, holo. †), *Rytigynia schumannii* Robyns var. *puberula* (K. Schum.) Robyns in B.J.B.B. 11: 159 (1928)). Several specimens from Lushoto District, Shume, e.g. *Msuya* 13, have very sparse bristly hairs on the midribs and margins and are undoubtedly *R. uhligii. Ruffo* 168 (Lushoto District, N. Kitivo Forest Reserve), with more densely hairy juvenile leaves, inflorescence axis and calyces, may genuinely represent this variety. Some other more pubescent but inadequate specimens from the W. Usambaras I have referred to *R. flavida.*

20. **R. adenodonta** (*K. Schum.*) *Robyns* in B.J.B.B. 11: 148 (1928); T.T.C.L.: 529 (1949); Verdc. in K.B. 42: 166 (1987). Type: Tanzania, Rungwe District, Uwarungu Mt., *Goetze* 1454 (B, holo. †, BR, iso.!)

Much-branched shrub or small tree 1.5–3.6 m. tall, entirely glabrous (save for domatia and insides of stipules) or with young shoots slightly to densely adpressed pubescent; older branches greyish, ± nodulose and verrucose, the bark often red-brown and fissured and flaky. Leaf-blades elliptic to broadly ovate-elliptic, 1.3–6 cm. long, 0.4–2.8 cm. wide but usually quite small in Flora area, abruptly shortly obtusely acuminate at the apex, ± rounded or cuneate at the base, scarcely discolorous, glabrous save for small pubescent domatia which may be almost or quite absent or, rarely, in Malawi, pubescent beneath and on midrib above; petioles 3–4 mm. long; stipules connate at the base, glabrous outside, densely long-villous inside, 2 mm. long, abruptly subulate-cuspidate, the appendage 1–2 mm. long. Flowers solitary or in lax 2–4-flowered cymes; pedicels rather thick, 3–5(–11) mm. long. Buds ± rounded to acuminate at the apex. Calyx-tube 1.5 mm. long; limb obsoletely dentate, truncate or teeth broadly triangular, 0.8 mm. long. Corolla creamy white; tube obconic, dilated from the base, 5–6 mm. long 3.5–5 mm. wide, with a ring of deflexed long hairs inside from the top of the tube; lobes oblong, reflexed, reaching about the middle of the tube, ovate, ± 3–3.5 mm. long, 2–2.5 mm. wide, obtuse or with short appendage to 0.5 mm. long. Ovary 3–4-locular; stigmatic club coroniform, 1 mm. wide, 3–4-lobed. Fruit purple-black, ± 8 mm. long.

var. **adenodonta**; Verdc. in K.B. 42: 166 (1987)

Young stems often hairy, with mostly abbreviated nodose lateral branches. Leaf-blades small, ± 2–2.5 × 1–1.5 cm., obtuse to acuminate.

TANZANIA.   Ufipa District: Malonje Hill, 9 Nov. 1933, *Michelmore* 760!; Mbeya Peak Forest Camp, Oct. 1959, *Procter* 1514!; Rungwe District: Umalila, Uwarungu Mt., Nov. 1899, *Goetze* 1454!
DISTR.   T 4, 7; Malawi and Zambia (Nyika Plateau)
HAB.   Dense bush and riverine forest; 2200–2550 m.
SYN.   *Vanguera adenodonta* K. Schum. in E.J. 30: 414 (1901); De Wild. in B.J.B.B. 8: 42 (1922)

var. **reticulata** (*Robyns*) *Verdc.* in K.B. 42: 167 (1987). Type: Malawi, Mt. Zomba, *Whyte* (K, holo.!)

Young stems usually glabrous, with side branches developed and leafy. Leaf-blades up to 7.5 × 4.5 cm., acuminate.

TANZANIA.   Iringa District: Mufindi, Kigogo, 31 Oct. 1947, *Brenan, Greenway & Gilchrist* 8246 & same locality, 18 Dec. 1961, *Richards* 15743! & Tapu, Nov. 1953, *Carmichael* 277! & Image Forest Reserve, Nov. 1959, *Procter* 1519!
DISTR.   T 7; Malawi and Zambia (Nyika Plateau)
HAB.   Upland evergreen forest, bamboo gorge forest; 1800–1950 m.
SYN.   *R. reticulata* Robyns in B.J.B.B. 11: 147 (1928)

NOTE.   I had at first treated the Iringa populations as a subspecies of *R. uhligii* differing in having rather longer wider corolla-tubes and longer lobes, thicker more discolorous leaves and often more developed calyx-teeth although one specimen in the same area *Polhill & Paulo* 1820 from Kigogo R., 1590 m. is nearer *R. uhligii;* there seems no doubt, however, that the populations are to be included in *R. adenodonta.* The situation in Malawi is less clear and the relationship with *R. uhligii* needs more study.

21. **R. pubescens** *Verdc.* in K.B. 42: 168 (1987). Type: Tanzania, Iringa District, Sao Hill, *Greenway* 6400 (K, holo.!, EA, iso.!)

A stiffly branched shrub 1.8–3 m. tall; bark on old shoots dark grey, often cracked and fissured to reveal a rusty undersurface; young shoots dark purplish, covered with short rusty hairs, thickened at the nodes; leafless older parts of side shoots often with approximate nodes. Branches and leaves often in whorls of 3 on main shoots but leaves paired on side shoots; leaves born on abbreviated shoots or slender side shoots, those on the main stems having fallen off; blades oblong, narrowly ovate-elliptic, ovate or almost round and perhaps not fully developed at flowering time, 1.5–6 cm. long, 0.8–3.5 cm. wide, rounded to acute or shortly acuminate at the apex, rounded to cuneate at the base, ± velvety on both surfaces with yellowish hairs but with an overall greyish look to the foliage; petioles 2 mm. long; stipule-bases 1.5 mm. long, densely brown pilose within and very obvious on older shoots but appendage ± obsolete. Cymes 1–3-flowered; peduncle ± suppressed or 2–3 mm. long and pedicels 5–6 mm. long, both pubescent; bracts scarcely 1 mm. long. Calyx-tube 1.2–1.5 mm. long; limb ± 0.6 mm. long, truncate. Buds clavate, obtuse, glabrous, limb-part about ½ the length of the tube. Corolla yellowish white or greenish yellow; tube 5 mm. long, hairy inside just below the throat but rest glabrous, glabrous outside; lobes elliptic, 3 mm. long, 1.5 mm. wide, acute but not appendaged. Ovary 3-locular. Young fruits globose, 7 mm. diameter, the pedicels up to ± 1.3 cm. long.

TANZANIA.   Mbeya District: Umalila, Inyala, Nov. 1975, *Leedal* 3066!; Iringa District: Sao Hill, 24 Nov. 1980, *Ruffo* 1609! & Ngowasi [Ngwazi], 3 Feb. 1987, *Lovett* 1423!
DISTR.   T 7; not known elsewhere
HAB.   Bushland clumps of *Apodytes, Byrsocarpus, Rhus, Erythrina* and *Parinari* in seasonally burnt *Hyparrhenia* grassland, sometimes on termite mounds; 1830–2135 m.

NOTE.   *Ruffo* 1609 with its slender side-shoots and subtending leaves in whorls of 3 has a very different facies from the type but seems to be conspecific. The species may grow taller since it is used for fuel and poles.

22. **R lichenoxenos** (*K. Schum.*) *Robyns* in B.J.B.B. 11: 156 (1928); Verdc. in K.B. 36: 539 (1982) & in K.B. 42: 169, fig. 1K (1987). Type: Tanzania, Uluguru Mts., Lukwangule Plateau [Mt. Lukwan], *Stuhlmann* 9120 (B, holo. †)

Shrub or small tree, 1.8–10 m. tall; young shoots with bright red-rusty velvety indumentum at length falling off to reveal reddish brown finely longitudinally ridged peeling bark or stems grey-brown pubescent only on younger parts, lenticellate. Leaves paired; blades elliptic to elliptic-oblong or ± lanceolate, 1.5–8 cm. long, 0.9–3 cm. wide, obtusely acuminate at the apex, rounded and sometimes a little unequal or cuneate at the base, very discolorous, shortly ferruginous pubescent above but not obscuring the dark surface, yellowish grey velvety beneath; petiole 2–3 mm. long; stipules (2.5–)6–7 mm. long, with a subulate apex from a broad triangular base, the base ± persistent, densely pilose within. Flowers in almost sessile, 1–3(–6)-flowered cymes; peduncle and pedicels 1–2(–3) mm. long, densely pubescent, sometimes becoming 5–6 mm. long in fruit; bracts hyaline, ovate, ± 2.5 mm. long, glabrous. Calyx-tube subglobose-ovoid, 1.5–2 mm. long, 3 mm. wide, densely ferruginous pubescent contrasting with the glabrous limb which is hyaline, the tubular part under 0.5 mm. tall, and the lobes broadly ovate, 0.5–1(–1.5) mm. long. Buds obtuse. Corolla greenish white or greenish yellow; tube cylindric-obconic to slightly funnel-shaped, 5–7 mm. long, sometimes distinctly narrower at the throat, somewhat sparsely pilose or glabrous, with a ring of deflexed hairs inside; lobes ovate, 2–3 mm. long, acute. Ovary 4–5-locular; style 6–7 mm. long; stigmatic club coroniform, 5-lobed. Fruit blackish when ripe, globose, 1–1.2 cm. in diameter, pubescent with pale and ferruginous hairs; pyrenes compressed-ellipsoid, 8 mm. long, 5 mm. wide, with a ± median notch.

subsp. **lichenoxenos**; Verdc. in K.B. 42: 169 (1987)

Corolla-tube sparsely pilose outside.

TANZANIA.   Morogoro District: Uluguru Mts., Lukwangule Plateau, 5 Feb. 1935, *E.M. Bruce* 779! & same locality, western slopes, 6 Dec. 1969, *Harris et al.* 3753! & same locality, above Chenzema towards Lukwangule, 1 Jan. 1975, *Polhill & Wingfield* 4657!
DISTR.   T 6; not known elsewhere
HAB.   Upland rain-forest; 1830–2400 m.

SYN. *Vangueria lichenoxenos* K. Schum. in E.J. 28: 70 (1899) & 493 (1900); De Wild. in B.J.B.B. 8: 58 (1922) subsp. **glabrituba** *Verdc.* in K.B. 42: 169 (1987). Type: Tanzania, Iringa District, Mufindi, *R.M. Davies* 924 (K, holo.!)

Corolla-tube glabrous outside.

TANZANIA. Iringa District: Mufindi, 3 Oct. 1934, *R.M. Davies* 924! & Mufindi, Kigogo Forest Reserve, 12 Oct. 1954, *Sangiwa* 54! & upper Ruhudje R. basin, Lupembe, Ditima [Nditima], 25 Oct. 1931, *Schlieben* 1396!
DISTR. **T** 7; not known elsewhere
HAB. Evergreen forest; 1830–2000 m.

23. **R. nodulosa** (*K. Schum.*) *Robyns* in B.J.B.B. 11: 155 (1928); T.T.C.L.: 530 (1949); Verdc. in K.B. 42: 169 (1987). Type: Tanzania, Uluguru Mts., near Nghweme [Nglewenu], *Stuhlmann* 8775 (B, holo.†)

Shrub 1–2 m. tall or branched tree 3–6 m. tall; stems grey or blackish grey, rugose; young shoots densely covered with short pale ferruginous pubescence; flowering shoots sometimes very short, corrugate and nodose. Leaf-blades narrowly ovate-elliptic, 0.8–5 cm. long, 0.4–2 cm. wide, acute to shortly acuminate at the apex, broadly cuneate at the base, discolorous, blackish above when dry, much paler beneath and densely pubescent on the midrib; petiole 1 mm. long, densely pubescent; stipules with pubescent triangular base 3 mm. long and subulate appendage 3–3.5 mm. long. Inflorescence 1–2(–3)-flowered; peduncle obsolete; pedicels 1–3 mm. long, ferruginous pubescent. Buds obtuse or shortly apiculate. Calyx-tube 1 mm. long; limb 1 mm. long, ± scarious, denticulate. Corolla greenish yellow, glabrous outside; tube ± cylindrical-urceolate, 6 mm. long, with a ring of deflexed hairs inside; lobes ovate, 2 mm. long, 1.5 mm. wide with very short appendages. Anthers 1.5 mm. long, half-exserted. Ovary 5-locular; style ± 6 mm. long, the stigmatic club coroniform, 5-lobed. Fruit not seen.

TANZANIA. Morogoro District: Uluguru Mts., NW. slopes of Lupanga, 14 Feb. 1970, *Harris & Pócs* 4130! & 12 Nov. 1932, *Schlieben* 2981! & without exact locality, 12 Oct. 1932, *Wallace* 182!
DISTR. **T** 6; not known elsewhere
HAB. Mossy forest with tree-ferns etc., mist-forest; 1350–2100 m.

SYN. *Vangueria nodulosa* K. Schum. in P.O.A. A: 92 (1895), *nomen*, & in E.J. 28: 71 (1899) & 493 (1900); De Wild. in B.J.B.B. 8: 60 (1922)

NOTE. *Akeroyd & Mayuga* 18 (Tanzania, ?Kilosa District, Chonwe Mts., Dec. 1968) may be variant of this species with leaves up to 8 × 2.8 cm., but it also resembles 42, *R. sp. F, Semsei* 2034.

24. **R. induta** (*Bullock*) *Verdc. & Bridson* in K.B. 42: 169 (1987). Type: Tanzania, Kondoa District, Kolo*, *B.D. Burtt* 1294 (K, holo.!, BR, EA, iso.!)

A much-branched shrub or small spreading tree, 3–11 m. tall, with brittle brownish or dark purple lenticellate branches and dense spreading yellow-brown pubescence when young but soon glabrous. Leaf-blades ovate-oblong to elliptic, 2.5–7.5 cm. long, 1.2–5 cm. wide, subacute, rounded or shortly obtusely acuminate at the apex, cuneate to ± rounded at the base, usually shortly softly pubescent all over but sometimes ± glabrous, drying blackish; petiole 1–3 mm. long, spreading pubescent; stipules with broad triangular base 3–5 mm. long, acuminate at the apex, becoming corky with basal part ± persistent. Cymes short, 2–4-flowered; peduncle 2–3 mm. long; pedicels 2–3.5 mm. long; bracts ovate, 1.5 mm. long, all parts spreading pubescent. Calyx-tube campanulate, ± 1 mm. long; limb-tube very short, with lobes triangular or narrowly triangular-oblong, ± 1 mm. long. Corolla greenish outside, cream inside; tube 6 mm. long, with some long hairs at the throat; lobes oblong, 3–5 mm. long, 1.8–2 mm. wide, acute but scarcely apiculate. Anthers exserted, 1.5 mm. long. Ovary 2–3-locular; style exserted, 2.5 mm. long, the stigmatic club 1 mm. long, cylindrical and 2–3-lobed. Fruits subglobose, didymous in dry state, 1.4–1.5 cm. long and wide, but plum-like when fresh, ± 2 cm. in diameter, with 2(–3) pyrenes, each ± 1.5 cm. long, 7 mm. wide; fruiting pedicel 0.8–1 cm. long.

KENYA. Machakos District: Kilungu Forest, between Nunguni and the dispensary, 9 Jan. 1972, *Mwangangi* 1945 A! (identification doubtful); Masai District: Emali Hill, 15 Mar. 1940, *V.G.L. van Someren* 128! & Leserin [Legeri] Hills, Ilgeri, 26 Feb. 1979, *Kuchar* 10692!
TANZANIA. Masai District: 6 km. E. of Loliondo, 6 Nov. 1964, *Gillett* 16326!; Mbulu District: Ufiomi Mt., 21 Jan. 1928, collector for *B.D. Burtt* 1241!

* Fide Brenan T.T.C.L.: 486 (1949).

DISTR. K 4, 6; **T** 2; not known elsewhere
HAB. *Podocarpus* forest, *Eucalyptus* plantation presumably relict of original woodland, forest edges, streamsides; 1770–2250 m.

SYN. *Canthium indutum* Bullock in K.B. 1932: 366 (1932); T.T.C.L.: 486 (1949)

25. **R. binata** (*K. Schum.*) *Robyns* in B.J.B.B. 11: 202 (1928); T.T.C.L.: 531 (1949); Verdc. in K.B. 42: 170 (1987). Types: Tanzania, Uzaramo District, Dar es Salaam, Mogo Forest Reserve [Sachsenwald], *Engler* 2162 & 2185 (B, lecto.†)*

Shrub or tree 2–6 m. tall, ?rarely scandent; stems glabrous, the older ones grey, sometimes ± reddish powdery beneath or with reddish brown epidermis which peels to reveal dark red-purple beneath. Leaf-blades elliptic, 5–14 cm. long, 2–6 cm. wide, shortly and broadly acuminate to very narrowly acuminate at the apex, the actual tip acute or narrowly rounded, rounded to cuneate at the base, slightly discolorous, drying olive-green, glabrous save for hairy domatia or rarely adpressed pilose on the lamina; petiole 2–6 mm. long; stipules forming a sheath 1–2.5 mm. long, densely pilose within, with narrow subulate appendages 4–6 mm. long. Inflorescences 2(–3)-flowered, often distinctly supra-axillary ± 5 mm. above the node; peduncle 1–2.5 cm. long; pedicels 0.5–1.8 cm. long; bracts 1.5 mm. long, apiculate. Buds glabrous, markedly acuminate, the acumen 2–3 mm. long. Calyx-tube 1.5 mm. long, the limb ± 1 mm. long, truncate or with minute teeth 0.5 mm. long. Corolla green, yellow-green or cream-yellow; tube 2–3 mm. long, with some deflexed hairs inside below throat; lobes narrowly lanceolate, 4–4.5 mm. long including the 1.5–2 mm. long appendage. Ovary 3–4-locular. Fruit subglobose, 8–10 mm. long and wide, with 1–3 pyrenes, the fruiting pedicels to 2.3 cm. long.

TANZANIA. Uzaramo District: Pugu Hills, near Kisarawe, Minaki, 27 Nov. 1968, *Harris & Walker* 2615! & near Kiserawe, 2 Nov. 1970, *Harris & Schlieben* 5344!; Kilwa District: Selous Game Reserve, Nunga Thicket, 11 Oct. 1978, *Vollesen* in M.R.C. 4336!
DISTR. **T** 6, 8; not known elsewhere
HAB. *Brachystegia microphylla* thicket; deciduous coastal bushland on sand, "dry forest"; 115–700 m.

SYN. *Vangueria binata* K. Schum. in E.J. 34: 333 (1904); De Wild. in B.J.B.B. 8: 46 (1922)

26. **R. pauciflora** (*Hiern*) *R. Good* in J.B. 64, Suppl. 2: 23 (1926); Robyns in B.J.B.B. 11: 197 (1928); F.P.S. 2: 462 (1952); Verdc. in K.B. 42: 170 (1987). Type: Sudan, Jur, Kuchuk Ali, *Schweinfurth* 1617 (K, lecto.!, BM, isolecto.!)

Shrub 2.4–5 m. tall; branches with paired straight supra-axillary spines ± 1 cm. long which are abbreviated undeveloped inflorescences; branchlets with bristly whitish or brown hairs, the purplish brown or grey bark eventually flaking to reveal a powdery surface. Leaf-blades elliptic to ovate, 1.5–7 cm. long, 1–4 cm. wide, acuminate at the apex, rounded to subcordate at the base, eventually subcoriaceous, discolorous, the venation above becoming ± impressed and surface slightly bullate, venation reticulate beneath, at first fairly densely pubescent, later with a few scattered hairs on the main nerves beneath and pubescent domatia but otherwise glabrous; petiole ± 2 mm. long, pubescent; stipule-bases 1–2 mm. long, at first fused, long white-pilose within, the appendages ± 2 mm. long, glabrous. Cymes 1–3-flowered, about half as long as the leaves; peduncle 1–2.5 cm. long, sparsely pubescent, the base thickened, being the part equivalent to a spine; pedicels 0.3–0.5(–1.5 in fruit) cm. long; bracts forming a small involucre ± 1 mm. long. Calyx-tube ovoid, 1.2 mm. long, glabrous, the limb scarcely developed, subtruncate. Corolla greenish white; tube ± 3 mm. long, glabrous outside, with a ring of deflexed hairs inside; lobes 5–8, narrowly triangular, 5 mm. long, 1.7 mm. wide, micropapillate inside, long-acuminate at the apex, the appendage ± 1.5 mm. long. Anthers half-exserted. Ovary 3–4-locular; style exserted 1.5 mm., the stigmatic club ± 1 mm. long. Fruits purple, globose, 0.7–1 cm. diameter, with (1–)3–5 pyrenes and 2–5-lobed in the dry state; pyrenes straw-coloured, ellipsoid, 7 mm. long, 4 mm. wide, 2.5 mm. thick.

UGANDA. W. Nile District; W. Madi, Abieso, May 1948, *Eggeling* 5778! & Koboko/Maracha [Maraca] counties, below Kadre Hill, June 1952, *Alonzie* 4! & Mt. Otzi, below summit, 7 June 1936, *A.S. Thomas* 1978! & War, May 1948, *Dale* U. 531!
DISTR. **U** 1; Sudan, Central African Republic, Cameroon and Angola (*fide* Good)
HAB. Wooded grassland; 1230–1500 m.

* From the citation of Robyns it would appear that these two Engler numbers constitute one sheet; all the syntypes have been destroyed.

SYN. *Vangueria pauciflora* Hiern in F.T.A. 3: 151 (1877)
　　Canthium pauciflorum Baillon in Adansonia 12:189 (1878) *non* Blanco (1837). Type: Sudan, Jur,
　　Kuchuk Ali, *Schweinfurth* 1617 (P, holo, BM, K, iso.!)

NOTE.　Both Hiern's and Baillon's epithets are based on Schweinfurth's manuscript name. The
fruits are said to be edible.

27. **R. mrimaensis** *Verdc.* in K.B. 42: 170 (1987). Type: Kenya, Kwale District, Mrima Hill,
Brenan & Gillett 14616 (K, holo.!, EA, iso.!)

Shrub 1–4 m. tall, with slender glabrous twigs bearing paired supra-axillary straight
ascending spines up to ± 2 cm. long; bark of young shoots dull purplish; apparently
sometimes flowering when leafless. Leaf-blades ovate, 2.7–7 cm. long, 1.3–4.5 cm. wide,
acuminate at the apex, the actual tip narrowly rounded, rounded at the base, drying
blackish, glabrous save for some bristly hairs on the midrib beneath and sparsely hairy
domatia; petiole ± 3 mm. long; stipule-bases broadly triangular, 1.5 mm. long, densely
setose within, the subulate appendage 2 mm. long. Flowers solitary in the leaf-axils or
abnormally a triad of flowers terminating a shoot; pedicels 0.8–1.2(–6 in fruit) mm. long.
Calyx-tube obconic, ± 2 mm. long; limb-tube ± 0.3 mm. long, with narrow acute triangular
teeth ± 1 mm. long. Buds acuminate, glabrous. Corolla greenish white; tube 4.5 mm. long,
with upper half inside with deflexed hairs; lobes narrowly triangular, 5 mm. long, 1.5 mm.
wide, narrowly acuminate, the subulate appendage ± 1 mm. long. Ovary 2-locular; style
exserted ± 1.5 mm.; stigmatic club broadly cup-shaped ± 1.5 mm. wide. Fruit 1.4 cm. in
diameter; pyrenes brown, almost straight, compressed, narrowly obovoid, obscurely
keeled at the apex, almost smooth.

KENYA.　Kwale District: N. bank of R. Mwachema mouth, Kaya Tiwi, 24 Aug. 1989, *Robertson & Luke*
　　5897! & Mrima Hill, 2 Feb. 1989, *Robertson et al.* in *MDE* 14!; Kilifi District: Kombeni R. valley, edge
　　of Kaya Fimboni, 2 Aug. 1989, *Robertson & Luke* 5868!
DISTR.　**K** 7; not known elsewhere
HAB.　Evergreen forest, particularly *Afzelia, Caesalpinia, Craibia, Garcinia, Gyrocarpus* and *Mimusops*,
　　also moist semi-deciduous forest; 10–200 m.

NOTE.　The shape of the calyx-tube and traces of ribbing hint at something anomalous, but now ripe
fruits are available there is no other reasonable position for it.
　　This taxon resembles the Asian *Canthium rheedii* DC. in some respects but differs in not having a
pubescent style. *C. rheedii* belongs to a group which includes the type of *Dondisia* DC. at present
considered a synonym of *Canthium*.

28. **R. celastroides** (*Baillon*) *Verdc.* in K.B. 42: 171, fig. 1H-J (1987). Types: Kenya,
Mombasa, *Boivin* (P, syntypes!)

Densely branched shrub or small tree 1.8–7.5 m. tall or sometimes scrambling, with
rather rough grey, brown or whitish grey bark; branches sparsely to densely pubescent on
young parts, often with ± yellowish hairs, persistent or becoming glabrous; branches
often in whorls of 3; spines frequently present, solitary, 0.7–1.3 cm. long. Leaves usually
paired but sometimes in whorls of 3; blades narrowly elliptic, ovate-lanceolate or
lanceolate, 2–5.5(–9) cm. long, 0.8–2.5(–4) cm. wide*, obtuse to narrowly acuminate at
apex, the tip itself usually obtuse, cuneate or rarely rounded at the base, mostly densely
pubescent and often ± velvety beneath when young or persistently so in one variant or
entirely glabrous save for domatia or only sparsely pubescent; petiole 0.5–2 mm. long;
stipule-bases 1–1.5 mm. long, connate, truncate, subscarious, pubescent to glabrescent
outside, densely villous inside, with subfleshy deciduous often ± reflexed appendage,
2.5–4(–6) mm. long. Inflorescence 2–4(–7)-flowered, not subumbellate, but with
secondary branches or rachis usually shortly developed; peduncle glabrous to pubescent,
obsolete or up to 3(–7) mm. long; bracts scarious, forming a cup ± 1 mm. long, glabrous to
pubescent; pedicels 2–9 mm. long, glabrous to pubescent, sometimes attaining 1.4–2 cm.
in fruit. Buds obtuse to distinctly but shortly acuminate, glabrous or very sparsely to
densely pubescent. Calyx-tube 1–1.5 mm. long, the limb truncate to shortly toothed,
0.25–0.5 mm. long or rarely calyx-lobes more developed, 0.5–0.8 mm. long. Corolla white,
cream, yellowish or greenish; tube ± campanulate, 1.5–2 mm. long, with a ring of deflexed
hairs inside; lobes oblong, (1.5–)2.5–3.5 mm. long, 1–1.5 mm. wide, glabrous to densely
pubescent outside, ± papillate inside, acute or with short subulate appendage 0.5–0.75

* A specimen from the Usambaras, Sigi to Longuza, has leaves up to 9 × 4 cm.

mm. long. Ovary 2(–5)-locular; style 2–2.5(–4) mm. long; stigmatic club coroniform, 3(–5)-lobed. Fruit black, subglobose, usually deeply lobed in dry state, 6–8(–9) mm. tall, 8.5–9.5 mm. wide, with 1–3(–5) pyrenes, glabrous; pyrenes oblong-segmentoid, ± 10 × 6 mm., with hilar notch just above middle.

var. **celastroides**; Verdc. in K.B. 42: 171, fig. 8A (1987)

Branchlets, leaves, inflorescences, etc. glabrescent to pubescent, but always some parts with sparse to dense pubescence; corolla entirely glabrous to densely pubescent.

KENYA. Kwale District: Mrima road, at bottom of Mrima Hill, 6 Sept. 1957, *Verdcourt* 1902!; Mombasa District: Mowesa, *R.M. Graham* BB383 in *F.D.* 1750!; Kilifi District: Mazeras, Apr. 1970, *R.M. Graham*, N824 in *F.D.* 2336!

TANZANIA. Tanga District: E. Usambara foothills, Sigi, Singali, 22 Apr. 1950, *Verdcourt* 164!; Pangani District: S. bank of R. Pangani, between Hale and Makinyumbe, 1 July 1953, *Drummond & Hemsley* 3120!; Morogoro District: Uluguru Mts., near Morogoro, 3 Jan. 1936, *E.M. Bruce* 409!

DISTR. **K** 4, 7; **T** 3, 5, 6, 8; Mozambique

HAB. Edges of *Brachystegia, Strychnos, Markhamia, Hyphaene*, etc., woodland, open forest, relict thickets in open grassy areas, grassland with scattered trees, also riverine thicket, bushland and lowland rain-forest; 0–750(–1140) m.

SYN. *Canthium celastroides* Baillon in Adansonia 12: 190 (1878)*
    Vangueria glabra K. Schum. in P.O.A. C: 384 (1895); De Wild. in B.J.B.B. 8: 52 (1922). Type: Tanzania, Tanga District, Amboni, *Holst* 2603 (B, holo. †, K, iso.!)
    V. microphylla K. Schum. in P.O.A. C: 385 (1895) & in E.J. 28: 494 (1900); De Wild. in B.J.B.B. 8: 59 (1922), pro parte. Types: Tanzania, Uzaramo [Usaramo] District, *Stuhlmann* 6242 (B, lecto. †) & Kenya, Kitui District, Ukambani, Malemba [Malemboa], *Hildebrandt* 2836 (B, syn. †, K, isosyn.!)
    V. oligacantha K. Schum. in E.J. 34: 334 (1904); De Wild. in B.J.B.B. 8: 60 (1922). Type: Tanzania, Usambara Mts., Lwengera, *Engler* 913 (B, holo. †)
    Plectronia amaniensis K. Krause in E.J. 43: 142 (1909). Types: Tanzania, E. Usambara Mts., Amani, *Zimmermann* 91 (B, syn. †, EA, isosyn.!) & *Warnecke* 347 (B, syn. †, BM, isosyn.!) & Kwamkuyo Falls, *Braun* 1138 (B, syn. †, EA, isosyn.!)
    Rytigynia oligacantha (K. Schum.) Robyns in B.J.B.B. 11: 172 (1928); T.T.C.L.: 531 (1949); K.T.S.: 471 (1961)
    R. glabra (K. Schum.) Robyns in B.J.B.B. 11: 174 (1928); T.T.C.L.: 530 (1949)
    R. microphylla (K. Schum.) Robyns in B.J.B.B. 11: 171 (1928); T.T.C.L.: 530 (1949); K.T.S.: 472 (1961)
    R. amaniensis (K. Krause) Bullock in K.B. 1932: 389 (1932); T.T.C.L.: 531 (1949); K.T.S.: 471 (1961)

var. **nuda** *Verdc.* in K.B. 42: 171, fig. 8B (1987). Type: Zanzibar, Kisimbani, *Faulkner* 3353 (K, holo.!)

Plant entirely glabrous save for domatia and insides of stipules and corolla-tube.

TANZANIA. Kilwa District: Selous Game Reserve, 10 km. NNW. of Kingupira, 18 Dec. 1975, *Vollesen* in *M.R.C.* 3110!; Zanzibar I., Mwera R., 15 Dec. 1929, *Vaughan* 1016! & Kidichi, 18 Dec. 1959, *Faulkner* 2438!

DISTR. **T** 6, 8; **Z**; not known elsewhere

HAB. Open places near cultivations, swamp margins, riverine forest and thicket and evergreen forest; 90–220 m.

NOTE. As circumscribed here *R. celastroides* is a very variable species and extreme specimens can be selected from the large collections now available which appear quite different from each other, there being great variation in indumentum, leaf-size and presence or absence of spines. *R. glabra* actually has sparse hairs on the stems and foliage although the buds are glabrous, but the range of indumentum on the buds varies from a few odd hairs to densely pubescent or entirely glabrous and cannot be used as a character. I have thought it best to erect a new variety for entirely glabrous specimens. *Greenway* 4589 (Tanzania, S. Pares, Kisiwani, 3 Feb. 1936) and *Hornby* 357 (Tanzania, Mpwapwa, 18 Feb. 1931) have the leaves very velvety beneath and may represent a distinct variant but no other material adequate to assess variation has been seen. *Robertson et al.* in MDE 123 (Kwale District, Mrima Hill, 5 Feb. 1989) a 20 cm. shrublet with 5–9 mm. long spines at each node and ovate leaves 0.5–2 cm. long is distinctive but perhaps just a juvenile state. *Robertson* 553 (Tanzania, Kilosa District, Ilonga, 14 Feb. 1967) is a woody herb 0.9 m. tall having most of the leaves in whorls of 3 and merges with *Fadogia*. Several other more shrubby specimens all from Morogoro District between 690 and 900 m. in grassland, open woodland or *Brachystegia* margins are similar with narrow velvety leaves in whorls of 3 or sometimes apparently more due to fastigiation on very abbreviated shoots; it may be an ecological variant but is distinctive and

* Index Kewensis erroneously gives = *Plectronia celastroides* Bak., a very different Seychelles species.

possible deserves a name. It is doubtless no coincidence that this species is mostly found in *Brachystegia* woodland. Very small-leaved specimens from the northern parts of the range in Kenya merge with *R. parvifolia* which may itself be only an extreme variant of *R. celastroides* although I have considered it advisable to recognise it as a distinct species. Whether it is a distinctive variety or whether there are two species joined by a hybrid swarm or some other explanation might be evident from field study.

29. **R. parvifolia** *Verdc.* in K.B. 42: 172 (1987). Type: Kenya, Kilifi District, 52 km. W. of Malindi, Kakoneni, *Spjut* 3956 (K, holo.!, BR, EA, iso.!)

Usually a stiff much-branched spiny shrub to 6 m. but with very variable branching; branches usually in whorls of 3, white or grey-brown, densely pubescent when young; spines short, up to 4–5 mm. long or sometimes absent (at least on pieces preserved), mostly placed just above the nodes; in some specimens the side-branches are reduced or very short and nodes telescoped, the internodes 0.3–1 cm. long. Leaf-blades elliptic, 1–2.5(–4) cm. long, 0.6–1.5(–2) cm. wide, rounded at the apex, ± cuneate at the base, not thin, with only 2–3 lateral nerves on each side of midrib, glabrous above and beneath or with curled pubescence on midrib and venation beneath and sometimes margins obscurely ciliate; barbellate domatia also present (but see note); petiole 1–2 mm. long, glabrous or pubescent; stipule-bases 1 mm. long, brown and scarious, pubescent, with reflexed appendage. 1.5–2 mm. long. Cymes 2–3-flowered; peduncle 1.5–2 mm. long; pedicels 1–10 mm. long; bracts brown and scarious, under 1mm. long. Buds acuminate. Calyx-tube 0.8–1 mm. long, the limb very short truncate or very slightly toothed. Corolla cream or greenish; tube 1.8–3 mm. long, with deflexed hairs inside on upper half; lobes 5–6, 2–3.5 mm. long, 1–1.5 mm. wide, acuminate with very short appendages, densely papillate inside. Anthers appendaged and papillate. Ovary 3-locular; style exserted ± 2 mm.; stigmatic club 1 mm. wide. Fruit 7–7.5 mm. long, 6–11 mm. wide, with 1–3 pyrenes, the pedicels up to 1.4 cm. long.

KENYA. Kilifi District: Marafa area, May 1959, *Rawlins* 726! & 24 km. S. of Garsen on road to Malindi, 10 Nov. 1957, *Greenway* 9489!; Lamu District: Utwani, Dec. 1956, *Rawlins* 276!
DISTR. **K** 7; Somalia
HAB. Bushland and scattered tree grassland, particularly of *Combretum*, *Commiphora*, *Lannea*, *Cordia*, *Salvadora*, *Haplocoelum*, etc., also *Brachystegia spiciformis* woodland; 30–180 m.

NOTE. The status of this taxon needs field study; it merges with *R. celastroides* and in the Sokoke Forest intermediates are common but extremes are so very diverse that I have maintained two species. Whether this reflects the truth I do not know, but the situation could be easily explained by introgression between two species once well separated by geographical isolation but now overlapping. *Gillett et al.* 25151 and *Friis et al.* 4946 (S. Somalia) are forms of this species.
    The problem has very recently been complicated by the collection of further specimens from the northern Kenya coast. *Luke* 1360 (Tana River District, Ozi, 20 Aug. 1988) and 1431 (Lamu District, Ras Tenewi, 20 Nov. 1988) found in degraded forest of *Mimusops*, *Afzelia*, *Tabernaemontana*, etc., at 10–30 m. are shrubs 3 m. tall with pale stems, pubescent when young; stipules with base 1.5 mm. long with recurved appendage to 2 mm.; leaves 1.5–5 cm. long, 1–3.5 cm. wide, glabrous save for domatia; petiole 2–3 mm. long; inflorescences 1-flowered; peduncle 2 mm. long with 2 pairs of bracts; pedicel 6 mm. long becoming 1.5 cm. long in fruit; calyx 1.5 mm. long including the very short annular truncate limb; corolla-tube 2 mm. long; lobes triangular 3.3 mm. long, 1.3 mm. wide including the appendage ± 0.7 mm. long; style 3.2 mm. long; stigmatic club 1 mm. long; fruit 7 mm. long. I suspect this population is best considered a variety of *R. parvifolia* but the limits of this, *R. celastroides* and 48, *R. sp. I* need reassessment in the field.

30. **R. monantha** (*K. Schum.*) *Robyns* in B.J.B.B. 11: 153 (1928); T.T.C.L.: 529 (1949), pro parte; Fl. Rwanda 3: 217, fig. 68.1 (1985); Verdc. in K.B. 42: 173 (1987). Type: Tanzania, Iringa District, Uhehe, Rungemba, *Goetze* 727 (B, holo. †)

Much-branched shrub or small tree 0.3–4.5 m. tall, with the older stems brown, purple-brown or reddish, with very finely fissured smooth bark or sometimes in exposed areas becoming rough and nodose, the dark very rough bark flaking to reveal a reddish or pale brownish powder; branching often ± horizontal, the young shoots green in life, densely shortly spreading pubescent, with sometimes bright rusty hairs, but in some areas (e.g. Malawi) ± glabrous. Leaf-blades ovate, oblong or elliptic, 1–6 cm. long, 0.5–4 cm. wide, narrowly acuminate at the apex, rounded to cuneate or sometimes subcordate at the base, typically markedly discolorous with dense adpressed yellowish or whitish pubescence above and densely ± coarsely velvety beneath but not entirely obscuring the ± reticulate venation but the indumentum is sometimes much sparser and some Malawi specimens have almost glabrous leaves; petiole 1–4 mm. long; stipules with hairy basal hyaline parts joined to form a sheath 1.5–2 mm. long; appendages compressed, lanceolate or narrowly falcate, drying dark, 3–10 mm. long or rarely pale and filiform. Inflorescences 1–2-flowered; peduncle 0.4–1.6 cm. long; pedicels (2–)3–9 mm. long, lengthening up to 1.9 cm. in fruit; bracts ovate, 1.5–3.5 mm. long, ± connate, acuminate. Buds clavate, acuminate, apiculate or with 5 distinct tails but very variable and sometimes (in Rwanda) ± blunt, usually spreading bristly pubescent but occasionally glabrous. Calyx-tube 1–1.5 mm. long, densely pubescent outside, densely hairy inside all over; lobes linear to triangular, 0.5–1.5 mm. long (long and linear in Malawi). Corolla white, pale yellow or greenish; tube 3–6 mm. long, ± glabrous inside, with few reflexed hairs outside or pubescent in upper half; lobes triangular to lanceolate, 2–4 mm. long, 1–2 mm. wide, acute, with apiculum or short appendage 0.5–1.5 mm. long, often creamy white inside, usually bristly pubescent outside or less often glabrous (particularly in Malawi material). Ovary 3–4-locular; style white, usually exserted 1–3 mm., sometimes with few hairs just beneath club, ± dilated at base; stigmatic club whitish, depressed subglobose or oblong-obconic, 1 mm. long, sulcate beneath, 3–4-lobed at apex. Fruit subglobose, 6–8 mm. long, 7–8 mm. wide, pubescent, with 1–4 pyrenes.

var. **monantha**; Verdc. in K.B. 42: 174 (1987)

Leaves ovate, oblong or elliptic, mostly densely pubescent, usually velvety beneath. Corolla with dense spreading hairs outside. Figs. 132/12, p. 754 & 145.

TANZANIA.   Bukoba District: Karagwe, Nyaishozi [Nyashozi], Dec. 1931, *Haarer* 2408!; Buha District: Kalinzi, 22 Nov. 1962, *Verdcourt* 3394!; Iringa District: Iheme, 24 Feb. 1962, *Polhill & Paulo* 1599! & near Sao Hill, 1 Aug. 1933, *Greenway* 3443!
DISTR.   T 1, 4, 7, 8; Rwanda, Burundi, Malawi and NE. Zambia
HAB.   *Brachystegia*, *Uapaca*, *Faurea* and *Parinari*, *Ozoroa* woodland, also *Isoberlinia*, *Commiphora*, *Combretum*, *Uapaca* dry scrub, thicket, open bushland, etc., often in rocky places and in ravines; also recorded from riverine forest; 1100–1800 m.
SYN.   *Vangueria monantha* K. Schum. in E.J. 28: 493 (1900); De Wild. in B.J.B.B. 8: 59 (1922)
         Rytigynia castanea Lebrun, Taton & Toussaint, Contr. Fl. Parc Nat. Kagera 1: 139 (1948); Fl. Pl. Lign. Rwanda: 598, fig. 203.1 (1982). Type: Rwanda, Akagera National Park, Mushushu to Uruwita [Muruhita], *Lebrun* 9672 (BR, holo.!, K, iso.!)
         R. sp. sensu Brenan in Mem. N.Y. Bot. Gard. 8: 452 (1954)
         R. sp. sensu White, F.F.N.R.: 420 (1962)

NOTE.   As circumscribed here this is an extraordinarily variable taxon occurring in a wide range of habitats, but I have been unable to subdivide it satisfactorily; *R. castanea* has a somewhat different facies but no constant characters to separate it. In Malawi and Zambia more extreme variants with contorted branching and small leaves occur, but even here the characters do not seem constant enough to separate off subspecies.

var. **lusakati** Verdc. in K.B. 42: 174 (1987). Type: Tanzania, Kigoma District, Ititie, *Azuma* 1042 (EA, holo.!)

Leaves lanceolate to oblong-lanceolate, glabrescent save for pubescence along the main nerves. Flowers glabrous or with a very few scattered scarcely visible hairs.

TANZANIA.   Kigoma District: Ititie, 24 Dec. 1963, *Azuma* 1042!
DISTR.   T 4; ? Malawi, ? Zambia
HAB.   "Floor of thin riverine forest and tall mixed forest";? ± 900 m.

31. **R. flavida** *Robyns* in B.J.B.B. 11: 198 (1928); T.T.C.L.: 531 (1949); Verdc. in K.B. 42: 176 (1987). Type: Tanzania, Rungwe District, Kyimbila, *Stolz* 173 (K, holo.! + fragment from B, iso.!, BM, W, iso.!)

Shrub to 4 m. (erect and very divaricately branched *fide* Robyns*); branches densely spreading yellowish pubescent when young but later glabrous, very dull purplish brown. Leaf-blades oblong, 2.5–8.5 cm. long, 1–3.5 cm. wide, acuminate at the apex, cuneate to rounded at the base, densely shortly adpressed yellowish pubescent on both surfaces; petiole ± 2 mm. long; stipules 1[–2.5 mm.]** long, pubescent outside, densely hairy inside, brownish membranous, with pubescent subulate appendage 1–4 mm. long. Inflorescences 1–2-flowered; peduncle very short; pedicels 1 cm. long, lengthening to 1.8 cm. in fruit. Calyx-tube 1.5 mm. long, densely pubescent, the limb with 5–6 triangular teeth ± 1 mm.

   * It is not evident where this information came from.
 ** See note.

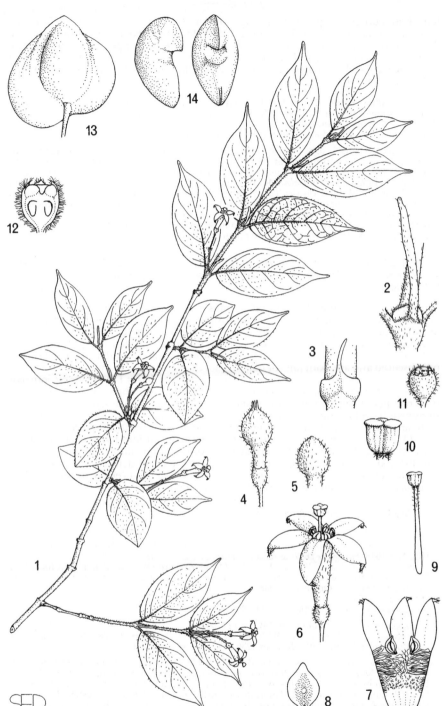

FIG. 145. *RYTIGYNIA MONANTHA* var. *MONANTHA* — **1**, flowering branch, × ⅔; **2**, young stipule, × 4; **3**, mature stipule, × 4; **4**, bud, × 4; **5**, top of untailed bud, × 4; **6**, flower, × 3; **7**, section through corolla, × 3; **8**, dorsal view of anther, × 8; **9**, style and stigmatic knob, × 3; **10**, stigmatic knob, × 8; **11**, calyx, × 4; **12**, section through ovary, × 6; **13**, fruit, × 3; **14**, pyrene (2 views), × 4. 1, 2, 6, from *Polhill & Paulo* 1599; 3, from *Procter* 940; 4, from *Procter* 381; 5, 7–12, from *Bridson* 280; 13, from *Peter* 37734; 14, from *Peter* 37611. Drawn by Sally Dawson.

long. Buds pubescent, up to 10 mm. long, apiculate. Corolla pubescent outside [greenish yellow]; tube subcylindrical, 5 mm. long, with a ring of deflexed hairs inside below the throat; corolla-lobes glabrous inside, pubescent outside, with appendages 2.5 mm. long. Anthers 2.75–3 mm. long, with distinct basal appendages. Ovary 4-locular; style 7–8 mm. long; stigmatic club 1.5 mm. long, 4-lobed. Fruit ellipsoid, 0.8–1.3 cm. long, 8(–10) mm. wide, pubescent, with 1–2(–3) pyrenes.

TANZANIA. Mbeya District: Poroto Mts., between Igogwe and Igali, Igali Pass, 2 Feb. 1979, *Cribb et al.* 11263!; Rungwe District: Kyimbila, ? Kondeland, 2 Apr. ? 1907, *Stolz* ? 173!
DISTR. **T** ? 3, 7; not known elsewhere
HAB. 'Steppe' [montane forest with *Hagenia*]; 1300[–2050] m.

NOTE. This species is virtually unknown; there is but one flower on *Cribb et al.* 11263 and the leaves are less acuminate and the indumentum more grey but I believe it is conspecific. *Paulo* 234 (Rungwe Forest, Jan. 1954) has much smaller leaves and less hairy corolla. Other material probably belonging here is *Shabani* 407 (Lushoto District, Gare footpath), *Shabani* 532 (? Lushoto) and *Magogo* 279 (Rungwe District, Tukuyu, Kiwira Forest) but the latter has glabrescent leaves and is stated to be a climber. Better material will solve the identity of all the specimens. Information derived from these specimens is in [   ]. There is some doubt about the number 173 since it could perhaps be 143; against this is the fragment from Berlin sent to Robyns by Mildbraed and accepted by him as 173. Stolz's early numbers are not in chronological sequence.

32. **R. bagshawei** (*S. Moore*) Robyns in B.J.B.B. 11: 152 (1928) & F.P.N.A. 2: 347 (1947); Verdc. in K.B. 42: 176 (1987). Type: Uganda, Toro District, Mpanga Forest, *Bagshawe* 1009 (BM, holo.!)

Erect branched shrub (2–)4.5 m. tall; stems slender, grey or brown with pale lenticels, densely pubescent with ± spreading yellowish hairs, but later glabrous and dull purplish brown. Leaf-blades elliptic to oblong-elliptic, 2.5–8.5 cm. long, 1–3.5 cm. wide, long-acuminate at the apex, the actual apex narrowly rounded, rounded to slightly subcordate at the base, discolorous, sparsely to densely shortly pilose above and beneath but surface not obscured and also with tufts of hair around the domatia in the nerve-axils beneath; petiole 2–3(–5) mm. long; stipules very hairy outside, the bases 1–2 mm. long, joined to form a sheath; appendages subulate (2–)5–7 mm. long, subpersistent. Cymes 1–2-flowered; peduncle 3–4 mm. long; pedicels 4–6 mm. long; bracts ovate, 1.5 mm. long. Buds densely pubescent save at base, with 5 long apical tails. Calyx-tube 1 mm. long, the limb subulate-dentate, teeth 0.7(–1) mm. long, densely pubescent. Corolla greenish white; tube (3–)4.5 mm. long, with a ring of deflexed hairs inside; lobes ovate-oblong to lanceolate, 2.5(–3) mm. long, (1–)1.7 mm. wide, with an apical filiform appendage (3.5–)5 mm. long. Anthers 1 mm. long, just exserted. Ovary 2-locular; style exserted 2.5 mm., the stigmatic club depressed, ± 2 mm. wide. Fruit ellipsoid to compressed-obovoid, ± 11 × 8 mm., glabrous or pubescent; pedicel (1.2)–2 cm. long.

var. **bagshawei**; Verdc. in K.B. 42: 176 (1987)

Leaf-blades, calyx-tube, etc., more densely pubescent.

UGANDA. Toro District: Mpanga Forest, Apr. 1906, *Bagshawe* 1009!; Kigezi District: Mulole [Murole] Hill, Apr. 1948, *Purseglove* 2687! & Kasatora Forest, Sept. 1947, *Dale* U. 518!
TANZANIA. Lushoto District: W. Usambara Mts., Mazumbai Forest Reserve, 17 Jan. 1973, *Ruffo* 1486! & Mkussu Forest Reserve, W. of Baga, 21 Jan. 1985, *Borhidi et al.* 85143! & E. Usambara Mts., Derema Tea Estate, 5 Oct. 1979, *Abdallah* 730!
DISTR. **U** 2; **T** 3; Zaire
HAB. Evergreen forest and thicket; 1000–2250 m.

SYN. *Vangueria bagshawei* S. Moore in J.B. 45: 42 (1907)
    V. bequaertii De Wild. in B.J.B.B. 8: 44 (1922) & in Pl. Bequaert. 2: 268 (1923). Type: Zaire, Ruwenzori, Lamia valley, *Bequaert* 4236 (BR, holo. seen by D.B.)

NOTE. The Tanzanian records are based on *Ruffo* 1486 and *Lovett* 175 (Lushoto District, West Usambaras, Mazumbai Forest Reserve) and three specimens very recently collected by Prof. Borhidi and his colleagues in the West Usambaras at Kitera R. Forest Reserve (85509), Mkussu Forest Reserve (85143) and Baga Forest Reserve (84549); also *Abdallah* 730 (Lushoto District, East Usambaras, Amani, Derema Tea Estate). It is remarkable that this species has not been previously collected at Amani despite some 90 years of botanical activity. *Abdallah* 739 (same locality as 730) only recently seen, has ± glabrescent leaves but is not otherwise similar to var. *lebrunii* having much smaller leaves; it may be distinct. *Iversen et al.* 85703 from the West Usambaras, summit of Ndamanyiru, 2200–2270 m., has narrow leaves but is probably a form of this from dry forest in a rocky wind-swept place; flowering material is needed. Var. *lebrunii* (Robyns) Verdc. differing in glabrescent leaves and ± glabrous calyx occurs in E. Zaire, Rwanda and Burundi.

33. **R. longicaudata** *Verdc.* in K.B. 42: 180, fig. 11 (1987). Type: Tanzania, W. Usambara Mts., 3 km. NE. of Bumbuli Mission on path to Mazumbai, *Drummond & Hemsley* 2474 (K, holo.!, BR, EA, iso.)

Erect or subscandent shrub 1.8–4 m. long or tall, the stems slender, glabrous, dark purple-brown with pale lenticels; branches held horizontally. Leaf-blades oblong to oblong-elliptic or -ovate, 2.5–8.5 cm. long, 0.8–3.7 cm. wide, acuminate at the apex, the actual tip mostly narrowly rounded, cuneate to rounded at the base, thin, the margins often crinkly, glabrous save for tufts of white hairs in the axils beneath or very young leaves with adpressed seta-like hairs but very soon glabrous; veins reticulate beneath on drying; petiole 3–7 mm. long; stipules joined to form a short hyaline sheath 1.5–2 mm. long, each with a subulate appendage 3–5 mm. long from just below the margin. Flowers solitary; pedicels 5 mm. long, lengthening to 1.8–2.3 cm. in fruit and thickening slightly towards the apex. Calyx-tube campanulate, ± 1.5 mm. long, the limb reduced to a narrow rim or with 5 very narrow linear lobes 0.5–1 mm. long. Buds with long tails 3–4 mm. long. Corolla dull yellow, cream or greenish white; tube 3.5 mm. long, sparsely pubescent outside, glabrous inside save throat sparsely pubescent; lobes narrowly triangular, 4–5 mm. long, 2 mm. wide, with subulate appendages 4.5 mm. long. Anthers brownish, curved, just exserted, ± 1.5–2 mm. long. Ovary 2–3-locular; style exserted, ± 4–5 mm. long; stigmatic club green, coroniform, 1–2 mm. wide, bilobed at apex. Fruit globose or didymous, ± 1–1.2 cm. long, 0.8–1.2 cm. wide, with 2–3 pyrenes; pyrenes ellipsoid-fusiform, ± 1 cm. long, 4–4.5 mm. wide, not markedly widened in apical third, dehiscing easily at the apex. Fig. 146, p. 830.

TANZANIA. Lushoto District: W. Usambara Mts., 6.5 km. E. of Lushoto, above Kwazinga, 27 Apr. 1953, *Drummond & Hemsley* 2255! & Mazumbai Forest Reserve, 8 Jan. 1977, *Polhill* 4714! & Shagayu, 23 July 1957, *Carmichael* 635!
DISTR. T 3; not known elsewhere
HAB. Undershrub in *Ocotea, Agauria* and similar upland rain-forest; 1750–1980 m.

34. **R. longituba** *Verdc.* sp. nov. ab omnibus speciebus adhuc descriptis calycis limbis carentibus, corollae tubo longiore 9 mm. longo differt; a speciebus *Hutchinsoniae* Robyns calycis limbo haud evoluto differt; ob stipulas lateraliter compressas *R. monanthae* (K. Schum.) Robyns similis foliis floribus ramisque glabris corollae tubo longiore valde differt; *R. longicaudatae* Verdc. probabiliter affinis sed stipulis lateraliter compressis corollae tubo longiore diversa. Type: Tanzania, Morogoro District, Nguru Mts., *Pócs* 88183/A (K, holo.!, NHT, SUA, UPS, VBI, iso.)

Glabrous shrub 1.8 m. tall; stems finely longitudinally wrinkled. Leaf-blades ovate, oblong-ovate or ± elliptic, 4–7.5 cm. long, 1.5–3.5 cm. wide, long-acuminate, the actual acumen rounded, broadly cuneate to rounded at the base, ± discolorous, sometimes with a few domatia beneath; petiole 2–5 mm. long; stipules lanceolate, 4 cm. long, 1 mm. wide, the limb held at right-angles to the base and in the plane of a radius of the stem. Flowers 1–2 on very reduced axillary ?viscid or at least shining nodules 2–3 mm. long with minute congested bracts; pedicels 5–6 mm. long. Calyx-tube 1.2 mm. long; limb reduced to a minute rim 0.2 mm. tall. Corolla white; buds only seen, with apical appendages ± 3 mm. long; tube ± 9 mm. long; lobes ± 6–7 mm. long including appendages.

TANZANIA. Morogoro District: Nguru Mts., between Chazi and Dikurura valleys, W. of Mogole Hill, 14 Sept. 1988, *Pócs* 88183/A!
DISTR. T 6; not known elsewhere
HAB. Rain-forest; 1400–1570 m.

NOTE. Two other sterile specimens have been seen, *Pócs* in *Borhidi* 89260 and 89286, both from the Nguru Mts.

35. **R. sp. E**

Shrub with slender greyish branches, shortly adpressed setulose-pubescent when young, later glabrous and with epidermis ± peeling; foliage very dense and delicate. Leaf-blades narrowly to broadly elliptic or ± lanceolate, 0.5–3 cm. long, 0.4–1.1 cm. wide, distinctly acuminate to a rounded tip, cuneate at the base, thin and pale green, glabrous or with a few scattered hairs above, slightly pubescent on venation beneath and with very conspicuous tufts of white hairs in the nerve-axils in lower half of the leaf; petiole 0.5–1 mm. long; stipules with tubular sheath 1.5–2 mm. long, with subulate appendages ± 1 mm.

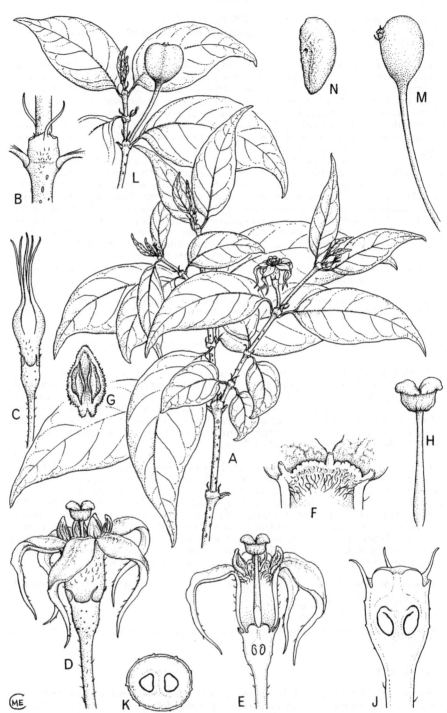

FIG. 146. *RYTIGYNIA LONGICAUDATA* — **A**, flowering branch, × 1; **B**, stipule, × 4; **C**, bud, × 4; **D**, flower, × 4; **E**, longitudinal section through flower, × 4; **F**, inside upper part of corolla-tube, × 8; **G**, anther, × 10; **H**, style and stigmatic knob, × 6; **J**, longitudinal section through ovary, × 10; **K**, transverse section through ovary, × 10; **L**, tip of fruiting branch, × 1; **M**, fruit, × 2; **N**, pyrene, × 2. All from *Drummond & Hemsley* 2474. Drawn by Mrs M.E. Church.

long, decurrent on to sheath for ± 1 mm., the sheath eventually splitting, white pubescent within. Flowers solitary; pedicel 5 mm. long, puberulous. Calyx-tube 1.5 mm. long, the limb under 0.5 mm. long, slightly undulate. Buds clavate, with tails, the limb-part sparsely pubescent. Corolla greenish yellow; tube cylindrical, 5–7 mm. long, glabrous outside, with deflexed hairs inside extending from throat to about halfway, the lower half glabrous; lobes oblong-elliptic, 3 mm. long, 2 mm. wide, with subulate tails 2.5 mm. long. Filaments extremely short; anthers just exserted, 1.5 mm. long. Ovary 2-locular; style ± 1 cm. long of which 3.5 mm. is exserted; stigmatic club obcoroniform, 1.3 mm. wide, 10-ribbed below, 2-lipped above. Disk cushion-shaped, 1.5 mm. wide. Fruit not known.

TANZANIA.   Lushoto District: W. Usambara Mts., Baga I Forest Reserve, 2 Mar. 1984, *Borhidi et al.* 84443!
DISTR.   **T** 3; known only from the above collection
HAB.   Evergreen forest; 1600–1800 m.

NOTE.   Very close to *R. longicaudata* Verdc. but the uniformly graceful small foliage, distinctly adpressed pubescent young stems give it a very different aspect; *R. bagshawei* (S. Moore) Robyns is much hairier with larger foliage.

36. **R. caudatissima** *Verdc.* in K.B. 42: 180, fig. 12 (1987). Type: Tanzania, Iringa District, Uzungwa Mts., Sanje, *D. Thomas* 3781 (K, holo.!, MO, iso.!)

Small tree to 4 m. with brown glabrous branches eventually covered with thin grey corky bark. Leaf-blades oblong to ovate-oblong, 2.5–8 cm. long, 1.2–3.2 cm. wide, acuminate at the apex, the actual tip narrowly rounded and minutely apiculate, truncate, rounded or usually subcordate at the base, slightly discoloured, glabrous save for pubescent domatia in axils and secondary axils of venation beneath; petioles 2–3 mm. long; stipular sheath cylindrical, 1.5–2 mm. long, ciliate at margin and with subulate or laterally compressed appendages 1–5 mm. long. Flowers solitary; pedicels 8–10 mm. long, becoming up to 2.5 cm. long in fruit. Calyx-tube campanulate, 2 mm. long; limb-tube 0.5 mm. long; lobes linear or subulate, 1–2(–4) mm. long, also sometimes one or 2 minute additional ones. Buds with very distinct tails ± 1 cm. long. Corolla yellow; tube 4–5 mm. long, glabrous outside, inside with a ring of deflexed hairs at the throat, otherwise glabrous; lobes oblong, 3–4 mm. long, 1.5–2 mm. wide, with tomentose margins, otherwise glabrous, terminated by linear-subulate deflexed appendages 6–12 mm. long. Ovary 4-locular; style 5 mm. long with a coroniform stigmatic club 2 mm. wide, ultimately 4-lobed. Fruit ellipsoid or subglobose, 10 mm. long, 8–10 mm. wide, with 1–4 pyrenes. Fig. 131/20, p. 752.

TANZANIA.   Iringa District: Uzungwa Mts., Sanje, *D. Thomas* 3781!
DISTR.   **T** 7; not known elsewhere
HAB.   Evergreen forest; 1100–1400 m.

37. **R. pseudolongicaudata** *Verdc.* in K.B. 42: 182 (1987). Type: Tanzania, Morogoro District, Nguru Mts., *Schlieben* 4189 (K, holo.!, BR, LISC, iso.!)

Shrub 2–3 m. tall or tree to 10 m.; stems slender, densely ferruginous hairy when young, later glabrous, the older parts grey and minutely fissured. Leaf-blades oblong to oblong-lanceolate, 2–9 cm. long, 0.9–3 cm. wide, acuminate at the apex, rounded at the base, sparsely pilose above and below mainly on the nerves to densely almost velvety hairy on both surfaces with pale brownish hairs; petiole 1.5–3 mm. long; stipules triangular, 2 mm. long, joined at base, with subulate hairy appendage 8 mm. long. Inflorescences 1-flowered; pedicels 3–5 mm. long, pubescent. Buds acuminate with 5 tails. Calyx-tube 1.2 mm. long, pubescent; lobes linear, 1.2 mm. long. Corolla greenish yellow or yellow; tube 6 mm. long, pubescent outside, glabrous inside; lobes 2.5 mm. long, pubescent outside, with subulate pubescent appendages 4 mm. long. Ovary 2-locular. Fruit creamy green, obovoid, flattened on one face, with a few hairs, subglobose ± 1.3 cm. in diameter or if consisting of 1 pyrene then 10 mm. long, 6 mm. wide; fruiting pedicels 1.6 cm. long.

TANZANIA.   Kilosa District: Ukaguru Mts., Mamiwa Forest Reserve, 16 Aug. 1972, *Mabberley & Alehe* 1487!; Morogoro District: Nguru Mts., W. slopes of Lukwangule Plateau, above Chenzema, *Harris et al.* 6076/AH!; Iringa District: Uzungwa Mts, Mwanihana Forest Reserve, above Sanje village, 10 Oct. 1984, *D. Thomas* 3780!
DISTR.   **T** 6, 7; not known elsewhere

HAB.　Evergreen montane moist forest with *Syzygium, Cyathea, Macaranga, Polyscias stuhlmannii*, etc.; 1650–2250(?–2400) m.

38. **R. hirsutiflora** *Verdc.* in K.B. 42: 183, fig. 13 (1987). Type: Tanzania, Iringa District, Uzungwa Mts., Sanje, *D. Thomas* 3811 (K, holo.!, MO, iso.!)

Shrub or small tree to 3 m.; young branches with dense short yellow hairs but soon merely pubescent or glabrescent; internodes often congested. Leaf-blades oblong or ± elliptic, 2–8 cm. long, 1–3.7 cm. wide, acuminate at the apex, cuneate to rounded at the base, slightly discolorous, densely yellow hairy when young but eventually pubescent mostly only on the nerves; domatia small, hidden by hairs on midrib; petiole 3–5 mm. long; stipule-sheath ± 2 mm. long, pubescent, with pubescent subulate appendages 4–6 mm. long. Flowers solitary; pedicels 2 mm. long, slightly accrescent in fruit to ± 5 mm. Calyx hairy; tube obconic, 1.5 mm. long; lobes ± 10, linear-triangular, 1–2.5 mm. long, shortly densely pilose. Buds with divaricate tails. Corolla white, hairy; tube campanulate, ± 2.5 mm. long, inside at throat with sparse deflexed hairs, otherwise glabrous; lobes oblong-elliptic, 4 mm. long, 2 mm. wide, with subulate subapical oblique slightly curved setulose-pilosulose appendages 3 mm. long. Ovary 2-locular; style 4 mm. long; stigmatic club subglobose, 1.2 mm. in diameter, sulcate basally, bilobed at apex. Fruits (only 1-seeded ones seen) narrowly obovoid-ellipsoid, 10 mm. long, 4–5 mm. wide, sparsely pubescent.

TANZANIA.　Iringa District: Uzungwa Mts., Sanje, *D. Thomas* 3811! & Uzungwa Mts., E. of W. Kilombero Forest Reserve, Udekwa village, Dec. 1981, *Rogers & Hall* 1454!
DISTR.　**T** 7; not known elsewhere
HAB.　Evergreen forest; 1100–1400 m.

39. **R. xanthotricha** (*K. Schum.*) *Verdc.* in K.B. 36: 539 (1981) & K.B. 42: 183 (1987). Types: Tanzania, E. Usambara Mts., Amani, *Engler* (1902) 609 & 610 (B, syn. †, K, fragments)

Scandent shrub with slender branches probably several m. long or ± erect shrub 2.5–3 m. tall; branches densely covered with somewhat bristly yellowish hairs when young but at length glabrous. Leaves paired; blades oblong-elliptic or narrowly elongate-elliptic, (1–)5–9.5(–12) cm. long, (0.4–)1.4–4.4(–5.5) cm. wide, acuminate at the apex, the acumen very narrowly rounded, rounded to subcordate at the base, rather densely covered with somewhat bristly yellowish ± adpressed hairs 1.5–2 mm. long, particularly on the nerves beneath and both sides of the midrib the hairs on this being set at right-angles; petioles 2–3 mm. long; stipule-bases joined to form an annulus 0.5 mm. tall with linear appendages 3–7(–11) mm. long. Flowers solitary, axillary, but sometimes that from terminal leaf-pairs appearing to be terminal; pedicels 0.8–1(–1.8 in fruit) cm. long, densely covered with patent bristly yellow hairs. Calyx-tube subglobose, ± 2 mm. long, white pilose; lobes filiform-subulate, 1–2 mm. long, bristly pubescent. Buds with 5 long tails. Corolla green or yellow, white pilose outside; tube 3 mm. long, pubescent below throat inside, glabrous beneath; lobes narrowly triangular, 3 mm. long, with long linear appendages 4–5 mm. long. Ovary 3-locular; style 3.5 mm. long; stigmatic club coroniform, 1.2 mm. wide, irregularly 3-lobed above. Fruit greenish white or becoming (?) black, ellipsoid, 0.8–1 cm. long, 8.5 mm. wide, with (1–)2–3 pyrenes, sparsely pilose with seta-like hairs.

TANZANIA.　Lushoto District: E. Usambara Mts., 1.6 km. NNE. of Amani on road to Monga, 28 July 1953, *Drummond & Hemsley* 3410! & Kwamkoro Forest Reserve, 19 Jan. 1961, *Semsei* 3175! & same locality, 10 Aug. 1961, *Mgaza* 413!
DISTR.　**T** 3; not known elsewhere
HAB.　Rain-forest; 850–950 m.

SYN.　*Plectronia xanthotricha* K. Schum. in E.J. 34: 335 (1904)
　　　Hutchinsonia xanthotricha (K. Schum.) Bullock in K.B. 1932: 389 (1932); T.T.C.L.: 501 (1949)

NOTE.　The corolla of this species was unknown to both K. Schumann and Bullock; despite the difference in habit the three specimens cited are undoubtedly conspecific with the fragments of syntypes preserved at K. This is undoubtedly closely related to *Hutchinsonia* but lacks the long corolla-tube. *H. cymigera* Bremek. with several-flowered inflorescences of small flowers is a *Rytigynia*; Bremekamp considerably altered the original description of *Hutchinsonia* to accommodate it (B.J.B.B. 22: 98 (1952)).

40. **R. griseovelutina** *Verdc.* in K.B. 42: 183 (1987). Type: Tanzania, Ulanga District, Sali, Ngongo Mt., *Cribb, Grey-Wilson & Mwasumbi* 11159 (K, holo.!)

Shrub to 3 m. tall; stems slender, densely spreading pubescent when young but soon glabrous and covered with brown rather ridged corky bark. Leaf-blades ovate or ± elliptic, 3.5–8 cm. long, 1.5–5.8 cm. wide, rounded to acute or mostly shortly acuminate at the apex, cuneate to ± truncate at the base, discolorous, green and adpressed pubescent above, softly grey velvety beneath with matted hairs which obscure the surface; petiole 4–6 mm. long; stipules with basal triangular parts 3 mm. long, connate, pubescent but becoming glabrous and persistent, bearing a filiform pubescent ± spreading apical appendage about 5 mm. long. Cymes 1–2-flowered, pubescent; peduncle 2–2.5 mm. long; pedicels 3–4 mm. long; bracts not developed. Calyx-tube obconic, 2 mm. long, with 8 minute lobes ± 0.5 mm. long. Corolla green; tube funnel-shaped, 3.5 mm. long, with a ring of tangled long hairs inside; lobes lanceolate, 5.5 mm. long including the apical 1.5–2 mm. long appendage, with scattered spreading hairs outside. Ovary 2-locular; style 6 mm. long, swollen below; stigmatic club coroniform, ± 1 mm. long, grooved. Fruit not known.

TANZANIA. Ulanga District: Sali, Ngongo Mt., 23 Jan. 1979, *Cribb, Grey-Wilson & Mwasumbi* 11159!
DISTR. **T** 6; not known elsewhere
HAB. Ridge forest; 1350 m.

41. **R. bugoyensis** (*K. Krause*) *Verdc.* in B.J.B.B. 50: 515 (1980); Fl. Pl. Lign. Rwanda: 597, fig. 204.1 (1982); Fl. Rwanda 3: 217, fig. 69.1 (1985); Verdc. in K.B. 42: 184 (1987). Type: Rwanda, Gisenyi [Kissenye], Bugoie [Bugoye] *Mildbraed* 1484 (B, holo. †, BM, iso.!, K, fragment!)

Shrub or small tree 1–6 m. tall, with slender spreading branches, perhaps sometimes ± scandent; older branches with ± lenticellate grey-brown bark, glabrous, bearing supra-axillary ± opposite paired spines, 0.6–2.5 cm. long; young shoots pubescent; leaves and flowers often borne on short condensed nodular lateral shoots bearing only one pair of leaves or apical portions of otherwise normal shoots can have the inter-nodes much condensed and similarly bear a pair of leaves. Leaf-blades ovate to elliptic or oblong-elliptic, 3–10.5 cm. long, 1.2–7 cm. wide, long-acuminate at the apex, the actual tip narrowly rounded, rounded to cuneate at the base, very thin, rather sparsely pilose with adpressed ± rough hairs on both surfaces and ciliate, with ± obscure barbellate domatia; petiole 0.3–1 cm. long, bristly pubescent; stipules connate at the base, triangular, 4 mm. long including the 1.5–2 mm. long filiform apex, deciduous. Inflorescences 2–3-flowered axillary cymes or flowers solitary; peduncles 4 mm. long and pedicels 3–10 mm. long, both ± densely pubescent. Calyx-tube hemispherical, 1.2–1.5 mm. long, ± glabrous or pilose; lobes linear-lanceolate, 1–3.5 mm. long, glabrous. Buds clavate, the tube glabrous to pilose, the apical swollen limb part ± spreading pilose, shortly tailed. Corolla greenish to yellowish cream; tube 3–6 mm. long, with most of inside pubescent with tangled hairs; lobes ovate-oblong, 1.5–3.5 mm. long. Ovary 2-locular; style 4.5–5 mm. long, thickened towards the base; stigmatic club coroniform, 0.5–0.8 cm. long, 2-lobed. Fruit black, compressed-obcordate, 1.2–1.3 cm. long, 1.1–1.2 cm. wide, broadly emarginate at the apex, glabrous or with a few bristly hairs at apex, crowned by calyx-limb, with 2 pyrenes; pedicels up to 1.5 cm. long.

subsp. **bugoyensis**; Verdc. in K.B. 42: 184 (1987)

At least corolla-lobes hairy outside and often tube as well; buds with less distinct tails. Fruiting pedicels up to 1.5 cm. long. Figs. 131/9, p. 752 & 147/1–12, p. 834.

UGANDA. Toro District: S. Ruwenzori, Kijomba ridge, Oct. 1940, *Eggeling* 4094! & near Mahoma R., 9 Sept. 1951, *Osmaston* 1189!; Kigezi District: Rubaya, Dec. 1947, *Purseglove* 2576!; Mbale District: Bugisu [Bugishu], Bubungi, July 1926, *Maitland* 1232!
KENYA. Northern Frontier Province: Marsabit, *T. Adamson* 122!; Meru Forest, Feb. 1932, *Honoré* in F.D. 2760!; N. Kavirondo District: Kakamega Forest, June 1933, *Dale* in F.D. 3118!
TANZANIA. Bukoba District: Kiamawa, Sept. 1935, *Gillman* 473!; Lushoto District: W. Usambara Mts., 2.4 km. NE. of Bumbuli Mission, on path to Mazumbai, 10 May 1953, *Drummond & Hemsley* 2464! & near Balangai, Mpalalu, 13 Mar. 1916, *Peter* 16105!
DISTR. U 2, 3; **K** 1, 4, 5; **T** 1, 3; Zaire, Rwanda and Burundi
HAB. Forest, *Agauria, Philippia* heath in W. Usambaras; (900–)1230–2400 m.

SYN. *Plectronia bugoyensis* K. Krause in Z.A.E. 1907–8, 2: 327 (1911)
    Vangueria butaguensis De Wild. in B.J.B.B. 8: 47 (1922) & in Pl. Bequaert. 2: 270 (1923). Type: Ruwenzori, Zaire, Butagu valley, *Bequaert* 3722 (BR, holo.)
    Rytigynia butaguensis (De Wild.) Robyns in B.J.B.B. 11: 145 (1928); F.P.N.A. 2: 347 (1947); K.T.S.: 472 (1961)
    Hutchinsonia bugoyensis (K. Krause) Bullock in K.B. 1932: 389 (1932)

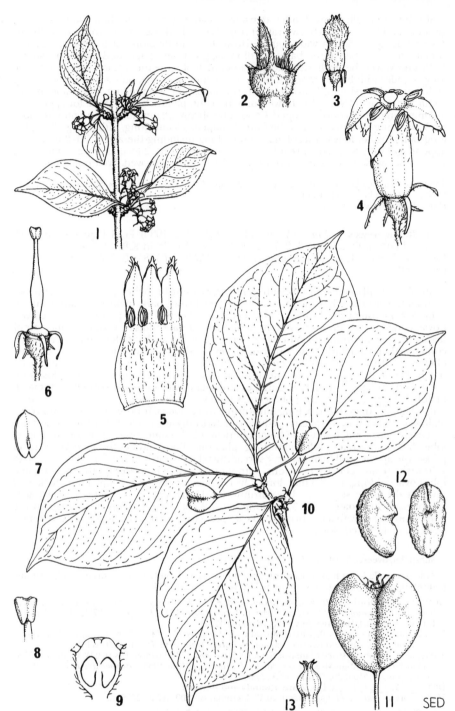

FIG. 147. *RYTIGYNIA BUGOYENSIS* subsp. *BUGOYENSIS* — **1**, flowering branch, × ⅔; **2**, stipule, × 4; **3**, flower bud, × 2; **4**, flower, × 4; **5**, portion of corolla, opened out × 4; **6**, calyx with style attached, × 4; **7**, dorsal view of anther, × 8; **8**, stigmatic knob, × 8; **9**, longitudinal section through ovary, × 8; **10**, fruiting branch, × ⅔; **11**, fruit, × 2; **12**, pyrene (2 views), × 2. Subsp. *GLABRIFLORA* — **13**, top of flower bud, × 2. 1, 4–10 from *Purseglove* 2576; 2, from *Paulo* 583; 3, 12, from *Snowden* 907/a; 11, from *Osmaston* 1189; 13, from *Richards* 6750. Drawn by Sally Dawson.

Canthium urophyllum Chiov., Racc. Bot. Miss. Consol. Kenya: 54 (1935), as *"urophyllam"*.* Type: Kenya, Fort Hall District, SW. Mt. Kenya, Mukarara [Mokkarara], *Balbo* 24 (TOM, holo.!)

NOTE. The 900 m. record is based on a sterile specimen, *Peter* 18163, the only one seen from the E. Usambaras.

subsp. **glabriflora** *Verdc.* in K.B. 42: 184 (1987). Type: Tanzania, Mt. Rungwe, *Richards* 6750 (K, holo.!)

Corolla glabrous outside; buds with more distinct tails at apex. Fruiting pedicels up to 3 cm. long. Fig. 147/13.

TANZANIA. Kilosa District: Ukaguru Mts., Mandege Forest Project, 24 May 1972, *Harris et al.* 6448!; Rungwe Forest, Jan. 1954, *Paulo* 233!; Songea District: Eastern Matagoro Hill, 27 Mar. 1956, *Milne-Redhead & Taylor* 9408!
DISTR. **T** 6–8; Malawi
HAB. Upland rain-forest; 1440–2400 m.

NOTE. Of the 8 sheets seen only 3 are in flower and confirmation is needed that the corolla is always glabrous.

42. **R. sp. F**; Verdc. in K.B. 42: 185 (1987)

Small slender tree; new shoots with spreading yellow-brown hairs but soon glabrous and dull purplish brown. Leaf-blades elliptic to oblong-elliptic, 2.5–9 cm. long, 1.2–3.8 cm. wide, narrowly acuminate at the apex, the actual tip rounded, broadly cuneate to rounded at the base, drying brown, bristly yellowish brown pubescent above and mainly on the nerves beneath, ± discolorous; petioles ± 2 mm. long; stipules with hairy base 2 mm. long and stoutly subulate bristly hairy appendage ± 3–4 mm. long, only visible at uppermost nodes, soon falling. Inflorescences 3–8-flowered, yellowish bristly pubescent; peduncle thick, 1–3 mm. long; bracts narrowly ovate, 1.5 mm. long; fruiting pedicels 1–1.2 cm. long and a few undeveloped pedicels 4 mm. long. Corolla not seen. Fruits subglobose, 9 mm. long and wide, with 2–5 pyrenes.

TANZANIA. Morogoro District: Chigurufumi Forest Reserve, Mar. 1955, *Semsei* 2034!
DISTR. **T** 6; known only from the above gathering
HAB. Presumably forest; altitude not known

NOTE. Flowering material is needed to identify this species but it does not appear to match any known species. The foliage and habit are similar to those of *Akeroyd & Mayuga* 18 treated here as perhaps a variant of *R. nodulosa*, but the inflorescence is different.

43. **R. decussata** (*K. Schum.*) *Robyns* in B.J.B.B. 11: 195 (1928); T.T.C.L: 531 (1949); Verdc. in K.B. 42: 185 (1987). Lectotype, see Robyns (1928): Tanzania, Uzaramo [Usaramo] District, *Stuhlmann* 6637 (B, lecto.†); Uzaramo, *Stuhlmann* 6836 (K, prob. isosyn.!)

Shrub or subshrub 0.6–1.5 m.** tall, with slender stems from a woody rootstock, densely spreading pubescent when young, becoming glabrescent and older stems covered with dark red-brown ± peeling bark. Leaves mostly paired but frequently in whorls of 3; blades elliptic, oblong or ± ovate, 1.7–8.5(–11) cm. long, 0.8–5(–6) cm. wide, shortly pubescent above, velvety woolly beneath with tangled adpressed grey hairs which almost or completely obscure the surface but main venation clearly evident when dry; petiole 4–7 mm. long; stipules joined to form a short sheath 1.5–2.5 mm. long, densely pilose within near node; appendages subulate, 2–3 mm. long. Cymes short, 1–5-flowered; peduncles (1–)2–4 mm. long; pedicels (0.5–)2–6 mm. long, both ± spreading pubescent; bracts and bracteoles ± obsolete. Calyx pubescent; tube campanulate or globose, 1–1.5 mm. long; limb-tube very short; lobes linear-lanceolate, 1.5–4 mm. long. Buds acuminate. Corolla white, cream or yellow; tube 2–4.5 mm. long, glabrous or with a few scattered hairs outside, with a ring of deflexed hairs inside; lobes triangular, 2.5–4 mm. long, 1–2 mm. wide, apiculate. Anthers purple-brown, half-exserted, with rather distinct connective appendages. Ovary 2-locular; style exserted ± 1.5 mm., the stigmatic club green, 1 mm. long. Fruit black, didymous, usually only one per cyme developing, ± 1 cm. long and wide, 4 mm. thick, with 2 pyrenes, crowned by calyx-teeth and pedicel lengthening to 9 mm.

---

* An obvious misprint — the type is labelled *urophyllum* correctly.
** A report by Semsei that it attains 4.5 m. seems unlikely.

KENYA. Kwale District: Shimba Hills Lodge, Mkomba R., 23 Feb. 1988, *Robertson & Luke* 5183!
TANZANIA. Handeni District: Kwamsisi road, May 1965, *Procter* 3006!; Kilosa District: Kibedya, Jan. 1931, *Haarer* 1983!; Lindi District: Lake Lutamba, 11 Jan. 1935, *Schlieben* 5860!
DISTR. **K** 7; **T** 3, 6, 8; Mozambique
HAB. *Acacia*, *Combretum* and *Brachystegia* woodlands, thicket, open grassland, etc.; 75–1000 m.

SYN. *Pachystigma decussatum* K. Schum. in P.O.A. C: 387 (1895)
     Vangueria longisepala K. Krause in E.J. 39: 534 (1907); De Wild. in B.J.B.B. 8: 58 (1922). Type: Tanzania, Dar es Salaam, Mogo Forest Reserve [Sachsenwald], *Holtz* 1085 (B, holo. †)
     Rytigynia sessilifolia Robyns in B.J.B.B. 11: 194 (1928). Type: Mozambique, Mt. Mkoto, *Stocks* 138 (K, holo.!)

NOTE. This is a peripheral species linking *Rytigynia*, *Fadogia* and *Tapiphyllum* and resembling *Fadogia* in habit. Since its position is likely to remain doubtful I have erected a subgenus *Fadogiopsis* Verdc. (K.B. 42: 185 (1987)) within *Rytigynia* rather than transfer it to *Fadogia* from which it differs in the ovary. *Welch* 450 (Morogoro District, 12 km. NE. of Kingolwira Station) has leaves in whorls of 3 and a very *Fadogia*-like habit.

44. **R. saliensis** *Verdc.* in K.B. 42: 185 (1987). Type: Tanzania, Ulanga District, Sali, *Cribb et al.* 11181 (K, holo.!)

Shrub to 3 m.; branches slender, glabrous or with very few curled hairs on apical parts, the older parts covered with chestnut to grey-brown finely longitudinally ridged corky bark, lenticellate. Leaf-blades oblong, thin, 4.5–12 cm. long, 2.3–5.5 cm. wide, rather shortly obtusely acuminate at the apex, rounded at the base, glabrous on both sides and without domatia; petiole 3–5 mm. long; stipules with rounded base, 2 mm. long, soon corky and persistent with a subulate appendage 6.5 mm. long also becoming corky and persistent. Inflorescence 5-flowered; peduncle ± 5 mm. long; pedicels ± 4 mm. long; bracts narrowly ovate, 1.5–2 mm. long. Buds acute. Calyx-tube 2–2.5 mm. long; limb-tube 0.4 mm. long, with lanceolate lobes 1.5–3 mm. long. Corolla white; tube urceolate, 3 mm. long, glabrous inside; lobes triangular-oblong, 3.5–4.5 mm. long, 2.5–3 mm. wide, finely papillate outside, acute or with slight appendage 0.3 mm. long. Ovary 3-locular; style subulate-conic, 3 mm. long; stigmatic club oblong, 1.2 mm. long. Fruits not known.

TANZANIA. Ulanga District: Sali, Muhulu Forest Reserve, 24 Jan. 1979, *Cribb et al.* 11181!
DISTR. **T** 6; not known elsewhere
HAB. Montane forest on ridges; 1425 m.

NOTE. I have referred this species to a separate subgenus *Sali* Verdc. (K.B. 42: 185 (1987)), which links *Rytigynia* with *Vangueria* and *Vangueriella*.

45. **R. fuscosetulosa** *Verdc.*, sp. nov. affinis *R. saliensis* Verdc. caulibus foliisque densissime fuscosetulosis, calycis lobis linearibus vel lineari-lanceolatis usque 5 mm. longis differt. Typus: Tanzania, Iringa District, Mufindi, Lulando Forest Reserve, *Cribb et al.* 11466 (K, holo.! & iso.!)

Shrub ± 1–2 m. tall; stems with minutely longitudinally fissured grey-buff bark, glabrescent; young shoots with spreading dense pale ferruginous bristly hairs. Leaf-blades elliptic-oblong, 7–10.5 cm. long, 3–6 cm. wide, obtusely acuminate at the apex, rounded at the base, harshly hairy on both surfaces with dense hairs similar to those on the shoots; petioles 5 mm. long, similarly bristly; stipules similarly bristly, ± 1 cm. long, the narrowly triangular base produced into a narrow acumen. Fascicles axillary, 2–6-flowered, ± sessile; peduncle and pedicels 1.5–6 mm. long; bristly; paired basal bracts ovate-triangular, 2 mm. long, 0.8 mm. wide; secondary linear, 2 mm. long. Calyx-tube 2 mm. long; lobes linear to linear-lanceolate, 2.5–5 mm. long, 0.6 mm. wide; all parts glabrous or bristly pubescent. Corolla pale green, glabrous save very few hairs at the throat; tube cupular, 2.5 mm. long; lobes triangular, 3.5 mm. long, 3 mm. wide at base, acute, venose. Anthers 1.6 mm. long (boiled up; 0.7 mm. dry). Style 2.5 mm. long, tapering from ± stout base; stigmatic club green, cylindrical, 10-ribbed, 1.2 mm. long, the actual stigmatic surface yellow, 3–4-lobed. Fruit drying black or chestnut, red in life, fleshy, 1.6 × 1.5 cm., glabrous to pilose, containing 3–4 pyrenes, or fruits reduced to one pyrene, 1.2 × 1 cm.; pyrenes straw-coloured, up to 1.4 cm. long, 8 mm. wide, 7 mm. thick, compressed at apex, the point of attachment ± beaked, situated ± 6 mm. from apex; testa-cells isodiametric.

TANZANIA. Iringa District: N. side of Uzungwa Scarp, Lubongolo Forest, 17 June 1979, *Mwasumbi* 11844! & Mufindi area, Lulando Forest Reserve, 19 Apr. 1987, *Lovett & Congdon* 2039! & 12 Apr. 1988, *Congdon* 203!

DISTR.   **T** 7; not known elsewhere
HAB.   Evergreen forest; 1450–1650 m.

SYN.   *Vangueriella sp. A* sensu Verdc. in K.B. 42: 197, fig. 2A (1987)

NOTE.   This species also is intermediate between several of the presently accepted genera. The pyrene and seed closely resemble *Vangueria* but the embryo is basally much more terete and the ovate-triangular cotyledons are distinctive; also the ovary is 3–4- not 5-locular. In foliage *R. fuscosetulosa* is remarkably similar to the W. African *Vangueriella campylacantha* (Mildbr.) Verdc., but *Vangueriella* is often spiny, has a 2-locular ovary and a mitriform stigmatic club. The present species seems best placed in *Rytigynia* subgen. *Sali* despite a few hairs at the corolla throat and very different indumentum since the inflorescence structure, bracts, calyx-lobes and habit are very similar.

Inadequately known species

46. R. sp. G

Apparently a shrub; stems purplish brown, lenticellate, the bark flaking off in very small pieces, almost powdery. Leaf-blades elliptic-oblong, 3–9 cm. long, 1.3–3.7 cm. wide, long-acuminate at the apex, rounded then abruptly cuneate at the base, glabrous, drying somewhat shining and bronzy green and with venation slightly raised on both surfaces; petioles up to 3 mm. long; stipule-bases forming a pale sheath 3 mm. long, white-hairy inside, the appendage obsolete or subulate ± 2 mm. long. Flowers and fruit quite unknown.

TANZANIA.   Iringa District: Mufindi, valley of Nyalawa R., below bridge which lies 1.6 km. below Idege Division Headquarters, Livalonge Tea Estate, 26 Aug. 1971, *Perdue & Kibuwa* 11248!
DISTR.   **T** 7; known only from the above gathering
HAB.   Presumably evergreen forest; 1500 m.

NOTE.   This distinctive sterile foliage has not been matched with any described species, but been misidentified as *R. schumannii* (*R. uhligii*) and may well be related to that.

47. R. sp. H

Shrub; stem glabrous with straw-coloured or brown flaking or peeling bark. Leaf-blades ovate-lanceolate, ± 4.5 cm. long, 2.4 cm. wide, narrowly acuminate at the apex, cuneate to ± rounded at the base, glabrous; venation reticulate but ± plane beneath, slightly impressed above when dry, the margin very obscurely wrinkled or crenate; petiole ± 4 mm. long; stipule-base 1–2 mm. tall, the appendage thick, subulate, ± 2 mm. long, deciduous. One fruiting inflorescence only seen; peduncle ± 9 mm. long; pedicel 6.5 mm. long; inflorescences probably 2-flowered. Fruit 8 mm. long, 7 mm. wide.

TANZANIA.   Tabora, 4 Feb. 1970, *Mungure* 221!
DISTR.   **T** 4; known only from the above gathering
HAB.   *Brachystegia* woodland; 1200 m.
NOTE.   Only one poor specimen has been seen; it could perhaps be related to *R. lewisii* Tennant from Zambia.

48. R. sp. I

Bushy shrub; stems glabrous or with sparse short pubescence, with youngest parts straw-coloured, later pale brown, the bark peeling in small flakes, somewhat powdery beneath; lateral branches sometimes in threes sometimes with 1 or all 3 reduced to spines, but with reduced cushion shoots below the spines and branches. Leaves opposite or in threes on expanded shoots or subfastigiate on the cushion shoots due to telescoping of nodes; blades elliptic, 2.5–5 cm. long, 1–3.2 cm. wide, obtuse or very obscurely rounded-acuminate at the apex, cuneate at the base, discolorous, glabrous save for rather diffuse domatia in the main nerve axils; petiole ± 2 mm. long, often pubescent; stipule-bases ± 1 mm. long, pubescent outside, hairy inside, appendages recurved or hooked, ± 2 mm. long, deciduous. Inflorescences 1–2-flowered, glabrous; peduncle 6–7 mm. long; pedicels 1.8–2 cm. long, the basal bracts forming a minute involucre 0.5 mm. long. Flowers unknown. Fruits preserved consisting of 1 pyrene, rounded reniform, 7 mm. long, 5 mm. wide.

KENYA. Tana River District: Ozi, 4 July 1973, *Homewood* 62! & 64!
DISTR. **K** 7; known only from the above gatherings
HAB. Unflooded coastal riverine forest on sandy soil; 6 m.

NOTE. The habit is that of *Meyna* but the fruits are quite different.

49. **R. sp. J**

Shoots said to be seedling specimens of a small tree. Youngest shoots very densely covered with soft spreading yellow-brown hairs, the previous season's growth with darker bark and similar but dark ferruginous hairs. Leaf-blades elliptic or oblong, 4–8.5 cm. long, 1.5–2.5 cm. wide, acute or shortly acuminate at the apex, rounded to ± cuneate at the base, discolorous, drying dark above, densely adpressed pubescent above and almost velvety beneath with rather bristly pale hairs, particularly noticeable at right-angles to the midrib, but surface nowhere obscured; petioles ± 3 mm. long, very densely spreading pilose; stipule-bases broad, ± 1 mm. tall, with subulate appendages 5–6 mm. long. Flowers and fruits unknown.

TANZANIA. Iringa District: Mufindi, Kigogo Forest, Mgololo path, 28 Oct. 1946, *Gilchrist* in *F.H.* 1701!
DISTR. **T** 7; known only from the above gathering
HAB. Presumably evergreen forest; 1500 m.

NOTE. The native name is given as Mtagamba (Kifuagi); very probably a *Rytigynia*, but not certain without fertile parts.

50. **R. sp. K**

Subshrubby ? ± erect herb 30–40 cm. tall, with several slender wiry stems from a ± slender woody rootstock; young stems rather densely pubescent with short curled hairs; at length glabrescent. Leaves paired or in whorls of 3; blades elliptic or oblong-elliptic, 1.2–3.8 cm. long, 0.7–1.6 cm. wide, acute to quite rounded at the apex, cuneate at the base, slightly discolorous, pubescent on both surfaces with short adpressed hairs; petioles short, ± 1 mm. long; stipules with broad base ± 0.7 mm. long and subulate appendage 2–3 mm. long. Inflorescences 1–several-flowered, borne in the upper axils; peduncles 2–3 mm. long; pedicels 2.5–7 mm. long, both pubescent like the stems. Open flowers not seen; buds subobtuse, densely pubescent. Calyx truncate. Fruits not seen.

TANZANIA. Lindi District: about 80 km. W. of Lindi, Mbemkuru, 22 Feb. 1935, *Schlieben* 6049!
DISTR. **T** 8; known only from the above gathering
HAB. Wooded grassland ("parkland"); 280 m.

NOTE. The habit and rather characteristic indumentum should render this easily recognisable. Schlieben describes the flowers as yellowish so presumably some of the gathering had open flowers.

51. **R. sp. L**

Tree to 4 m.; shoots slender, glabrous with peeling white epidermis revealing brown bark and with pale lenticels. Leaves oblong-elliptic to elliptic, 4.5–12 cm. long, 2.5–5 cm. wide, distinctly acuminate at the apex, cuneate to ± rounded at the base, thin to very thin, entirely glabrous save for the domatia, patches of white pubescence in the axils of nerves beneath; stipule-sheath under 1 mm. long with ring of erect white hairs projecting from apex, the appendage slightly curved and 5 mm. long, 0.5 mm. wide.

KENYA. Kwale District: NE. side of Gongoni Forest Reserve, 1 June 1990, *Robertson & Luke* 6309!
DISTR. **K** 7; not known elsewhere
HAB. Moist lowland forest; 30 m.

NOTE. *Luke & Robertson* 2399 (W. Gongoni Forest, 3 June 1990) can hardly be different but the cymes are 1–3-flowered with peduncles 3 mm. long and pedicels 1–3 mm. long, buds 2–3 mm. long, glabrous and distinctly acuminate; leaves smaller, up to 6 × 2.5 cm. and finely puberulous as are the stems. A third specimen, *Luke & Robertson* 2327 (the same plant as *Robertson & Luke* 4864 formerly misidentified as *R. binata* — Kwale District, Jego, 27 May 1990), is small-leaved but glabrous save for the domatia. All these are from the same type of forest and must be conspecific and can hardly be anything but a population of *R. celastroides* with distinctive characters but I hesitate to include them without more evidence. True *R. binata* has a different calyx, different leaf-colour, bud-shape, etc. and is, I am sure, not the same.

100a. GENUS ? NOV. A

Tree 10 m. tall; stems slender, grey-brown, obscurely lenticellate, glabrous save for dense bristly hairs within the stipule-bases. Leaves narrowly oblong-elliptic, 1.5–4.5 cm. long, 0.8–1.8 cm. wide, acuminate or in a few small leaves rounded or even emarginate at the apex, the actual acumen somewhat blunt, cuneate at the base into a short petiole, glabrous, lacking even domatia, paler beneath; stipules very characteristic, subulate, 3–3.5 mm. long, widened at the base, bearing a relatively large club-shaped colleter at the apex, hairy within. Flowers fasciculate or sometimes solitary in the axils of fallen leaves which usually bear young leafy shoots; pedicels 4–5 mm. long; basal paired bracts 2–3 mm. wide from tip to tip; bracteoles 1.5 mm. long. Calyx-tube globose, 1.8 mm. tall, 2.2 mm. wide; lobes narrowly triangular, 2 mm. long, 1 mm. wide, thickened at the apex. Corolla pale green inside, the tube darker, 7 mm. long, glabrous outside but with a dense ring of deflexed bristle-like hairs at the middle inside and tangled hairs above this to the throat; basal 2 mm. glabrous; lobes lanceolate, 5 mm. long, 2 mm. wide, scarcely apiculate. Ovary 5-locular, the locules strongly convex outside; style 9 mm. long; stigmatic club subcylindrical, 1 mm. long, widest above, constricted near to the base which is hollowed out. Anthers ovoid, 1.5 mm. long, the apical apiculae inturned. Fruit not seen.

TANZANIA. Morogoro District: N. Nguru Mts., Kanga Mt., 2 Dec. 1987, *Lovett & Thomas* 2713!
HAB. Moist evergreen forest on steep slopes; 1200 m.

NOTE. The pollen grains are triporate with fine reticulation as in very many Vanguerieae.

101. VANGUERIELLA

Verdc. in K.B. 42: 189 (1987)

Vangueriopsis subgen. *Brachyanthus* Robyns in B.J.B.B. 11: 249 (1928)

Shrubs or small trees 2–10 m. tall or, sometimes, scandent shrubs, glabrous to densely hairy, unarmed or branches with short opposite often recurved spines. Leaves opposite, petiolate; stipules connate into a sheath at base, triangular or oblong, villous inside, produced into a linear or subulate subpersistent or soon deciduous appendage. Flowers small or medium-sized in few–many-flowered dichasial axillary and opposite pubescent or rarely glabrous cymes; bracts small or scarcely developed. Buds lanceolate or cylindrical, ± 4–8 mm. long, acute, ± obtuse, the apex filiform-apiculate, glabrous or tomentose. Calyx-lobes developed but short, rounded-ovate, linear or linear-subulate, usually spreading, often connate at the base. Corolla-tube campanulate or cylindrical, glabrous at the throat but (save in 5 species) with a ring of deflexed hairs in the middle inside; lobes oblong, usually shorter or equalling the tube or in one species filiform-apiculate and much longer. Anthers small, half-exserted, ovate or ovate-lanceolate, glabrous or subpapillose; filaments very short. Ovary 2-locular; style shortly exserted, cylindrical to narrowly obconic; stigmatic club coroniform or mitriform usually widened at base, rarely cylindrical, bilobed at the apex. Disk glabrous. Fruit usually comparatively large, ± didymous, containing 2 often verrucose pyrenes and crowned by the calyx-limb.

A genus of about 17 species mainly in W. Africa as far as Angola; only the inadequately known *V. rhamnoides* occurs in East Africa and belongs to sect. *Stenosepalae* (Robyns) Verdc.

1. **V. rhamnoides** (*Hiern*) Verdc. in K.B. 42: 197 (1987). Type: Angola, Pungo Andongo, Tunda Quilombo, *Welwitsch* 5350 (LISU, holo., BM, K, iso.!)

Straggling slender shrub or scandent, 3–3.6(–9) m. tall; branches in whorls of 3 and said to bear rather stout spines in whorls of 3; branchlets chestnut-brown, lenticellate, ± longitudinally ridged, with ± adpressed bristly hairs but soon glabrescent. Leaves ? paired or in whorls of 3; blades oblong, 6–15 or more cm. long, 3–9 cm. wide, sharply acuminate at the apex, rounded to truncate at the base, with rather sparse short stiff hairs above, finely adpressed pubescent beneath particularly on the raised venation; petiole 5–8 mm. long, adpressed hairy; stipule-bases triangular with ± thin margins, 3–4 mm. long, with subulate appendage 5–7.5 mm. long, eventually deciduous. Inflorescences axillary ± many-flowered distichous cymes; peduncle ± 5 mm. long; pedicels 4–6 mm. long,

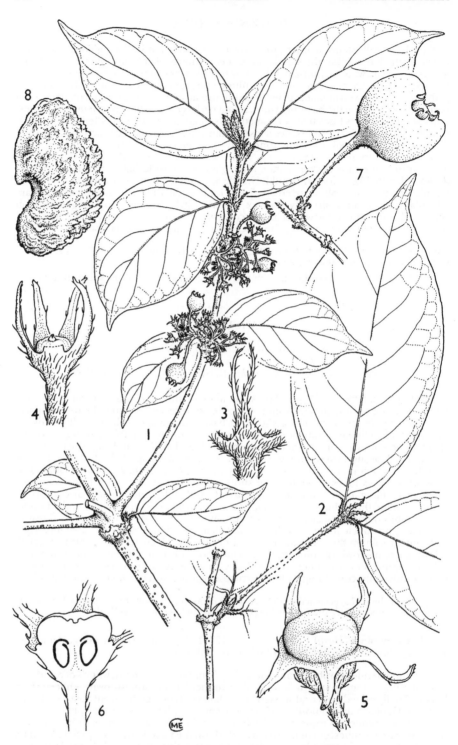

FIG. 148. *VANGUERIELLA RHAMNOIDES* — **1**, fruiting branch, × ⅔; **2**, shoot with spines, × ⅔; **3**, stipule, × 3; **4**, immature calyx with one lobe removed, × 10; **5**, calyx, × 8; **6**, longitudinal section through ovary, × 8; **7**, fruit, × 1; **8**, pyrene, × 3. 1, 4–6, from *Chandler* 2150; 2, 3, 7, 8, from *Synnott* 657. Drawn by Mrs M.E. Church.

lengthening and becoming much thicker in fruit, all axes adpressed pubescent. Calyx-tube campanulate, 1.5 mm. long, adpressed pubescent; lobes linear-lanceolate, 2–2.5 mm. long, thin; corolla not seen; disk depressed hemispherical. Ovary 2-locular. Immature fruits subglobose, ± 8 mm. diameter, crowned with persistent calyx-lobes, with few obscure scattered hairs. Ripe fruits compressed, bilobed, up to 2 cm. wide (*fide* collector), 1.6 cm. long and wide when dry, on pedicels up to 3 cm. long, bluntly warted in dry state due to rugae on pyrenes. Figs. 132/13, p. 754 & 148.

UGANDA. Bunyoro District: Budongo Forest, near R. Nyabisabu, 13 Oct. 1971, *Synnott* 657! & E. Boundary, Budongo Forest, Nyakafunjo Nature Reserve, 4 Aug. 1973, *Synnott* 1501!; Mengo District: Entebbe–Kampala road, km. 16, Kanjansi Forest, *Chandler* 2150!
DISTR. U 2, 4; Zaire S. Sudan and Angola
HAB. Evergreen forest; 1050–1170 m.

SYN. *Canthium rhamnoides* Hiern, Cat. Afr. Pl. Welw. 1: 472 (1898)
       Vangueriella sp. sensu Verdc. in K.B. 42: 196 (1987)

NOTE. It is extraordinary that no other material of this plant appears to have been collected between two such important places in Uganda so it may be genuinely rare.

102. **MULTIDENTIA**

Gilli in Ann. Naturhist. Mus. Wien 77: 21, t. 1 (1973); Bridson in K.B. 42: 641–654 (1987)

Canthium Lam. sect. *Granditubum* Tennant in K.B. 22: 438 (1968)

Shrubs, small trees to 12 m. tall or sometimes pyrophytic subshrubs, mostly glabrous, unarmed; stems often lenticellate. Leaves mostly restricted to new growth, paired or rarely ternate, petiolate; blades chartaceous to coriaceous, typically glaucous beneath, with a conspicuous network of finely, or less often coarsely, reticulate tertiary nerves; domatia present as tufts of hair or absent; stipules sheathing at base, pubescent within, provided with a linear, somewhat keeled lobe. Flowers (4–)5(–6)-merous, usually medium-sized, borne in pedunculate cymes, usually at nodes from which the leaves have fallen, except in pyrophytic species; bracteoles linear to lanceolate, small. Calyx chartaceous or coriaceous; limb cupular, truncate, repand, dentate or cupular below and lobed above. Corolla normal or coriaceous and drying wrinkled; tube cylindrical, subequal to lobes or occasionally much longer, glabrous outside, with a ring of deflexed hairs inside and usually rather sparsely pubescent at throat; lobes reflexed, rounded or obtuse, thickened towards apex. Stamens set at throat; anthers partly or fully exserted, oblong-ovate or oblong, with dark coloured connective tissue on dorsal face except for margin, apiculate or not. Ovary 2-locular, each locule with one pendulous ovule; style slightly longer than or less often up to twice as long as corolla-tube, slender; stigmatic knob ± as broad as long or less often elongate, ribbed, hollow to mid-point, apex cleft ± to mid-point when mature. Disk glabrous. Fruit a 2-seeded drupe, large, subglobose, laterally compressed, often somewhat didymous, often lenticellate or rarely small and didymous (in the West African *M. pobeguinii* (Hutch. & Dalz.) Bridson), crowned by persistent calyx-limb; pyrenes thickly woody, broadly ellipsoid, sometimes curved, truncate at point of attachment, with line of dehiscence extending from point of attachment to apex then arching on either side back towards point of attachment but stopping short of it, very strongly rugulose (except in *M. pobeguinii*) see fig. 132, p. 754. Seeds with endosperm entire; testa finely reticulate; embryo slightly curved; radicle erect; cotyledons about ⅓ the length of the embryo, set perpendicular to the ventral face of the seed.

A genus of 11 species (2 poorly known) restricted to tropical Africa, 8 of which occur in the Flora area.

Pyrophytic subshrub up to 0.35 m. tall; inflorescence borne at
     leafy nodes, very often supra-axillary with peduncle set at
     an acute angle to stem; calyx-limb with very distinct lobes
     up to 6 mm. long   . . . .   8. *M. concrescens*
Shrubs or small trees more than 1 m. tall; inflorescence mostly
     borne on leafless nodes, truly axillary; calyx-limb repand
     to dentate or sometimes with lobes up to 2 mm. long:

Leaves coriaceous, often (but not always) drying yellow-
green, oblong-elliptic, oblong or rotund; stems
frequently thick and covered with pale corky bark;
corolla leathery and wrinkled when dry . . . . .   7. *M. crassa*
Leaves papery or subcoriaceous, usually drying blackish,
brown or dark green above, elliptic or occasionally as
above; stem not as above; corolla occasionally leathery
when dry:
    Tertiary nerves very coarsely reticulate; corolla-tube 0.3–
    1.4 cm. long:
      Corolla-tube 0.8–1.4 cm. long, greatly exceeding lobes;
      style somewhat longer than corolla-tube; midrib
      cream (when dry) . . . .   2. *M. fanshawei*
      Corolla-tube 3–5 mm. long, ± equal to lobes; style up to
      twice as long as corolla-tube; midrib not cream
      (when dry) . . . .   3. *M. exserta*
    Tertiary nerves finely to moderately coarsely reticulate;
    corolla-tube 2–3 mm. long:
      Leaves with 8–12 main pairs of lateral nerves; tertiary
      nerves finely reticulate and very conspicuous;
      stipules with lobe 6–13 mm. long . . . . .   1. *M. dichrophylla*
      Leaves with 5–7 main pairs of lateral nerves; tertiary
      nerves finely to moderately finely reticulate,
      conspicuous or somewhat obscure; stipules with
      lobe 3–7 mm. long:
        Inflorescence lax; calyx papery; corolla not thick;
        leaves not strongly discolorous . . . . .   4. *M. kingupirensis*
        Inflorescence not markedly lax; calyx subcoriaceous
        to coriaceous; corolla thick, often drying wrinkled;
        leaves strongly discolorous:
          Leaves distinctly acuminate; shrub or tree 3.5–13
          m. tall; calyx subcoriaceous; limb 1.5–2 mm.
          long . . . .   5. *M. sclerocarpa*
          Leaves shortly acuminate or sometimes obtuse;
          shrub 1–5 m. tall; calyx coriaceous, often
          drying yellow-green; limb 2–3 mm. long   6. *M. castaneae*

1. **M. dichrophylla** (*Mildbr.*) *Bridson* in K.B. 42: 646 (1987). Type: Cameroon, Baja Highlands, between Lom and Kunde, *Mildbraed* 9201 (B, holo.†, K, iso.!)

Small tree or scandent shrub, 4.5–12 m. tall; young stems lenticellate. Leaf-blades elliptic, 6.5–12.5 cm. long, 2.7–6.2 cm. wide, distinctly acuminate at apex, acute to rounded and often somewhat unequal at base, papery, strongly discolorous (in dry state); lateral nerves in 8–12 main pairs; tertiary nerves finely reticulate; domatia pubescent or absent; petiole 4–11 mm. long; stipules with sheath-like base, 3–5 mm. long, bearing a linear lobe 6–13 mm. long, pubescent inside. Flowers 5-merous, borne in rather lax pedunculate 30–100-flowered cymes; peduncles 1.5–3 cm. long, sparsely pubescent; pedicels 1–3 mm. long, pubescent; bracteoles linear or lanceolate, up to 2 mm. long. Calyx-tube ± 1 mm. long, sparsely pubescent; limb repand to shortly dentate, ± 1.25 mm. long. Corolla-tube 3 mm. long, 2.5 mm. wide, pubescent at throat, then with a ring of deflexed hairs; lobes triangular, 2.5 mm. long, 1 mm. wide at base, thickened towards apex, acute. Style 4 mm. long, glabrous; stigmatic knob as broad as long, ± 1 mm. in diameter. Fruit broader than long, laterally compressed, somewhat didymous (when dry), 1.8 cm. long, 2.5 cm. wide, 1.3 cm. thick, lenticellate; pyrene not known.

UGANDA. Mengo District: Kivuvu, Feb. 1918, *Dummer* 4095! & ? Kasii, Mar. 1917, *Dummer* 3116!
DISTR. U 4; Cameroon, Zaire and Sudan
HAB. Forest; 1220 m.

SYN. *Plectronia dichrophylla* Mildbr. in N.B.G.B. 9: 204 (1924)
    P. bicolor De Wild., Pl. Bequaert. 3: 179 (1925). Type: Zaire, Irumu, *Bequaert* 2932 (BR, holo.!)
    Vangueriopsis sillitoei Bullock in K.B. 1940: 57 (1940). Type: Sudan, Yei R., Lado, *Sillitoe* 377 (K,
    holo.!)
    Canthium bicolor (De Wild.) Evrard in B.J.B.B. 37: 459 (1967)

2. **M. fanshawei** (*Tennant*) *Bridson* in K.B. 42: 647 (1987). Type: Malawi, Ntchisi Mt., *Robson* 1668 (K, holo.!, EA!, LISC!, PRE!, SRGH, iso.)

Shrub or tree 2–3(–12) m. tall, erect or sometimes scandent, or a tree 14 m. tall, glabrous; stems covered with reddish brown rather powdery bark, lenticellate. Leaves tending to be restricted to apex of branches; blades elliptic to broadly elliptic or occasionally rotund or oblong-elliptic, 4–12 cm. long, 2.5–5(–7) cm. wide, acute to subacuminate or acuminate at apex, acute to cuneate or sometimes obtuse at base, stiffly papery, drying blackish brown with nerves pale above and glaucous beneath; lateral nerves in 4–5(–6) main pairs; tertiary nerves coarsely reticulate, either apparent or obscure; domatia present as hairy tufts; petiole 6–10 mm. long; stipule-base sheathing, 2–3 mm. long, bearing a linear-subulate lobe 2–3 mm. long, pubescent within. Flowers 5-merous, borne in many-flowered pedunculate cymes; peduncle 1–2.5 cm. long, glabrous to sparsely pubescent; pedicels 2–6 mm. long, sparsely pubescent to pubescent; bracts and bracteoles linear, 1–2 mm. long. Calyx-tube 1–1.5 mm. long, glabrous; limb 1–2 mm. long, repand to slightly dentate. Corolla white; tube 0.8–1.4 cm. long, slightly pilose at throat, with a ring of deflexed hairs set at ± ⅓ of the way from the top inside; lobes oblong-elliptic to narrowly obovate, 5–9.5 mm. long, 2.2–5 mm. wide, acute. Anthers erect or sometimes spreading but not reflexed. Stigmatic knob well exserted, ovoid, 2.25–2.5 mm. long, 1.5–2 mm. wide, ridged. Fruit subglobose, somewhat bilobed, 1.8–2.5 cm. long, 1.9–2.7 cm. wide; pyrene almost semicircular in outline, 1.7 cm. long, 1.3 cm. wide, 1.1 cm. thick, gradually rounded at point of attachment, strongly rugulose and grooved.

TANZANIA. Iringa District: Mufindi, 8 June 1986, *Paget-Wilkes* 95! & Nyumbanyitu Mt., 25 Sept. 1958, *Ede* 63!; Rungwe District: Kyimbila, *Stolz* 2152!
DISTR. T 7; Zaire (Shaba), Malawi and Zambia
HAB. Bamboo forest; 1980 m.

SYN. *Canthium fanshawei* Tennant in K.B. 22: 438 (1968)

3. **M. exserta** *Bridson* in K.B. 42: 647, fig. 3 (1987). Type: Tanzania, Kilwa/Lindi District, Machinga, *Anderson* 826 (K, holo.!, EA, iso.)

Shrub or small tree 1–7 m. tall. glabrous; young stems pale brown, lenticellate. Leaves restricted to apices of branches and young lateral shoots, mature or less often immature at time of flowering; blades elliptic to broadly elliptic or broadly ovate, 4.5–14 cm. long, 2.5–9.5 cm. wide, acute to subacuminate at apex, obtuse or truncate at base, papery, strongly discolorous when dry, glaucous beneath; lateral nerves in 5–6 main pairs; tertiary nerves apparent, coarsely reticulate; domatia present as tufts of hair, sometimes spread along the midrib; petioles 5–9 mm. long; stipules 2–3 mm. long, sheath-like at base, then somewhat triangular, terminating in a linear lobe 4–7 mm. long, pubescent inside. Flowers 5-merous, borne in pedunculate cymes, up to 30-flowered, arising at nodes from which the leaves have fallen; peduncles 1.6–2.5 cm. long, sparsely puberulous or puberulous; pedicels 2–6 mm. long, puberulous; bracts and bracteoles inconspicuous. Calyx-tube 1–1.5 mm. long; limb ± 1 mm. long, truncate, repand or shortly toothed. Corolla-tube 3–5 mm. long, pubescent at throat, then with a ring of deflexed hairs inside; lobes 3.5–5.5 mm. long, 1–2 mm. wide, thickened towards apex, but not apiculate, puberulous at margins. Style long-exserted, up to twice as long as corolla-tube; stigmatic knob 1–1.25 mm. in diameter. Fruit globose, up to 3.5 cm. in diameter or broadly elliptic in 1-seeded fruit, not lenticellate; pyrene not known in detail, ± 2.5 cm. long, 1.5 cm. wide. Fig. 131/6, p. 752.

subsp. **exserta**

TANZANIA. Kilwa District: Selous Game Reserve, Nahomba valley, 18 Jan. 1977, *Vollesen* in M.R.C. 4348!; Lindi District: western slopes of Rondo, 21 Nov. 1966, *Gillett* 17962! & Mlinguru, 21 Dec. 1937, *Schlieben* 5760!
DISTR. T 8; Mozambique and Zimbabwe
HAB. Thicket or forest; 240–500 m.

SYN. [*Canthium stuhlmannii* sensu T.T.C.L.: 489 (1949), pro parte, quoad *Gillman* 1476; Vollesen in Opera Bot. 59: 67 (1980), *non* Bullock]

NOTE. Subsp. *robsonii* Bridson from Mozambique (Tete Province), southern Malawi and eastern Zambia can be distinguished from subsp. *exserta* by the shorter style, 6–7.5 mm. as opposed to 9–10 mm., shorter corolla-tube and shorter, broader corolla-lobes.

4. **M. kingupirensis** *Bridson* in K.B. 42: 649, fig. 4 (1987). Type: Tanzania, Kilwa District, Kingupira, *Vollesen* in *M.R.C.* 4277 (K, holo.!, C, DSM, iso.)

Shrub 3–4 m. tall, glabrous; young stems lenticellate. Leaves restricted to apices of main branches and new shoots; blades elliptic, 6.5–10.5 cm. long, 3.2–4.5 cm. wide, acuminate at apex, acute at base, papery, somewhat discolorous when dry; lateral nerves in 6 main pairs; tertiary nerves apparent but not markedly discolorous ; domatia absent; petiole 0.8–1.2 cm. long; stipules sheath-like, 2–4 mm. long, bearing a linear lobe 4–5 mm. long, pubescent inside. Flowers 5-merous, borne in lax pedunculate 20–40-flowered cymes arising at nodes from which the leaves have fallen; peduncles 1–3 cm. long, sparsely pubescent; pedicels 2–7 mm. long, pubescent; bracteoles inconspicuous. Calyx-tube 1 mm. long, sparsely pubescent; limb 1–2 mm. long, wider than tube, rather irregularly lobed for about half its length. Corolla greenish white; tube 3 mm. long, 1.5–3 mm. wide, sparsely pubescent at throat and with a ring of deflexed hairs a short way from the top inside; lobes triangular-oblong, 3.5 mm. long, 2 mm. wide at base, thickened towards apex, acute, puberulous at margins. Stigmatic knob shortly exserted, slightly longer than wide, 1 mm. long. Fruit (immature) 1.8 cm. long and wide; pyrene not known. Fig. 131/7, p. 752.

TANZANIA. Kilwa District: Selous Game Reserve, Kingupira, *Ludanga* in *Rodgers* 828!, 31 Jan. 1971, *Ludanga* 1171! & 20 Feb. 1977, *Vollesen* in *M.R.C.* 4480!
DISTR. T 8; not known elsewhere
HAB. Riverine thicket; 120–125 m.

SYN. [*Canthium sclerocarpum* sensu Vollesen in Opera Bot. 59: 67 (1980), *non* (K. Schum.) Bullock]

5. **M. sclerocarpa** (*K. Schum.*) *Bridson* in K.B. 42: 650 (1987). Type: Tanzania, Lushoto District, E. Usambara Mts., Sigi to Longuza [Longusa], *Peter* 56025 (B, neo.!)

Shrub or tree 3.5–13 m. tall, glabrous; young branches lenticellate. Leaves restricted to new growth at apices of branches; blades narrowly elliptic to elliptic, 6–11 cm. long, 2.5–4.5 cm. wide, distinctly acuminate at apex, acute at base; papery to subcoriaceous, strongly discolorous when dry; lateral nerves in 6–7 main pairs; tertiary nerves apparent, moderately finely reticulate; domatia present as tufts of hair; petiole 0.8–1.2 cm. long; stipules sheath-like, 2–5 mm. long, bearing a caducous linear somewhat keeled lobe 4–7 mm. long, pubescent inside. Flowers 5-merous, borne in pedunculate 15–40-flowered cymes arising from nodes at which the leaves have been shed; peduncle (0.5–)1.5–3 cm. long, sparsely pubescent on either side; pedicels 3–6(–9) mm. long, with crisped hairs on either side; bracteoles linear, up to 4 mm. long. Calyx subcoriaceous; tube ± 1.5 mm. long, sparsely pubescent; limb 1.5–2 mm. long, distinctly wider than tube, lobed to ± halfway; lobes ovate, rounded, scarcely spreading. Corolla lime-green, fleshy, obtuse to acute in bud, wrinkled when dry; tube 3 mm. long, 3 mm. wide, pubescent at throat, then with a ring of deflexed hairs $\frac{1}{3}$ of the way from top; lobes triangular, 4 mm. long, 2 mm. wide at base, thickened towards apex, acute, puberulous at margins. Stigmatic knob shortly exserted, slightly longer than wide, 1.5 mm. long, bifid at apex when mature. Fruit 1.5 cm. long (*fide* K. Schumann).

KENYA. Kwale District: Mkongani North Forest, 14 July 1987, *Luke & Robertson* 530!
TANZANIA. Lushoto District: E. Usambara Mts., Kisiwani–Mashewa, 25 May 1943, *Greenway* 6689! & Longuza, 7 Apr. 1922, *Soleman* in *Herb. Amani* 5983! & Korogwe, 2 Jan. 1958, *Tanner* 3943!
DISTR. K 7; T 3; not known elsewhere
HAB. Forest; 100–550 m.

SYN. *Plectronia sclerocarpa* K. Schum. in E.J. 34: 334 (1904)
Canthium sclerocarpum (K. Schum.) Bullock in K.B. 1932: 375 (1932); T.T.C.L.: 489 (1949)

NOTE. K. Schumann compared this species to *Plectronia zanzibarica*, a *Keetia*, while Bullock compared it to *Canthium afzelianum* (sensu auctt., *non* Hiern) also a *Keetia*, and mentioned several specimens outside the Flora area, none of which bears any true relationship to *M. sclerocarpa* (see Bridson in K.B. 42: 651 (1987)). Schumann's description in no way conflicts with that above; the details concerning the brown colour of the nerves on drying and the fact that this endemic species is the only *Multidentia* known from the E. Usambara Mts. leaves little doubt that the name is correctly applied. Schumann originally cited *Engler* 389 & 394 from Sigi between Muheza and Longuza, but since it is unlikely that duplicate collections exist a neotype was chosen.

6. **M. castaneae** (*Robyns*) *Bridson & Verdc.* in K.B. 42: 651 (1987). Type: Tanzania, Dar es Salaam, ? Rudju, *Holtz* 3080 (B, holo.†)

Erect glabrous shrub or small tree, 1–5 m. tall, sparsely divaricately branched; young branches lenticellate; bark greyish, somewhat scaly and eventually longitudinally fissured. Leaves restricted to new growth at apices of branches; blades oblong to broadly elliptic, 5.5–10 cm. long, 3–5 cm. wide, shortly acuminate or sometimes obtuse at apex, acute at base (? or rounded), stiffly papery, chestnut to blackish brown above, glaucous to pale brown beneath when dry; lateral nerves in 5–6 main pairs; tertiary nerves coarsely reticulate and sometimes rather difficult to observe beneath; domatia present as tufts of hair; petiole 0.6–1 cm. long; stipules sheath-like, 1–3 mm. long, bearing a caducous linear somewhat keeled lobe 3–5 mm. long. Flowers (4–)5-merous, borne in pedunculate 15–35-flowered cymes arising at nodes from which the leaves have been shed; peduncle 1–1.5 cm. long, glabrous or sparsely pubescent on either side; pedicels 2–4 mm. long, pubescent on either side; bracteoles linear or subulate, up to 2 mm. long. Calyx usually coriaceous, drying yellow-green or sometimes greyish; tube ± 1.5 mm. long; limb distinctly wider than tube, 2–3 mm. long, lobed to ± the middle; lobes oblong or lanceolate to ovate, obtuse to rounded, eventually spreading. Corolla green, rather thick, wrinkled when dry, blunt in bud; tube 2–3 mm. long, villous at the throat, then with a ring of deflexed hairs $\frac{1}{3}$ to halfway from the top; lobes oblong or triangular, ± 4 mm. long, thickened and subobtuse at apex. Stigmatic knob shortly exserted, slightly longer than wide, 1.5 mm. long, bifid at apex. Fruit almost square in outline, 2.3 cm. across, 1.2 cm. thick; calyx-limb persistent; pyrene ellipsoid, ± 2–3 cm. long, 1.2–2 cm. wide, 1.1 cm. thick, the walls thickly woody with longitudinal ridges.

TANZANIA. Uzaramo District: 33 km. on Dar es Salaam–Kilwa road, 8 May 1972, *Harris* 6369! & Kazimzumbwi Forest Reserve at the Two Hippo Dams, 2 km. WSW. of Kisarawe, 12 Nov. 1977, *Mwasumbi & Wingfield* 11459!; Rufiji District: Utete, Kibiti, 18 Dec. 1968, *Shabani* 240!
DISTR.   **T** 6; not known elsewhere
HAB.   Coastal forest; 60–150 m.

SYN.   *Vangueriopsis castaneae* Robyns in B.J.B.B. 11: 258 (1928); T.T.C.L.: 538 (1949)
        Canthium stuhlmannii Bullock in K.B. 1932: 376 (1932); T.T.C.L.: 489 (1949). Type: Tanzania, Uzaramo [Usaramo], *Stuhlmann* 7007 (B, holo.†)

NOTE.   Robyns' careful and detailed description of *V. castaneae* mentions all parts except the fruit which was unknown to him; it in no way deviates from the above description. In fact the note on the chestnut-colour of the dry leaves above and thick rounded flower buds leave little doubt that the name is correctly applied.
        Bullock described *C. stuhlmannii* from an inadequate specimen, but the fruit size is more in accordance with *Multidentia* than any *Canthium*. Brenan noted that *Gillman* 1476 (corollas fallen) from Lindi District was perhaps this species; more weight is given to the assumption that Bullock's taxon is correctly placed in *Multidentia* since the *Gillman* specimen also proved to be a *Multidentia* (*M. exserta* subsp. *exserta*). So far *M. castaneae* is the only *Multidentia* recorded from the coastal forests of Uzaramo and Rufiji Districts; Bullock's taxon is highly likely to be conspecific.
        I decided not to neotypify either of the above epithets since there is a slight chance that isotypes could still be extant and in both cases the type locality is not precisely known.

7. M. crassa (*Hiern*) *Bridson & Verdc.* in K.B. 42: 652, figs. 1F, M & 2A–C (1987). Type: Sudan, Jur, Kurschuk Ali, *Schweinfurth* 1707 (K, lecto.!, BM, isolecto.!)

Shrub or small tree 0.6–6 m. tall, the main trunk with almost black or rough grey bark, which peels to expose a reddish undersurface; shoots usually rather stout, with thick often powdery very pale white-grey or straw-coloured wrinkled corky bark; sap copious, sometimes apparently milky. Leaves mostly restricted to apices of branches; blades oblong-elliptic to oblong or sometimes round, 3–27.5 cm. long, 1.7–15.5 cm. wide, obtuse or ± acute at the apex or sometimes tapering-acute, cuneate to ± rounded at the base, rather fleshy or coriaceous or sometimes papery, often drying yellow-green and mostly often markedly paler and glaucous beneath, glabrous or occasionally densely velvety tomentose on lower surface; lateral nerves in 5–7 main pairs; tertiary nerves finely reticulate, sometimes drying pale yellowish above and dark beneath; domatia present as tufts of hair; petioles 0.3–2.5 cm. long; stipules with broad basal part 3.5–5 mm. long and subulate apex 1.5–7.5 mm. long, pubescent within. Flowers 5-merous, borne in axillary many-flowered divaricate cymes, all parts of inflorescence mostly pubescent and sometimes densely woolly tomentose; peduncle 1–5 cm. long; pedicels 1–3 mm. long; bracts linear, 3–5 mm. long. Calyx-tube obconic, 1.5–2 mm. long, glabrous or sometimes pubescent; tubular part of limb short, 1–3 mm. long, almost truncate or undulate or with short triangular teeth ± 0.5–1 mm. long. Corolla greenish yellow, fleshy, thick, wrinkled when dry; tube 2.5–4.5 mm. long, pilose at throat, then with a ring of deflexed hairs $\frac{1}{3}$ of the way from the top; lobes narrowly triangular, 3 mm. long. Stigmatic knob distinctly

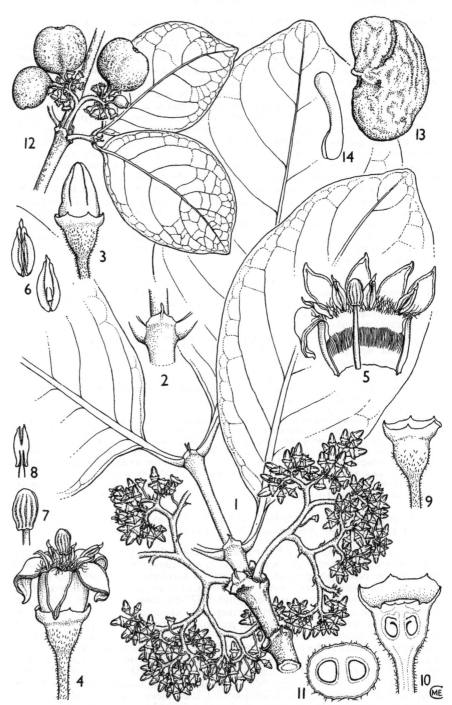

FIG. 149. *MULTIDENTIA CRASSA* var. *CRASSA* — **1**, flowering branch, × ⅔; **2**, stipule, × 1; **3**, flower bud, × 4; **4**, flower, × 4; **5**, corolla opened out with style and stigmatic knob, × 4; **6**, anther (2 views), × 6; **7**, stigmatic knob, × 6; **8**, longitudinal section through stigmatic knob, × 6; **9**, calyx, × 4; **10**, longitudinal section through ovary, × 6; **11**, transverse section through ovary, × 10; **12**, part of branch with fruits, × ⅔; **13**, pyrene, × 2; **14**, embryo, × 2. 1–11, from *Richards* 25838; 12, from *B.D. Burtt* 3591; 13, 14, from *Angus* 185. Drawn by Mrs M.E. Church.

exserted, green, ellipsoid, 2 mm. long, ridged. Fruit green mottled brown, yellow, dull red or brown spotted with white, fleshy and edible, depressed globose, 2.2–3 cm. long, 2.6–3.7 cm. wide (probably at least 3.5–4 cm. wide in life), lenticellate; pyrenes ellipsoid, 1.8–2.2 cm. long, 1.1–1.7 cm. wide, 1.1–1.5 cm. thick, the walls ridged, exceedingly thick and woody, up to 5–7 mm. thick, with a deep lateral semicircular groove on either side above the point of attachment, meeting on apical edge.

var. **crassa**

Leaf-blades glabrous even when young or with tomentose domatia beneath. Fig. 149.

UGANDA.   W. Nile District: Paidha [Payida], Dec. 1947, *Dale* U472!; Acholi District: Kitgum, Lira road, Mar. 1935, *Eggeling* 1696!; Teso District: Serere, May 1932, *Chandler* 576!
KENYA.   Trans-Nzoia District: Lugari, *Bogdan* in *E.A.H.* 10414!
TANZANIA.   Biharamulo District: Lusahungu, 15 Oct. 1960, *Tanner* 5334!; Mpanda District: 19 km. N. of Kasoge, Belengi, 18 Aug. 1959, *Harley* 9382!; Dodoma District: Manyoni, Kazikazi, 14 May 1932, *B.D. Burtt* 3591!
DISTR.   U 1, 3; K 3; T 1, 4, 5, 7, 8; Burundi, Zaire, Sudan, ? Ethiopia (*fide* von Breitenbach & Cufodontis), Central African Republic, Mozambique, Malawi, Zambia, Zimbabwe (Victoria Falls) and Angola
HAB.   *Brachystegia, Combretum, Terminalia*, etc., woodland, thicket, grassland, particularly on burnt hillsides and rocky places; 900–2100 m.

SYN.   *Canthium crassum* Hiern in F.T.A. 3: 145 (1877), pro parte, excl. *Schweinfurth* 1695; Bullock in K.B. 1932: 379 (1932), pro parte, excl. syn. *C. platyphyllum*; T.C.L.: 488 (1949); I.T.U., ed.2: 339 (1952); F.P.S. 2: 430 (1952); K.T.S.: 428 (1961); F.F.N.R.: 404, fig. 68J (1962), pro parte, excl. syn. *C. randii*; E.P.A.: 1008 (1965)
    Craterispermum orientale K. Schum. in P.O.A. C: 387 (1895); T.C.L.: 493 (1949). Type: Tanzania, Shinyanga District, Usule [Usula] to Usiha, *Fischer* 302 (B, holo.†, K, iso.!)
    Canthium opimum S. Moore in J.L.S. 37: 308 (1906); K. Krause in Wiss. Ergebn. Schwed. Rhod.-Kongo-Exped. 1911–12, 1 (Nachtr.): 15 (1921). Type: Angola, Malange, Kahella, *Gossweiler* 1239 (BM, holo.!, K, iso.!)
    C. dictyophlebum S. Moore in J.B. 57: 87 (1919). Type: Zaire, Lubumbashi [Elisabethville], *F.A. Rogers* 10085 (BM, holo.!, K, iso.!)
    Plectronia opima (S. Moore) Mildbr. in N.B.G.B. 9: 204 (1924)

var. **ampla** (*S. Moore*) *Bridson & Verdc.* in K.B. 42: 653 (1987). Type: Zambia, Chilanga, *F.A. Rogers* 8446 (BM, holo.!, K, iso.!)

Leaf-blades velvety beneath, with short matted hairs.

TANZANIA.   Ufipa District: Namwele, 24 Feb. 1950, *Bullock* 2578!; Mbeya District: Ruaha National Park, Magangwe Hill, 10 Dec. 1970, *Greenway & Kanuri* 14758! & Magangwe airstrip, 11 Mar. 1972, *Bjørnstad* 1569!
DISTR.   T 4, 7; Cameroon, Mozambique, Malawi and Zambia
HAB.   *Brachystegia* woodland, eroded fireswept hillsides; 1200–1800 m.

SYN.   *Canthium amplum* S. Moore in J.B. 57: 87 (1919)
    Plectronia buarica Mildbr. in N.B.G.B. 9: 203 (1924). Types: Cameroon, Baja Highlands, Buar, *Mildbraed* 9330 & 9459 (B, syn.†, K, isosyn.!)

NOTE.   Some specimens with a sparser indumentum on the lower leaf surface do occur and one from Zambia, *Silungwe* s.n., has densely tomentose inflorescence-axes and calyx-tube.
    Bullock cited *Canthium platyphyllum* as a synonym of this species, noting "*C. platyphyllum* is a further form with a lanate tomentum on the inflorescence and young branchlets; this falls away and the plant develops a rusty-red coloured bark." It is in fact a synonym of *Vangueriopsis lanciflora*. Sterile material of both *V. lanciflora* and *Ancylanthos rubiginosus* Desf. (not from Flora area) can resemble *M. crassa* var. *ampla* but usually have a stronger indumentum, and rusty-red bark.

8. M. concrescens (*Bullock*) *Bridson & Verdc.* in K.B. 42: 653 (1987). Type: Tanzania, Njombe, *Lynes* D.p.108 (K, holo.!)

Pyrophytic erect subshrub (7.5–)15–35 cm. tall; stem simple or slightly branched near the base, arising from a thick woody rootstock, glabrous, dark, wrinkled, with elongate lenticels. Leaves opposite or in whorls of 3, drying the bright yellow-green of an aluminium accumulator, mostly paler and often glaucous beneath, oblong, elliptic, oblong-elliptic or obovate to oblanceolate, 2.5–16(–21.5) cm. long, 0.8–8.5(–12.5) cm. wide, very shortly acutely or obtusely acuminate or less often merely acute or rounded at the apex, cuneate at the base, sometimes quite coriaceous, glabrous to rather sparsely, shortly pubescent on both surfaces; lateral nerves in 5–7 main pairs; tertiary nerves apparent and coarsely reticulate; domatia absent; petioles 0.3–1(–2) cm. long; stipules with broad basal (usually sheath-like) part 2–3 mm. long, hairy inside and along margin, with a glabrous subulate lobe 0.3–1.2 cm. long. Flowers 5(–6)-merous; inflorescences slightly to distinctly supra-axillary, divaricately cymose, usually 10–many-flowered, the axes, etc., mostly pubescent; peduncle 2–4.5 cm. long, forming an acute angle with main stem; pedicels 1–11 mm. long; bracts and bracteoles linear-lanceolate, 3–8 mm. long. Calyx-tube obconic, 2–4 mm. long, glabrous or pubescent; tubular part of limb 0.5–2.5 mm. long, drying distinctly wrinkled; lobes oblong to linear-lanceolate or narrowly triangular, (1–)2.5–9 mm. long, acute or subacute at the apex. Corolla greenish or greenish white, the lobes white inside, thickened, often drying wrinkled; tube 3–5 mm. long, pilose at throat, then with a ring of deflexed hairs set ⅓ down from top; lobes reflexed, narrowly triangular, 4–5 mm. long, 2 mm. wide, appearing to have lateral margins broadening towards apex in the dry state. Style 6–8.5 mm. long; stigmatic knob green, oblong-ellipsoid, 1.5–2 mm. long, 1–2 mm. wide, dividing at the apex. Fruit yellowish green becoming red, globose, 1.8–2.2 cm. in diameter; calyx-limb persistent; pyrenes ellipsoid, 1.1–1.4 cm. long, 8–9 mm. wide, 8 mm. thick, with deep irregular longitudinal ridges. Fig.132/14, p. 754.

TANZANIA.   Mbeya District: Mbozi, 29 Aug. 1933, *Greenway* 3639!; Iringa District: 48 km. S. of Iringa, 25 Feb. 1962, *Polhill & Paulo* 1612!; Songea District: Matengo Hills, Litembo, 15 Oct. 1956, *Semsei* 2608!
DISTR.   **T** 4, 7, 8; Malawi and Zambia
HAB.   Grassland, open *Brachystegia, Uapaca* woodland, sometimes on rocky hills; 1080–2400 m.

SYN.   *Pygmaeothamnus concrescens* Bullock in K.B. 1933: 471 (1933) & in Hook., Ic. Pl. 33, t. 3242 (1934); T.T.C.L.: 525 (1949); F.F.N.R.: 418 (1962)
    Multidentia verticillata Gilli in Ann. Naturhist. Mus. Wien 77: 21, t. 1 (1973). Type: Tanzania, Songea District, Matengo Highlands, upper Luaita valley, between Mbinga and Ndenga [Ndengo], *Zerny* 285 (W, holo.!)

NOTE.   Apart from the type of *Multidentia verticillata* none of the fairly extensive material at Kew shows leaves in whorls of 3.

103. VANGUERIA

A.L. Juss., Gen. Pl.: 206 (1789); Robyns in B.J.B.B. 11: 273 (1928); Verdc. in K.B. 36: 547 (1981)

Unarmed or rarely spiny deciduous shrubs or small trees with glabrous to densely velvety foliage. Leaves paired, often quite large; stipules broad at base, hairy inside, connate into a short sheath at the base, produced into a linear or subulate appendage at the apex, deciduous or ± persistent. Inflorescences axillary divaricately branched cymes at opposite sides of nodes, usually many-flowered, often at leafless nodes and sometimes flowering when leaves are not fully developed; branches of cymes often rather elongate with regularly spaced shortly petiolate flowers. Calyx-tube hemispherical or depressed campanulate; limb divided into 5 triangular, oblong, linear or ligulate lobes. Corolla small, usually yellow or greenish yellow, often distinctly apiculate in bud (due to lobe appendages); corolla-tube shortly cylindrical or campanulate, glabrous or hairy outside, with a ring of deflexed hairs inside and densely hairy at the throat; lobes 5, narrowly triangular, mostly reflexed, sometimes with distinct narrow appendage at the apex. Anthers ovate or oblong, apiculate, shortly exserted. Ovary 5-locular; ovule solitary in each locule, pendulous; style filiform-cylindric, glabrous, shortly exserted; stigmatic club cylindrical, very shortly 5-lobed at the apex. Disk annular, slightly raised. Fruit large, indehiscent and fleshy, ± globose, often lobed or angular in dry state, (4–)5(–6)-locular, containing (4–)5(–6) pyrenes, sometimes bearing remains of calyx-limb, glabrous, often edible.

Robyns recognised 27 species from tropical and S. Africa and Madagascar, but they were distinguished on minute characters such as indumentum of leaves and corolla, length of calyx-lobes and degree of development of corolla-lobe appendages hence nature of bud apex. Various combinations of these characters are treated as distinct species. It is clear that too many species have been recognised in the past, but it is not easy to decide which characters are most important. Different classifications result from the choice made. Robyns' recognition of every combination of characters as distinct obviates this difficulty but leads to great difficulties with numerous intermediate specimens. Many will probably consider that not enough reduction has been made in this present treatment. The genus is rare in W. Africa. The inflorescences are very attractive to many kinds of insects.

Two subgenera are recognised — subgen. *Vangueria*, species 1–7, and subgen. *Itigi* Verdc., species 8.

1. Flowers in sessile fascicles, under 1 cm. long, borne on
leafless twigs or while leaves are ± immature; calyx-
lobes linear, 2–5.5 mm. long; leaves glabrous
(subgen. *Itigi*) . . . .   8. *V. praecox*
Flowers in at least shortly pedunculate inflorescences
exceeding 1 cm., often not precocious; calyx-lobes
ovate, triangular or linear; leaves glabrous or hairy
(subgen. *Vangueria*) 2
2. Calyx-lobes triangular-oblong to narrowly oblong, 1.5(–2)
mm. long 3
Calyx-lobes more elongate, oblong to linear (1.2–)3–8 mm.
long 4
3. Leaves glabrous or almost so; corolla ± glabrous (both
occasionally somewhat pubescent) . . . . . .   1. *V. madagascariensis*
Leaves mostly densely pubescent to velvety; corolla
glabrous to densely pubescent . . . .   2. *V. infausta*
4. Inflorescences short and/or graceful, under 2.5 cm. long,
rather few-flowered; leaves glabrous or sometimes
ciliate and ± sparsely hairy above 5
Inflorescences longer and coarser, usually over 2 cm. long,
often with more flowers; leaves glabrous to velvety,
usually thicker; buds over 2 mm. wide and 5 mm. long   6
5. Leaves immature at time of flowering; corolla-buds 6 mm.
long (Pemba I.) . . . .   4. *V. sp. A*
Leaves mostly mature at time of flowering, mostly rather
thin; buds 1–2.2 mm. wide, mostly under 5 mm. long
(**K** 7; **T** 3, 6, 8) . . . .   7. *V. randii*
6. Leaves glabrous or nearly so . . . .   3. *V. apiculata*
Leaves densely pubescent to velvety 7
7. Young stems with short indumentum; widespread   5. *V. volkensii*
Young stems with rather long hairs; E. Usambaras, very
poorly known species . . . .   6. *V. bicolor*

1. **V. madagascariensis** *Gmelin*, Syst. Nat., ed. 13, 2: 367 (1791); Webster, Food Plants of
the Philippines: 196, t. 47b (1924); Verdc. in K.B. 36: 548 (1981) & Fl. Masc. 108. Rubiacées:
108, t. 32 (1989). Type: Madagascar, *Commerson* (P, holo.)

Shrub or small tree, 1.5–15 m. tall, often multi-stemmed and sometimes with a
spreading crown; stems glabrous, longitudinally ridged, with pale to dark bark, mostly
smooth and unpeeling but peeling or powdery in one variant. Leaf-blades narrowly to
broadly elliptic or elliptic-lanceolate, 8–28 cm. long, 3.2–15 cm. wide, acute to shortly
acuminate at the apex, cuneate to rounded or less often ± subcordate at the base, entirely
glabrous or sometimes very young leaves pilose beneath and adult ones sparsely
pubescent; petiole 0.8–1.8 cm. long; stipules with broad base 3–5 mm. long, and narrow
apex 0.4–1.8 cm. long, glabrous or pubescent. Inflorescence pubescent; branches 1–4.5
cm. long, 7–10-flowered; main peduncles ± 1 cm. long; pedicels ± 2 mm. long, save those
of central flower of inflorescence which measure 4 mm. Calyx-tube 1.2–3 mm. long; lobes
triangular-oblong to narrowly oblong, 0.5–1.5 mm. long, ± pubescent. Corolla greenish
yellow, yellow or cream, glabrous or rarely with few hairs; tube 3–4.5 mm. long; lobes
3.5–4.5 mm. long, with appendages up to 0.5 mm. long; buds slightly to usually markedly
acuminate or apiculate due to the corolla lobe-appendages. Style 7–8 mm. long; stigmatic
club cylindrical, yellow, 1.2–1.5 mm. long. Fruits green to brownish, subglobose, 2.5–5 cm.
in diameter, with 4–5 pyrenes, each ± 2 cm. long, 1.2 cm. wide, 8 mm. thick, with thick
woody walls 1–2 mm. thick. Seeds ± 1.6 cm. long, 6 mm. wide, 4.5 mm. thick, narrowed at
one end. Figs. 131/22, p. 752; 132/15, p. 754 & 150.

UGANDA. Karamoja District: Mt. Kadam, Amaler, Jan. 1936, *Eggeling* 2537!; Teso District: Serere,
Kyere, Feb. 1933, *Chandler* 1123!; Masaka District: Sese Is., Nkose I., 22 Jan. 1956, *Dawkins* 850!
KENYA. Northern Frontier Province: Moyale, 23 Apr. 1952, *Gillett* 12919!; Elgon, Mar. 1931, *E.J. & C.
Lugard* 571!; Masai District: NW. Mt. Suswa, 13 Feb. 1949, *Vesey-FitzGerald* 123!
TANZANIA.   Mbulu District: Lake Manyara National Park, Endabash R., 8 Nov. 1963, *Greenway &
Kirrika* 10993!; Moshi District: Kilimanjaro, Mashati, Jan. 1929, *Haarer* 1751!; Manyoni District:
Singida, Oct. 1935, *B.D. Burtt* 5271!; Zanzibar I., Muyuni, 19 Oct. 1930, *Vaughan* 1649!; Pemba I.,
Mawani, 19 Dec. 1930, *Greenway* 2761!

FIG. 150. *VANGUERIA MADAGASCARIENSIS* — 1, flowering branch, × ⅔; 2, stipules, × 2; 3, flower bud, × 6; 4, flower, × 6; 5, corolla opened out with style and stigmatic knob, × 6; 6, stigmatic knob, × 12; 7, calyx, × 8; 8, longitudinal section through ovary, × 12; 9, transverse section through ovary, × 12; 10, fruit, × ⅔; 11, pyrene, × 1. 1, 3–9, from *Haarer* 1752; 2, from *Greenway & Kanuri* 13468; 10, from *Lindsay* s.n.; 11, from *Richards* 20808. Drawn by Mrs M.E. Church.

DISTR. U 1–4; **K** 1, 3, 4, 6; **T** 1–8; **Z**; **P**; Ghana, Nigeria, Cameroon, Zaire, Central African Republic, Sudan, Ethiopia, Malawi and South Africa (Transvaal); also cultivated in Madagascar, Mascarenes, Zaire, India, Singapore, N. Australia, Ghana and Trinidad

HAB. Evergreen forest, riverine forest and woodland, bushland, grassland with scattered trees, sometimes on rock outcrops and termite mounds; 0–2130 m.

SYN. *Vavanga chinensis* Rohr in Skr. Naturh.-Selsk. Kjøbenhavn 2: 207 (1792). Type: specimen grown at St. Croix I., W. Indies, introduced by Rohr from Guadeloupe (? C. holo.)

    V. edulis Vahl in Skr. Naturh.-Selsk. Kjøbenhavn 2: 208, t. 7 (1792). Type as for *V. chinensis, nom. illegit.*

    Vangueria edulis (Vahl) Vahl in Symb. Bot. 3: 36 (1794); Willd., Sp. Pl. 1: 926 (1798); Robyns in B.J.B.B. 11: 287 (1928)

    V. edulis Lam., Tab. Encycl. 2: 235 (1819) & t. 159 (1792) without specific epithet. Type: "ex India and Madagascar", Mauritius, *Commerson* (P, holo.)

    V. floribunda Robyns in B.J.B.B. 11: 285 (1928). Type: South Africa, Transvaal, Barberton, *F.A. Rogers* 18214 (K, holo.!)

    V. acutiloba Robyns in B.J.B.B. 11: 286 (1928); T.T.C.L.: 537 (1949); K.T.S.: 478 (1961). Type: Tanzania, Moshi District, Kilema, *Volkens* 1687 (K, holo.!)

    V. venosa Robyns in B.J.B.B. 11: 290 (1928); F.P.S. 2: 466 (1952), *non* Sond., *nom. illegit.* Type: Ethiopia, Djeladjeranne, *Schimper* 653 (P, holo., K, W, iso.!)

    Canthium maleolens Chiov. in Miss. Biol. Borana, Racc. Bot.: 228, fig. 71 (1939). Type: Ethiopia, Moyale, *Cufodontis* 731 (FT, holo., W, iso.!)

    Vangueria robynsii Tennant in K.B. 22: 443 (1968). Type as for *V. venosa*

NOTE. Most material from Lindi District, **T** 8 (e.g. W. Rondo slopes, 21 Nov. 1966, *Gillett* 17952! & Rondo Forest Reserve, 3 Dec. 1964, *P. Mason* 1!) has peeling bark on the branches which reveals a powdery layer. A similar bark is demonstrated by *Boaler* 571, a specimen of *V. infausta*, from Nachingwea. Logically speaking the various variants I have treated under *V. infausta* cannot be kept separate; specimens of *V. madagascariensis* with some pubescence on either leaves and/or corolla are not uncommon and hairy and glabrous variants have been found growing together, e.g. *Brummitt & Polhill* 13617 and 13617A, Iringa District, 8 km. SW. of Iringa on Mbeya road. However, when one compares typical *V. infausta* and typical *V. madagascariensis* and considers that all but a very few specimens are easily named and that the rest are abundantly distinct, combining the two would be an impracticality and unacceptable to most African botanists.

    Robyns treats *V. venosa* as published by Richard but it appears only in synonymy.

2. **V. infausta** *Burchell*, Travels S. Afr. 2: 258, fig. on p. 259 (1824); DC., Prodr. 4: 454 (1830); Sond. in Fl. Cap. 3: 13 (1865); Hiern in F.T.A. 3: 147 (1877); K. Schum. in P.O.A. C: 384 (1895); Robyns in B.J.B.B. 11: 307 (1928); Breitenb., Indig. Trees S. Afr.: 1130, fig. on p. 1129 (1965); Palmer & Pitman, Trees S. Afr. 3: 2079, figs. on p. 2080 (1972); van Wyk, Trees Kruger Nat. Park 2: 571, t. 702 (1973); Palmer, Field Guide Trees S. Afr.: 293 (1977); Coates Palgrave, Trees S. Afr.: 873 (1977); Verdc. in K.B. 36: 549 (1981); Fl. Pl. Lign. Rwanda: 610, fig. 207.3 (1982); Fl. Rwanda 3: 230, fig. 73.2 (1985). Type: South Africa, Cape Province, Sensavan, *Burchell* 2629 (K, holo.!)

    Shrub or small tree 1.5–8 m. tall, very similar to the last species save for the indumentum; trunk smooth, grey or eventually rough and ridged; branches ridged, the young parts densely ferruginous, pubescent or velvety. Leaf-blades elliptic, oblong-elliptic, ovate or sometimes almost round or lanceolate, 4–30 cm. long, 2.5–18 cm. wide, rounded, subacute or ± acute to shortly acuminate at the apex, rounded to cuneate or rarely ± subcordate at the base, often discolorous, deeply pubescent to usually densely softly velvety on both surfaces, the hairs often yellowish or ferruginous in dry state; petiole 0.3–1 cm. long, similarly hairy; stipules with base 2–4 mm. long and apical part 0.3–1.2 cm. long, hairy. Inflorescences densely hairy, sometimes extensive and much branched but typically rather short, branches 1.5–3.5 cm. long, each 5–10-flowered; peduncles 6–8 mm. long; pedicels 1–2.5 mm. long, but those of the central flower of each dichasial element ± 3.5 mm. long. Calyx-tube 0.75–1.2 mm. long; lobes obtusely triangular to narrowly oblong, 1–1.25(–2) mm. long. Corolla green or yellow-green, typically densely spreading hairy outside but glabrous in some variants; tube 3–4.5 mm. long; lobes 3–4 mm. long with appendages up to 0.5 mm. long or practically obsolete. Style 4.5–6 mm. long; stigmatic club yellow, 1 mm. long. Fruit green, usually ripening to dull orange-brown or purplish, depressed subglobose, 1.5–4.7 cm. in diameter; pyrenes up to 1.3–2 cm. long, 0.6–1.3 cm. wide, 5–8 mm. thick.

KEY TO INFRASPECIFIC VARIANTS

1. Leaves usually small, often not developed at flowering time; buds rounded at apex, densely spreading hairy outside . . . . subsp. **infausta**

Leaves usually larger, developed at flowering time; buds
   apiculate (subsp. *rotundata*) 2
2. Corolla pubescent to hairy   . . . .   var. **rotundata**
   Corolla ± glabrous   . . . .   var. **campanulata**

subsp. **infausta**; Verdc. in K.B. 36: 549 (1981)

Leaves usually toward the lower end of the scale of size given and often not fully developed at
flowering time, always densely velvety. Inflorescences not extensively branched. Buds rounded at
apex and not at all apiculate. Corolla densely spreading hairy outside. Fruit up to 3.5 cm. in diameter.

TANZANIA.   Tabora District: S. of Tabora, Igigwa [Kawewe's Country], 20 Oct. 1928, *Carnochan* 61!;
Singida District: NW. Iramba Plateau, 10 Dec. 1933, *Michelmore* 802!; Mbeya District: Kinyangesi
[Kenyangeti], 27 Nov. 1970, *Greenway & Kanuri* 14670!; Iringa District: Ruaha National Park road
turn-off on Iringa–Idodi road, 27 Nov. 1970, *Greenway & Kanuri* 14683!
DISTR.   T ?1, ?2, 4–7; Rwanda, Mozambique, Malawi, Zambia, Zimbabwe, Namibia and South Africa
HAB.   Woodland (usually *Brachystegia* or *Commiphora* mixed woodland), rocky places; 500–2500 m.

SYN.   *V. tomentosa* Hochst. in Flora 25: 238, adnot. (1842); Robyns in B.J.B.B. 11: 308 (1928); T.T.C.L.:
   538 (1949), pro parte; F.F.N.R.: 424 (1962). Type: South Africa, Natal, Durban, *Krauss* 219 (?B,
   holo.†, K, iso.!)

NOTE.   Drummond (Kirkia 10: 276 (1975)) sinks *V. lasioclados* K. Schum. into *V. infausta*, but White
(F.F.N.R.: 424 (1962)) maintains them separate though admitting close relationship; it certainly
does not appear identical but is probably no more than a variety. *V. velutina* Hook., described from
Madagascar, is probably also a form of *V. infausta* and probably not a native of Madagascar.

subsp. **rotundata** (*Robyns*) Verdc. in K.B. 36: 549 (1981). Type: Tanzania, Rungwe District, Kyimbila,
Stolz 432 (K, holo.!, BR, EA, K, iso.!)

Leaves attaining dimensions towards the upper end of the scale of size given and often more
developed at flowering time, densely pubescent to densely velvety. Inflorescences often extensively
branched. Buds acute to distinctly apiculate at the apex. Corolla glabrous to densely spreading hairy
outside. Fruit up to 4.7 cm. in diameter.

var. **rotundata**; Verdc. in K.B. 36: 550 (1981)

Corolla pubescent to densely spreading hairy outside.

UGANDA.   W. Nile District: Leya R., 25 Mar. 1945, *Greenway & Eggeling* 7255!; Karamoja District:
Loyoro [Loyoru], 3 Nov. 1939, *A.S. Thomas* 3151! & Mt. Morongole, *J. Wilson* 714!
KENYA.   SE. slopes of Elgon, 3 May 1953, *Padwa* 3!; Nairobi District: 14.4 km. on Nairobi–Mbagathi
road, 8 Nov. 1932, *C.G. Rogers* 33!; Machakos District: Kibwesi, 15 Sept. 1961, *Polhill & Paulo* 464!
TANZANIA.   Mbulu District: Pienaars Heights, Dauar, 6 Jan. 1962, *Polhill & Paulo* 1067!; Pare District:
Same, 13 Dec. 1974, *Wingfield* 2844!; Rungwe District: Mpuguso, 10 Feb. 1976, *Cribb, Grey-Wilson &
Mwasumbi* in G.W. 10672!
DISTR.   U 1, 3, ?4; K ?2, 3–7; T 1–3, ?5, 6, 7, ?8; Mozambique and Malawi
HAB.   Dry evergreen forest, fringing forest, woodland, *Acacia* bushland, grassland with scattered
trees, rocky thickets; 30–2100 m.

SYN.   *V. rotundata* Robyns in B.J.B.B. 11: 300, figs. 27, 28 (1928); T.T.C.L.: 538 (1949); K.T.S.: 479 (1961)
   [*V. tomentosa* sensu I.T.U., ed. 2: 358 (1952), pro parte; F.P.S. 2: 466 (1952); K.T.S.: 479 (1961), *non*
   Hochst.]
   [*V. campanulata* sensu K.T.S.: 478 (1961), pro parte, *non* Robyns]

var. **campanulata** (*Robyns*) Verdc. in K.B. 36: 550 (1981). Type: Kenya, Nairobi Government Farm,
Linton 21 (K, holo.!)

Corolla glabrous or almost so save for slight pubescence above.

KENYA.   Nairobi District: Thika Road House, 22 Oct. 1950, *Verdcourt* 363! & 17 Feb. 1951, *Verdcourt*
437! & Bahati, 7 Apr. 1933, *C.G. Rogers* 644!; Masai District: Siyabei [Syabei] Gorge, Nov. 1940, *Bally*
1302!
TANZANIA.   Moshi District: Ngare Nanyuki R., 12 Dec. 1965, *Richards* 20818!; Kilosa District: Ukaguru
Mts., 3 km. E. of Mandege, Mt. Kwamba, 2 Feb. 1976, *Cribb, Grey-Wilson & Mwasumbi* in G.W. 10524!;
Morogoro District: E. Uluguru Mts., Tawa, Oct. 1930, *Haarer* 1937!
DISTR.   K 4, 6; T 2, 6, 8; ? Cameroon, Malawi
HAB.   Evergreen forest margins, *Brachystegia, Julbernardia* woodland; 480–1680 m.

SYN.   *V. campanulata* Robyns in B.J.B.B. 11: 293 (1928); K.T.S.: 478 (1961)
   V. sp. sensu T.T.C.L.: 537 (1949)

NOTE.   (on species as a whole). The species separated by Robyns on shape of bud apices and
indumentum of corolla certainly cannot be upheld but there is a tendency for blunt buds, short
inflorescences, smaller leaves, etc., to predominate in southern Africa. Some South African
specimens have distinctly pointed buds. Many specimens identical with subsp. *infausta* save for ±

acute buds. There is much variation in the amount of indumentum on the corolla. Of 5 duplicates of the type number of *V. rotundata* at Kew some have quite sparse corolla indumentum. I have compromised by recognising infraspecific variants. A few specimens with very sparse leaf indumentum are intermediate with *V. madagascariensis* and a similar specimen has been seen from Nigeria but lacks corollas. *V. infausta* sensu lato is also recorded from Zaire and the Central African Republic. Material from Uganda, Ankole District (e.g. Rwampara, Kigarama Hill, Oct. 1932, *Eggeling* 652, Mbarara, Gayonza, *A.S. Thomas* 4456, & from same locality, *Maitland* 859) has uniformly small fruits, 1.5–1.8 cm. in diameter, with pyrenes 6 mm. long, 5 mm. thick, was at first considered to be a distinct subspecies, but small-fruited specimens occur sporadically elsewhere; nevertheless it is distinctive.

3. **V. apiculata** *K. Schum.* in P.O.A. C: 385 (1895); Z.A.E. 1907–8, 2: 326 (1911); Robyns in B.J.B.B. 11: 283 (1928); T.T.C.L.: 537 (1949); Cufod. in Phyton 1: 145 (1949); I.T.U., ed. 2: 358 (1952); F.P.S. 2: 466 (1952); K.T.S.: 478 (1961); Drummond in Kirkia 10: 276 (1975); Verdc. in K.B. 36: 550 (1981); Fl. Pl. Lign. Rwanda: 610, fig. 207.1 (1982); Fl. Rwanda 3: 230, fig. 73.1 (1985). Type: Kenya, S. Kavirondo District, Karachuonyo, *Fischer* 294 (B, holo.†, K, iso.!)

Shrub, subscandent shrub or small spreading tree 1.8–12 m. tall, the stems sometimes several, virgate; branching often horizontal; bark grey, brown or red-brown, ± smooth or finely ridged; young shoots mostly dark plum-coloured and lenticellate, quite glabrous. Leaf-blades elliptic, oblong, ovate or ovate-lanceolate to lanceolate, 3–15(–17) cm. long, 1.5–6(–8) cm. wide, distinctly acuminate at the apex, rounded, cuneate or occasionally subcordate at the base, often discolorous and venose beneath when dry, quite glabrous or rarely slightly pubescent beneath; petiole 0.7–1(–1.4) cm. long; stipules with filiform part 3–9.5 mm. long from a short broad base 1.5–2 mm. long. Inflorescences typically short, 1–3.5 cm. long including peduncle, fairly lax to very condensed, the rhachis, etc., almost glabrous to usually pubescent or densely shortly hairy; peduncle 0.5–1 cm. long, similarly hairy, with paired bracts 3 × 2 mm. near apex; pedicels 2–4 mm. long, pubescent. Calyx-tube subglobose, glabrescent to densely shortly hairy, 1.5 mm. in diameter; lobes oblong to linear, sometimes slightly spathulate, (1.2–)3–7 mm. long, glabrous or ciliolate. Buds distinctly apiculate at the apex, the 5 appendages often quite long. Corolla glabrous, greenish white, or green to yellow; tube 4–5 mm. long; lobes narrowly to very narrowly triangular, 4–5 mm. long. Style 6–8 mm. long; stigmatic club 1.2 mm. long. Fruit green turning brown, subglobose or sometimes irregularly ellipsoid where only one pyrene has developed, 1.7–2.2 cm. long, 1.4–2.2 cm. wide, glabrous; pyrenes 0.9–1.7 cm. long, 4–6 mm. wide, 3–5 developing. Fig. 131/10, p. 752.

UGANDA. Acholi District: Gulu, 26 Mar. 1932, *Hancock* 1116!; Busoga District: Lolui I., 16 May 1964, *G. Jackson* U 93!; Mengo District: 8 km. on Kampala–Bombo road, Mar. 1938, *Chandler* 2186!
KENYA. Turkana District: Murua Nysigar [Muruanysiga] Hills, 18 Feb. 1965, *Newbould* 7273!; Elgon, 11 Apr. 1931, *Lugard* 234A!; S. Kavirondo District: Kisii, Sept. 1933, *Napier* in C.M. 6711!
TANZANIA. Bukoba District: Lubafu, July 1931, *Haarer* 2073! & Nyakato, 28 Oct. 1934, *Gillman* 203!; Arusha National Park, Ngurdoto Crater, 16 Apr. 1968, *Greenway & Kanuri* 13467!; Ufipa District: Sumbawanga, 15 Jan. 1950, *Bullock* 2230A & B!
DISTR. U 1–4; K 1–3, 5, 6; T 1–4; Rwanda, Zaire, ? Somalia, Sudan, Ethiopia, Mozambique, Malawi and Zimbabwe
HAB. Evergreen forest (including *Juniperus* and *Podocarpus* forest), riverine, marsh-side and lake-side forest, bushland and thicket, grassland with scattered trees, often on termite mounds or rocky outcrops; 900–2190 m.

SYN. *V. longicalyx* Robyns in B.J.B.B. 11: 278 (1928). Type: Zimbabwe, Chirinda outskirts, *Swynnerton* 64 (K, holo.! & iso.!)
V. sp. nr. randii sensu T.T.C.L.: 537 (1949)

NOTE. A few specimens are intermediate with *V. madagascariensis* Gmelin or possibly are genuine hybrids between the two, e.g. Arusha District, Lake Momela to Engare Nanyuki, 12 Mar. 1914, *Peter* 55454! and Ufipa District, Malonge Plateau, Molo village, 1 Jan. 1962, *Richards* 15846!; the inflorescences are longer and the calyx-lobes shorter. *Chandler* 1015 (Uganda, Mbale District, Bugishu, Bufumbo, Nov. 1932) has the foliage and inflorescences of *V. madagascariensis* but linear calyx-lobes to 2.5 mm. — it is almost certainly a hybrid. *Snowden* 412 (Uganda, Bukedi, Buwalasi, 17 Oct. 1916) on the other hand is merely a large-leaved form of *V. apiculata* with leaves to 9 cm. wide. *Polhill* in E.A.H. 12088 (Kiambu District, Kamiti Forest, Oct. 1960) has short inflorescences but large leaves and short calyx-lobes; this and similar specimens are treated as variants of *V. madagascariensis*. Despite these few intermediates the two species are I think adequately distinct. The specimen cited in T.T.C.L. as near *V. randii* (Bukoba District, Bugolora [Bugorora], *Pitt-Schenkel* 262!) seems to be a form with small leaves some of which are slightly pubescent beneath; it differs from the true *V. randii* in the less apiculate buds and more acuminate leaflets. The fruits are edible.

4. **V. sp. A**

Tallish tree; stems brown, somewhat powdery in places beneath the ± peeling epidermis, flowering when ± leafless or leaves only present at apical node, glabrous save for brownish hairs within the stipule-bases at upper nodes. Leaves seen elliptic-lanceolate, 2–3 cm. long, 1–1.2 cm. wide, narrowed to the rounded apex, ± rounded at base, glabrous; petiole 6 mm. long; stipules with linear appendage 2 mm. long from a short broad base. Inflorescences small, axillary at many leafless nodes, 1–1.5 cm. long, 1.5–2 cm. wide; peduncle 5 mm. long; paired bracts ± ovate, 1.5 mm. long; pedicels 0.5–1 mm. long. Calyx-tube 1 mm. long, glabrous; lobes linear-oblong, 3–4 mm. long, 0.8 mm. wide, obtuse, glabrous. Corolla white; buds acuminate, 6 mm. long including very short tails; tube 3.5 mm. long; lobes 2.5 mm. long. Fruits not seen.

TANZANIA. Pemba I., Ngezi Forest, 22 Jan. 1933, *Vaughan* 2063!
DISTR. **P**; not known elsewhere

NOTE. It seems necessary to notice this for phytogeographical reasons. It comes near to *V. acutiflora* and *V. randii* but has a quite different facies from both. It certainly would not be wise to describe it from one sheet.

5. **V. volkensii** *K. Schum.* in P.O.A. C: 384 (1895); Robyns in B.J.B.B. 11: 298 (1928); T.T.C.L.: 538 (1949); Verdc. in K.B. 36: 551 (1981); Fl. Pl. Lign. Rwanda: 610, fig. 207.2 (1982); Fl. Rwanda 3: 200, fig. 73.3 (1985). Type: Tanzania, Kilimanjaro, Marangu, *Volkens* 247 (B, holo.†)

Spreading shrub or small tree 2.4–9(–15) m. tall, with dark grey smooth or slightly fissured bark; shoots densely ferruginous velvety when young, later dark and lenticellate. Leaf-blades ovate-oblong or elliptic to ovate, 3–17(–26) cm. long, 1.5–10(–14) cm. wide, acuminate at the apex, cuneate to rounded or subcordate at the base, densely pubescent to velvety all over with yellowish hairs; petiole 0.5–1.3 cm. long; stipules with base 2 mm. long and filiform part 0.5–1.2 cm. long. Inflorescences short, 1–4 cm. long, very densely shortly hairy; peduncle 0.6–2 cm. long, with bracts as in last species; pedicels 0.75–3 mm. long, densely shortly hairy. Calyx-tube subglobose, densely shortly hairy, 1–1.5 mm. in diameter; lobes linear or oblong, 1.5–6(–8) mm. long, densely hairy. Buds distinctly apiculate, often the 5 appendages very marked. Corolla bright green or yellow-green, greenish cream inside, sparsely to densely hairy outside, sometimes only on the lobes, the tube ± glabrous; tube 3.5–5.5 mm. long; lobes narrowly triangular, 4–4.5 mm. long. Fruit eventually turning brown, subglobose, 2–2.5 cm. long or sometimes asymmetric and ellipsoid if reduced to 1 pyrene by abortion; pyrenes 4–5 normally, up to 1.6 cm. long, 7–8 mm. wide, 6 mm. thick.

KEY TO INFRASPECIFIC VARIANTS

1. Buds pubescent . . . .   a. var. **volkensii**
   Buds glabrous   2
2. Inflorescence-axes and calyx densely shortly hairy;
      inflorescences 2–2.5 cm. long . . . .   b. var. **fyffei**
   Inflorescence-axes sparsely hairy or glabrescent;
      inflorescences 3–4 cm. long . . . .   c. var. **kyimbilensis**

a. var. **volkensii**; Verdc. in K.B. 36: 551 (1981)

Buds distinctly pubescent. Rachis, peduncle, pedicels, calyx-tube and -lobes all densely shortly pubescent. Inflorescences 1–3.5 cm. long.

UGANDA. Karamoja District: Mt. Kadam, Apr. 1959, *J. Wilson* 735!; Mbale District: Bubulo, 15 May 1938, *Tothill* 2704!; Masaka District: Buyaga [Buwaga], Nov. 1925, *Maitland* 862!
KENYA. Northern Frontier Province: Moyale, 28 Apr. 1952, *Gillett* 12949!; NE. Elgon, May 1955, *Tweedie* 1315!; Nairobi District: Karura Forest, May 1939, *Bally* in C.M. 9214!
TANZANIA. Arusha District: Ngongongare to Lake Momela, 12 Mar. 1914, *Peter* 55455!; Lushoto District: Usambara Mts., Balangai, 12 Mar. 1916, *Peter* 16071!; Iringa District: Mt. Image, Tapu, Nov. 1953, *Carmichael* 278!
DISTR. **U** 1–4; **K** 1–7 (Teita); **T** 1–3, 7, 8; Zaire, Rwanda, Ethiopia and Sudan
HAB. Evergreen forest, particularly *Juniperus* and *Podocarpus*, especially at edges, scrub, thickets in grassland, often in rocky places or on termite mounds, occasionally riverine; 1140–2300 (?–2400) m.

SYN. *V. linearisepala* K. Schum. in E.J. 33: 351 (1903); Robyns in B.J.B.B. 11: 296 (1928); T.T.C.L.: 537
      (1949); I.T.U., ed. 2: 358 (1952); K.T.S.: 479 (1961). Type: Tanzania, W. Usambara Mts., Kwai,
      Albers 276 (B, holo.†)
      V. bicolor K. Schum. var. *crassiramis* K. Schum. in E.J. 34: 333 (1904). Type: Tanzania, W.
      Usambara Mts., near Kwai, *Engler* (1902) 1238 (B, holo.†)

b. var. **fyffei** (*Robyns*) *Verdc.* in K.B. 36: 551 (1981). Type: Uganda, Masaka District, Malabigambo
Forest Reserve, *Fyffe* 53 (K, holo.!)

Buds entirely glabrous or with very sparse hairs at top of tube. Rachis, peduncle, pedicels,
calyx-tube and -lobes all densely shortly hairy. Inflorescences 2–2.5 cm. long.

UGANDA. Masaka District: Malabigambo Forest, *Fyffe* 53!
DISTR.   U 4; known only from the above gathering
HAB.   Evergreen forest; 1150 m.

SYN. *V. fyffei* Robyns in B.J.B.B. 11: 295 (1928)

NOTE. This may be merely a state of *V. volkensii* and not a genuine variety; it is certainly not as
       distinctive as the following variant.

c. var. **kyimbilensis** (*Robyns*) *Verdc.* in K.B. 36: 552 (1981). Type: Tanzania, Rungwe District,
Mwakaleli, *Stolz* 2289 (K, holo.! & iso.!, BR, EA, iso.!)

Buds entirely glabrous. Rachis, peduncle, pedicels and particularly calyx-tube and -lobes sparsely
hairy or glabrescent. Inflorescences 3–4 cm. long.

TANZANIA. Rungwe District: Mwakaleli, 13 Dec. 1913, *Stolz* 2289!
DISTR.   T 7; known only from the above gathering
HAB.   Bamboo forest; 1500 m.

SYN. *V. kyimbilensis* Robyns in B.J.B.B. 11: 294 (1928); T.T.C.L.: 537 (1949)

NOTE. (on species as a whole). As mentioned in I.T.U., ed. 2: 358 (1952) this taxon could be treated
       as a hairy variant of *V. apiculata*, but since the two seem to be easily distinguished over a wide area I
       have maintained them at specific rank. One's suspicions are, however, raised by such sheets as
       Wilson 735 (Mt. Kadam, 2100 m., Apr. 1959) which consists of a mixture of the two, presumably
       collected side by side. Observations are needed throughout its range.

6. **V. bicolor** *K. Schum.* in E.J. 34: 332 (1904); Robyns in B.J.B.B. 11: 291 (1928); T.T.C.L.:
537 (1949); Verdc. in K.B. 36: 552 (1981). Type: Tanzania, E. Usambara Mts., Derema
[Nderema], *Scheffler* 202 (B, holo.†, BR, EA, K, iso.!)

Small divaricately branched tree to 8 m. Youngest shoots densely pale ferruginous
hairy with rather long hairs 1–1.5 mm. long, older parts dull purplish grey, slightly striate,
glabrous. Leaf-blades oblong-elliptic, 4–11 cm. long, 2–4.5 cm. wide, subacute to shortly
acuminate at the apex, rounded to rounded-cuneate at the base, distinctly discolorous,
fairly densely ferruginous hairy above but not velvety, nor obscuring the surface, densely
softly velvety tomentose beneath with matted yellowish grey hairs; lateral nerves slightly
impressed in dry state; petioles 5–7 mm. long, hairy like the stems; stipules with
base 0.5–1.5 mm. long and filiform part 6–7.5 mm. long, mostly obscured by long bristly
hairs. Inflorescences ± 2 cm. long, densely ferruginous hairy; peduncle thick, bracteolate,
up to 7 mm. long; pedicels 2 mm. long; bracts ovate, 3 × 2 mm. Calyx-tube ovoid, 1 mm.
long; lobes linear-lanceolate, 6–8 mm. long, 1.5 mm. wide. Buds shortly apiculate,
glabrous. Corolla green, 1 cm. long. Unripe fruits subglobose, 1.2 cm. long and wide.

TANZANIA. Lushoto District: E. Usambara Mts., Derema [Nderema], Jan. 1900, *Scheffler* 202!
DISTR.   T 3; known only from the type
HAB.   Rain-forest; 800 m.

NOTE. Presumably related to *V. volkensii*, but the isotypes seen have no complete flowers. The
       young stems have longer hairs, the leaf-blades are more softly tomentose beneath and the lateral
       nerves faintly impressed. No further material has ever turned up in what was a well-collected area
       now much degraded. Robyns' reference to a peduncle 7 cm. long is an error.

7. **V. randii** *S. Moore* in J.B. 40: 252 (1902); Robyns in B.J.B.B. 11: 281 (1928); F.F.N.R.:
424 (1962); Drummond in Kirkia 10: 276 (1975); Coates Palgrave, Trees of S. Afr.: 873
(1977); Verdc. in K.B. 36: 552 (1981). Type: Zimbabwe, Bulawayo, *Rand* 123 (BM, holo.!)

Shrub or small tree or occasionally scandent, 1.2–7 m. tall, with slender branches;
shoots brown, red-brown or purplish brown, usually with sparse to dense lenticels and
sometimes rather peeling epidermis, glabrous; in some forms some side branches are

modified into spines. Leaves thin or very thin; blades elliptic to oblong-elliptic, 2–15.5 cm. long, 0.8–6.6 cm. wide, bluntly subacute to obtusely acuminate or distinctly narrowly acuminate at the apex, cuneate to almost rounded at the base, glabrous or with both bristly hairs and much shorter hairs on the upper and particularly lower surface and usually also bristly cilia on the margins; stipules with base 1–2 mm. long, with filiform apex 0.2–1.1 cm. long; petiole 3–7 mm. long. Inflorescences graceful, 1–3.5 cm. long, the rachis, etc. glabrescent to shortly pubescent; ultimate branches 3–8-flowered; peduncle 0.4–1.5 cm. long, with paired ovate-oblong bracts 1–2 mm. long; pedicels 1–5 mm. long. Calyx-tube subglobose, 1 mm. long, glabrous to pubescent; lobes linear or oblong-linear, 2–6(–7) mm. long. Buds acuminate or acute, usually 5-tailed, glabrous. Corolla white, green or golden green; tube 2–3 mm. long; lobes narrowly triangular, 2.5–3 mm. long, ± obtuse to distinctly appendaged and apiculate at the apex. Style slender, 3.5–4.5 mm. long; stigmatic club cylindrical, 0.75 mm. long, minutely 5-lobulate. Fruit yellow, subglobose, 1.4–2 cm. in diameter on stalks 0.5–3 cm. long; pyrenes 3–5, 1.3–1.5 cm. long.

subsp. **vollesenii** *Verdc.* in K.B. 36: 553 (1981). Type: Tanzania, Kilwa District, Selous Game Reserve, Kingupira, *Vollesen* in *M.R.C.* 4271 (K, holo.!, C, EA, MRC, iso.)

Shrub 0.5–5 m. tall, often ± scandent, sometimes with branches modified into spines, either 1 or 2 at a node. Leaf-blades rounded to acuminate at apex, usually more rounded at base, glabrous. Inflorescences shorter and less slender, the peduncles ± 3 mm. long; ultimate branches of inflorescence 3-flowered. Calyx-tube glabrous; lobes very narrow, 3–6(–7) mm. long, 0.4–0.9 mm. wide. Corolla-tube wider; buds apiculate, mostly with 5 tails. Fruit-stalks 1–2 cm. long.

TANZANIA. Ulanga District: Selous Game Reserve, Ngwina road, 6 Feb. 1971, *Rees* T 110!; Kilwa District: Kingupira, 23 Jan. 1977, *Vollesen* in *M.R.C.* 4370! & 23 km. SW. of Kingupira, 1 Jan. 1976, *Vollesen* in *M.R.C.* 3173!
DISTR. **T** 6, 8; not known elsewhere
HAB. Groundwater forest, deciduous thicket, often riverine, also on termite mounds in wooded grassland; 100–390 m.

NOTE. *Vollesen* in *M.R.C.* 4252 (Rufiji District, Kichi Hills, 22 Dec. 1976, from deciduous coastal thicket on sand, 450 m.) has the short inflorescences of subsp. *vollesenii* but small leaves 3.3 × 1.4 cm. pilose above and strongly ciliate as in subsp. *acuminata*.

subsp. **acuminata** *Verdc.* in K.B. 36: 553 (1981). Type: Tanzania, Tanga District, Kange Gorge, *Faulkner* 1828 (K, holo.! & iso.!)

Shrub 1.2–3 m. tall, apparently never scandent nor with branches modified into spines. Leaf-blades very distinctly narrowly acutely acuminate, glabrous or sometimes markedly ciliate and/or bristly pubescent on lamina with understorey of much smaller hairs; stipules up to 1.1 cm. long. Inflorescences very slender, with peduncles to 1.5 cm. long; ultimate branches of inflorescences 3–5-flowered. Calyx-tube glabrous; lobes 2–3 mm. long, up to 0.5 mm. wide. Corolla-tube slender; buds apiculate, mostly with 5 tails. Fruit-stalks 2–3 cm. long.

KENYA. Kwale District: Shimba Hills, Kwale Forest, 23 Mar. 1968, *Magogo & Glover* 437! & Muhaka Forest, 14 Apr. 1977, *Gillett* 21053! & Jadini, Diani Forest, 11 July 1972, *Gillett & Kibuwa* 19895!
TANZANIA. Lushoto District: E. Usambara Mts., Maneno Mbangu-Kwamtili, 1 Oct. 1918, *Peter* 25158!; Tanga District: Kange Gorge, 19 Apr. 1956, *Faulkner* 1852!; Pangani District: Bushiri Estate, 7 Apr. 1950, *Faulkner* 574!
DISTR. **K** 7; **T** 3, 6; ?Somalia
HAB. Evergreen forest, often on coral or limestone; (?0–)30–300(–1200) m.

NOTE. I at first considered *acuminata* to be a subsp. of *V. chartacea* Robyns (type: South Africa, Natal, Mt. Edgecombe, *Medley Wood* 11586! (K, holo.!)) but later merged that species with *V. randii* and recognised 4 subspecies; subsp. *randii* from Zimbabwe and Zambia has a pubescent calyx-tube, shorter calyx-lobes, glabrous leaves and short peduncles; subsp. *chartacea* (Robyns) Verdc. has glabrous leaves and calyx-tube and peduncles to about 8 mm. *V. esculenta* S. Moore is scarcely distinct, differing only in the short calyx-lobes. The East African populations referred to subsp. *acuminata* are distinctly different in detail from those of subsp. *chartacea* in Natal but undoubtedly all belong to the same species. The variation in indumentum in subsp. *acuminata* is extraordinary but even in one locality material varies from having nearly glabrous leaves to leaves with strongly ciliate margins and densely bristly pilose and puberulous blades; subsp. *chartacea* has glabrous leaves, calyx-lobes usually broader, leaf-blades acuminate, stipules shorter and inflorescence-branches mostly with more numerous flowers and occurs in Natal and possibly in the Transvaal (but material from there is dubiously named). The Natal, Kenya, Tanzania coastal distribution is a quite common one with some coastal forest species recurring in Chirinda, etc.

8. **V. praecox** *Verdc.* in K.B. 36: 554, fig. 12 (1981). Type: Tanzania, Iringa District, Ruaha National Park, Kimiramatonge [Kirimatonge] Hill, *Greenway & Kanuri* 14793 (K, holo.!, EA, iso.)

FIG. 151. *VANGUERIA PRAECOX* — **A**, flowering branch, × ²/₃; **B**, flower, × 3; **C**, longitudinal section through flower, × 3; **D**, transverse section through ovary, × 4; **E**, pyrene, × 2; **F**, transverse section through pyrene, × 2; **G**, fruiting branch with mature leaves, × ²/₃. A–D, from *Greenway & Kanuri* 14793; E–G, from *Renvoize & Abdulla* 2309. Drawn by Christine Grey-Wilson.

Much-branched many-stemmed deciduous shrub or slender small tree 3–6 m. tall; branches ± horizontal; bark greyish brown, purplish or blackish, the young shoots ± smooth, with rather pale peeling epidermis, the older more ridged, obscurely lenticellate; flowers mostly appearing when leaves are very young or almost absent; current season's shoots sometimes ± herbaceous and drying dark but soon covered with bark; some lateral branches have very congested internodes. Leaves terminating short lateral branches, dark green; blades elliptic or somewhat obovate, 5.5–14.5 or more cm. long, 2.2–10.5 cm. wide, acuminate at the apex, the acumen usually rounded at the tip, cuneate at the base, glossy above, glabrous, at first thin but eventually almost coriaceous and ± paler beneath; venation usually drying pale; petioles 2–3 mm. long and where the leaves appear to come direct from the apex of an old shoot the base of the petiole appears adnate and decurrent, forming a shield 6 mm. long drying black which contrasts strongly with the pale corky bark; stipules with base 1.5–6 mm. long and wide and subulate part 3–7 mm. long, at first herbaceous but soon becoming corky and hairy within; the persistent corky bases of paired stipules terminate old branchlets and appear as a mouth from which new leaves and inflorescences emerge, both drying black and contrasting. Inflorescences glabrous very short 3–4-flowered cymes 7 mm. long, almost sessile or peduncles only 1–2 mm. long; pedicels 0.5–1.5 mm. long; bracts linear-lanceolate, 2–4 mm. long. Buds ± acuminate, glabrous. Calyx-tube depressed-subglobose, 2 mm. tall, 2.5–2.7 mm. wide; lobes linear-oblong, 2–5.5 mm. long, recurved. Corolla green or yellow-green; tube 3–4 mm. long; lobes narrowly triangular, 2.5 mm. long, slightly appendaged. Style 3.5 mm. long; stigmatic club cylindrical, 0.7 mm. long. Fruits yellow or greenish yellow to orange or brown, fleshy, subglobose, almost sessile, 2.2–2.8 cm. long, 2.3–2.6 cm. wide, containing 4–5 pyrenes; pyrenes buff, kidney-bean-shaped, ± 1.5 cm. long, 7 mm. wide, 9 mm. thick, with a hilum-like notch on inner margin, rather rugose. Figs. 132/16, p. 754 & 151.

TANZANIA.   Dodoma District: Manyoni, Kazikazi, 16 May 1932, *B.D. Burtt* 3551! & near Dodoma, 12 Dec. 1925, *Peter* 33079!; Iringa District: Ruaha National Park, summit of Mpululu Mt., 21 May 1968, *Renvoize & Abdallah* 2309! & Kinyantupa track, 21 Oct. 1970, *Greenway & Kanuri* 14597!
DISTR.   T 5, 7; not known elsewhere
HAB.   Mixed woodland on granite slopes and bushland on sandy soils, also valleys with impeded drainage (mbugas); 710–1300 m.

NOTE.   The fruit is sugary and edible.

103a. GEN. ? NOV. B

Tree to 10 m. with pendulous branches or scandent shrub 5–6 m. tall; stems slender with reddish brown bark flaking in minute pieces; youngest parts drying straw-coloured, glabrous. Leaf-blades oblong, 5.2–13.5 cm. long, 2.5–6 cm. wide, acuminate at the apex, rounded at the base, thin; petiole 0.5–1.3 cm. long; stipules triangular, ? 3 mm. long, inadequately preserved. Inflorescences axillary, clearly branched; peduncle ± 1 cm. long, pubescent; pedicels up to 2 mm. long, pubescent. Flowers unknown. Remains of calyx-lobes oblong-linear, 3 mm. long. Fruits globose, ± 3 cm. long (in flattened dried state), containing several pyrenes; pyrenes elongate, 2 cm. long, 9 mm. wide, 6.5 mm. thick, dorsally ± keeled, apically compressed, somewhat rugose. Testa cells isodiametric.

TANZANIA.   Lushoto District: Usambara Mts., Sigi R. valley near Chemka, 30 July 1974, *Baagøe et al.* 139! & Amani–Sigi Forest Reserve, along the Sigi and Nguruwe rivers and the N. slopes above, 10 May 1987, *Borhidi et al.* 87388!
DISTR.   T 3; not known elsewhere
HAB.   Rain-forest of *Cephalosphaera, Leptonychia usambarensis, Sorindeia madagascariensis* and *Pachystela msolo*; 420–600 m.

NOTE.   Despite the well-collected nature of the area no other material has been seen; its generic placing will remain dubious until flowers are available. The large fruit suggests *Vangueria*.

104. MEYNA

Link in Jahrb. Gewächskunde 1(3): 32 (1820); Robyns in B.J.B.B. 11: 226 (1928)

Shrubs, small trees or less often lianes of very characteristic habit, usually freely armed with paired spines (modified branches) below which is a normal node bearing a pair of leaves or a pair of very short contracted nodose branches forming spurs bearing apical fascicles of flowers and a pair of leaves, so that each node on the main stem appears to have a whorl of 4 leaves; stipules transverse or broadly triangular, adnate to the petioles, produced into a subulate tip, densely hairy inside, persistent. Cymes fasciculate, usually reduced so that flowers appear to emerge straight from the short spurs, or cymes few-flowered with short common peduncles or sometimes borne at the end of the spines; flowers small, usually many per spur, pedicellate, the buds distinctly apiculate. Calyx-tube short, ± hemispherical or depressed obconic; lobes short, linear-subulate or lanceolate, erect or spreading. Corolla usually green; tube shortly cylindrical, sometimes a little swollen at the base, hairy at the throat and with a ring of deflexed hairs inside below the throat; lobes 4–5, ovate-triangular, usually with an apiculate appendage at the apex. Stamens with very short filaments; anthers small, ± just exserted. Ovary 4–5-locular; style slender or rather thickened, always shortly exserted; stigmatic club crown-shaped or subglobose, slightly to distinctly marginally 4–5-lobed or depressed subpeltate with divaricate lobes, the lobes overtopping the middle. Fruit subglobose, rather large, on mostly accrescent pedicels, with 4–5 pyrenes and appearing 4–5-lobed in dry state and possibly slightly so in life.

A genus extending from East Africa to India and Indo-China. Robyns recognised 10 species but they appear to me to be very poorly delimited; it is likely that the African species will eventually prove to be a variant of the Indian *M. spinosa* Link. The genus is very close to *Canthium sensu stricto* and close to *Rytigynia*.

M. tetraphylla (*Hiern*) *Robyns* in B.J.B.B. 11: 232 (1928); Cufod. in Nuov. Giorn. Bot. Ital., n.s. 55: 88 (1948); F.P.S. 2: 443 (1952); E.P.A.: 1008 (1965). Type: Sudan, Jur, Abu Qurun, *Schweinfurth* 1856 (K, holo.!, BM, iso.!)

Shrub or small tree 2.5–9 m. tall with ± scandent branches or a liane 6–20 m. long; stems often forming dense tangles; bark on branches dark plum-coloured, ± longitudinally fissured; spines slender, 0.4–2.6 cm. long, supra-axillary, sometimes one of the pair developed as a side branch. Leaves pseudoverticillate, appearing to be in whorls of 4; blades ovate-elliptic, elliptic or elliptic-oblong, 2–7(–12) cm. long, 1.2–3.4(–6) cm. wide, mostly obtuse at the apex, ± cuneate at the base, glabrescent to pubescent and ciliate and domatia absent or feebly developed, not markedly discolorous; petioles 3–6 mm. long, pubescent or glabrescent; stipules with subulate tip 1–2 mm. long. Flowers in short fascicles on very abbreviated shoots, sometimes appearing when the leaves are fully developed; pedicels 1–6(–8 in fruit) mm. long, hairy or glabrous; buds glabrous or sparsely hairy. Calyx-lobes triangular to subulate, 0.5–0.7 mm. long. Corolla green; tube 1.25–2(–3) mm. long; lobes 2–3 mm. long, with appendage 0.5 mm. long; stigmatic club crown-shaped, 0.5–0.8 mm. long; exserted part of style 1–1.5 mm. long. Fruit brown, subglobose, 1.7 cm. long, 1.9–2 cm. wide, ± bluntly angularly 5-lobed; pyrenes 4–5, segment-shaped, 1.6 cm. long, 8–9 mm. wide, 4–6 mm. thick. Seeds soft, compressed-fusiform, 1 cm. long, 3 mm. wide, narrowed at both ends, margins thickened.

subsp. **tetraphylla**; Verdc. in K.B. 36: 540 (1981)

Buds sparsely spreading hairy. Leaf-blades ± pubescent. Pedicels densely spreading hairy.

UGANDA. W. Nile District: Madi, 13 Feb. 1863, *Grant*!; Karamoja District: Matheniko, base of Moroto Mt., Apr. 1954, *Philip* 542! & Dodoth, Kidepo National Park, *Synnott* 979!
KENYA. Turkana District: Oropoi, Feb. 1965, *Newbould* 6999! & Upe, Apelapong, June 1955, *Philip* 725!; W. Suk District: Sigor, *Kenya Nat. Museum 2nd. 1974 Exped.* 144!; Baringo District: near R. Ndau, above Kibiugoi, *Hivernel* 14!
DISTR. U 1; K 2, 3; Sudan and S. Ethiopia
HAB. Thicket in gullys, woodland; 1100–1650 m.

SYN. *Vangueria tetraphylla* Hiern in F.T.A. 3: 152 (1877); De Wild. in B.J.B.B. 8: 25 (1922)
[*Randia dumetorum* sensu Hiern in F.T.A. 3: 94 (1877), quoad spec. *Grant, non* Lam.]
Canthium tetraphyllum Baillon in Adansonia 12: 192 (1878). Type as for *M. tetraphylla*

subsp. **comorensis** (*Robyns*) *Verdc.* in K.B. 36: 540 (1981). Type: Comoro Is., Moely, *Boivin* (P, holo.)

Buds glabrous. Leaf-blades practically glabrous or ± pubescent. Pedicels glabrous to pubescent. Figs. 131/3, 21, p. 752; 132/17, p. 754 & 152.

KENYA. Kitui District: Mutha Plains, 24 Jan. 1942, *Bally* 1636!; Teita District: Voi R. crossing, Kandecha, 16 Dec. 1966, *Greenway & Kanuri* 12777!; Kwale District: Diani Forest, 13 July 1972, *Gillett & Kibuwa* 19843!

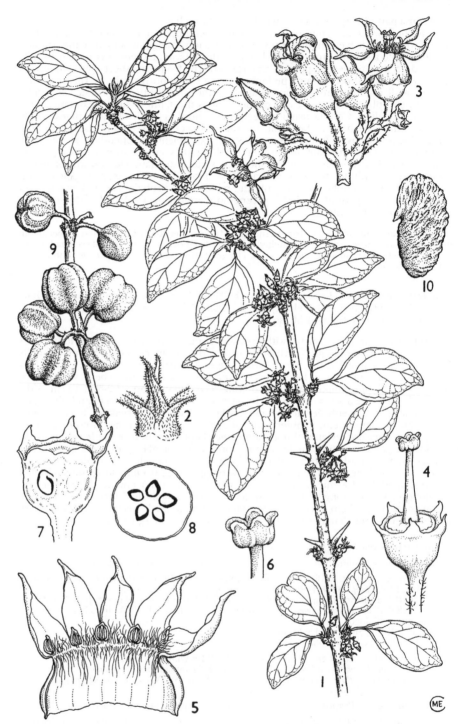

FIG. 152. *MEYNA TETRAPHYLLA* subsp. *COMORENSIS* — **1**, flowering branch, × ²⁄₃; **2**, stipule, × 6; **3**, inflorescence, × 4; **4**, flower with corolla removed, × 8; **5**, corolla opened out, × 8; **6**, stigmatic knob, × 18; **7**, longitudinal section of ovary, × 12; **8**, transverse section of ovary, × 12; **9**, fruits on branch, × ²⁄₃; **10**, pyrene, × 2. 1–8, from *Greenway & Kanuri* 14619; 9, 10, from *Bally* 1636. Drawn by Mrs M.E. Church.

TANZANIA. Dodoma District: 13 km. N. of Saranda, Luwila [Ruwiri] Gorge, Dec. 1935, *B.D. Burtt* 5437!; Uzaramo District: Dar es Salaam, Kunduchi, 22 Jan. 1969, *Mwasumbi* 10442!; Iringa District: Great Ruaha R., Trekimboga [Trikamboga] track, 27 Oct. 1970, *Greenway & Kanuri* 14619!
DISTR. **K** 4, 7; **T** 5–8; ? Somalia and Comoro Is.
HAB. Coastal bush on coral, bushland with scattered trees, dry or moist evergreen forest and woodland, often riverine or by waterholes; 0–1200 m.

SYN. *M. comorensis* Robyns in B.J.B.B. 11: 233 (1928)
   [*M. tetraphylla* sensu T.T.C.L.: 505 (1949), *non* (Hiern) Robyns sensu stricto]

105. **CANTHIUM**

Lam., Encycl. Méth. Bot. 1: 602 (1785)

[*Plectronia* sensu auctt. div., *non* L.]

Shrubs or small trees, sometimes scandent; spines absent or if present straight. Heterophylly sometimes apparent; leaves either deciduous and restricted to brachyblasts or apex of stem, or evergreen and spaced along the branches, opposite, paired, petiolate, papery to subcoriaceous, domatia absent, pit-like or present as tufts of hair; stipules shortly sheathing, apiculate or aristate, glabrous or pubescent within. Flowers ♂ or ♀ and ♂, 4–5-merous, pedicellate or not, borne in pedunculate to subsessile few–many-flowered cymes; inflorescence-branches present or reduced; bracteoles inconspicuous. Calyx-tube broadly ellipsoid to ovoid; tubular part of limb ± obsolete or somewhat reduced, scarcely equalling the disk, bearing dentate, triangular or linear lobes. Corolla white or yellowish, glabrous or rarely pubescent outside; tube broadly cylindrical, ± equal to lobes or sometimes longer or shorter, with or without a ring of deflexed hairs inside, often pubescent at throat; lobes reflexed, obtuse, acute or shortly apiculate. Stamens set at throat; anthers subsessile or borne on short filaments, with or without darkened connective tissue on dorsal face. Style slender, shortly exceeding the corolla-tube, glabrous; stigmatic knob ± spherical, point of attachment distinctly recessed, 2(–3)-lobed; ovary 2(–3)-locular. Disk glabrous. Fruit a 2(–3)-seeded drupe or often 1-seeded by abortion, small to moderately large, strongly dorsiventrally flattened or not, strongly or scarcely indented at apex; pyrenes ellipsoid to ovoid or obovoid, often flattened on ventral face, thinly woody, slightly to distinctly triangular or truncate at point of attachment, with a shallow crest extending to apex, usually rugulose. Seeds flattened on ventral face, very shortly crested around apex; endosperm entire; testa finely reticulate; embryo ± straight to curved with radicle erect; cotyledons small, perpendicular to ventral face of seed.

A diverse and poorly defined genus of very approximately 50 species, widely distributed throughout eastern and southern tropical Africa, South Africa, Seychelles, Madagascar, Asia and Malesia.
   Multidentia, Pyrostria, Psydrax and *Keetia* were formerly included in *Canthium* in Africa.

KEY TO SUBGENERA

1. Spines paired, always borne above brachyblasts   . . .   subgen. 1. **Canthium**
                                                                      (p.862)
   Spines absent or, if present, paired or ternate but not
      strictly associated with brachyblasts   2
2. Flowers unisexual; inflorescence fasciculate or
      subumbellate, ♀ flowers always fewer than ♂ flowers;
      corolla-tube with a ring of deflexed hairs inside;
      stipules lacking conspicuous tuft of hair inside   . . .   subgen. 4. **Bullockia**
                                                                      (p.880)
   Flowers ♂; inflorescence cymose or sometimes reduced
      and appearing subumbellate, occasionally solitary to
      few-flowered; corolla-tube lacking a ring of deflexed
      hairs inside; stipules often conspicuously hairy inside   3

3. Spines always absent; stems often lenticellate; leaves always deciduous, flowers typically borne at naked nodes; stipules sheathing when young, but usually quickly becoming corky, triangular to ovate, sometimes deciduous; calyx-limb reduced to a rim; anthers with little or no darkened connective on dorsal face; pyrenes usually triangular at point of attachment · · · · · subgen. 2. **Afrocanthium** (p.864)

Spines present, usually in threes or sometimes absent; stems not markedly lenticellate; leaves deciduous or not strictly so, flowers occasionally borne at naked nodes; stipules sheathing, only occasionally becoming corky; calyx-limb lobed almost to base; anthers with all but marginal area covered with darkened connective on dorsal face; pyrenes ± truncate at point of attachment . . . . subgen. 3. **Lycioserissa** (p.876)

Subgen. 1. **Canthium**

Shrubs or small trees, sometimes scandent; spines usually present, situated above brachyblasts, straight. Leaves deciduous, restricted to brachyblasts and apex of stem, opposite, paired, petiolate, papery to chartaceous, never large, discolorous or not; tertiary nerves apparent; domatia present as small tufts of hair; stipules triangular, apiculate or aristate, pubescent within. Flowers ⚥, 4(or 4–5 in Africa)-merous, pedicellate, borne on pedunculate to subsessile few to several-flowered cymes; inflorescence-branches present or reduced; bracteoles inconspicuous. Calyx-tube broadly ellipsoid to ovoid; limb with tube rather reduced, scarcely equalling the disk, bearing dentate or triangular-linear lobes. Corolla ?whitish, glabrous or rarely pubescent outside; tube broadly cylindrical, ± equal to lobes or sometimes shorter, with a ring of deflexed hairs inside (apparently absent in one Indian species) and pubescent at throat; lobes reflexed, acute or shortly apiculate. Anthers with central area covered with darkened connective on dorsal face. Style slender, shortly exceeding the corolla-tube, glabrous; stigmatic knob ± spherical, point of attachment recessed $\frac{1}{3}$–$\frac{1}{2}$ of the way from the base, 2-lobed; ovary 2-locular. Fruit a 2-seeded drupe, small to moderately large, strongly dorsiventrally flattened, scarcely indented at apex; pyrenes ellipsoid, flattened on ventral face, thinly woody, slightly (or distinctly in African species) triangular at point of attachment, with a shallow crest extending to apex, rugulose. Seeds narrowly obovoid, flattened on ventral face, very shortly crested around apex; embryo ± straight.

Only three species have been considered as *Canthium* sensu stricto, the type *C. coromandelicum* (Burm. f.) Alston (Fig. 131/23, p. 752), *C. travancoricum* Bedd. and the African species. They occur in central and southern India, Sri Lanka and Eastern tropical Africa (Somalia to Zimbabwe). Species 1.

1. **C. glaucum** *Hiern* in F.T.A. 3: 134 (1877), pro parte, excl. specim. *Kirk* ex Tete; Bullock in K.B. 1932: 359 (1932); E.P.A.: 1009 (1965). Type: Somalia, Tula (Tola) R., *Kirk* (K, lecto.!)

Shrub, sometimes scandent, 1–5 m. tall; young branches glabrous or rarely pubescent (not in Flora area), lenticellate or not, armed with spines up to 1.5(–1.8–2.1) cm. long, situated above very abbreviated lateral branches, apparently not developing into lateral branches; bark light to dark grey or brown. Leaves restricted to lateral spurs at apex of stem; blades narrowly to broadly elliptic, 2.4–6 cm. long, 1.4–3 cm. wide, acute to subacuminate or sometimes obtuse at apex, acute to cuneate at base, strongly or moderately discolorous, glabrous or glabrous above and sparsely pubescent at least on nerves beneath or rarely pubescent on both faces; lateral nerves usually in 3 main pairs; tertiary nerves apparent; domatia present as small tufts of hair; petiole 1–1.6 mm. long, glabrous, sparsely pubescent or rarely pubescent; stipules triangular, apiculate, 1–4 mm. long, densely pubescent within. Flowers (4–)5-merous, borne in subsessile to pedunculate, 2–8-flowered cymes; peduncles up to 1 cm. long; pedicels 0.3–1 cm. long, glabrous or sparsely pubescent; bracteoles inconspicuous. Calyx-tube 1 mm. long, glabrous or rarely pubescent; limb divided, often almost to base, into triangular or narrowly triangular lobes, 0.5–1 mm. long. Corolla glabrous or rarely pubescent outside; tube 1–1.25 mm. long, with a ring of deflexed hairs at throat and pubescent immediately

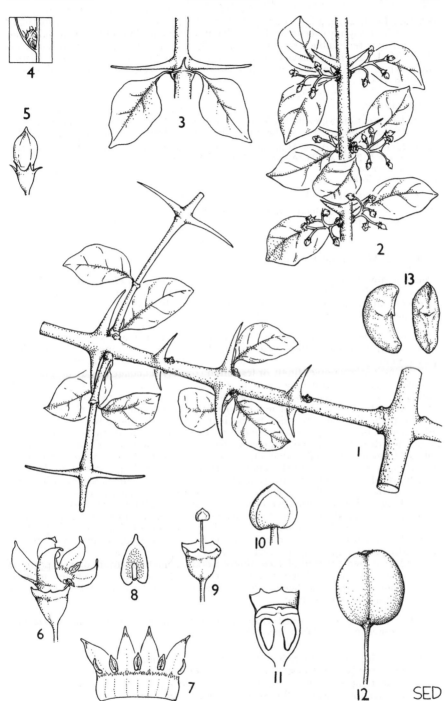

FIG. 153. *CANTHIUM GLAUCUM* subsp. *GLAUCUM* — **1**, vegetative branch, × 1; **2**, flowering shoot, × 1; **3**, node with stipule, × 2; **4**, domatium, × 6; **5**, flower bud, × 6; **6**, flower, × 6; **7**, part of corolla opened out, × 6; **8**, dorsal view of anther, × 14; **9**, calyx with style and stigma, × 6; **10**, stigmatic knob, × 14; **11**, longitudinal section of ovary, × 12; **12**, fruit, × 2; **13**, pyrene (2 views), × 2. 1, 13, from *Kuchar* 17410; 2, 4, 7–11, from *R.M. Graham* 2352; 3, from *Gillett & Hemming* 586; 5, from *Gisau SOK* 22; 6, from *Hemming* 392; 12, from *Langridge* 30. Drawn by Sally Dawson.

above it; lobes triangular-lanceolate, 2–2.5 mm. long, 1.5 mm. wide, scarcely apiculate to apiculate. Style slightly exceeding the corolla-tube; stigmatic knob ± spherical, 0.75 mm. wide. Fruit pale red, edible, ± square in outline, scarcely indented at apex, hardly tapered at base, 0.9–1.2 cm. long and wide; pyrene ellipsoid, flattened on ventral face, 7–12 mm. long, 4–6 mm. wide, triangular at point of attachment, crested at the apex, rugulose. Fig. 132/18, p. 754.

subsp. **glaucum**

Stems light grey to light greyish brown, often with a glaucous bloom; leaf-blades entirely glabrous, glaucous beneath; pedicels 3–5 mm. long when flowering, 6–9 mm. long when fruiting. Fig. 153.

KENYA. Kilifi District: Giriama, Mfuranje, *Gisau SOK* 22! & Malindi, Arabuko, Mar. 1930, *R.M. Graham* in *F.D.* 2352! & Kilifi, Aug. 1965, *Langridge* 30!
DISTR. **K** 7; Somalia
HAB. ?Coastal forest

SYN. *Plectronia glauca* (Hiern) K. Schum. in P.O.A. C: 386 (1895); Chiov., Fl. Somala 2: 245 (1932)

NOTE. Subsp. *frangula* (S. Moore) Bridson (fig. 132/18, p. 754) with usually darker stems and longer pedicels occurs in southern tropical Africa; glabrous and densely pubescent varieties have been described.

Subgen 2. **Afrocanthium** *Bridson*

Subgen. nov. a subgen. *Canthio* spinis carentibus, a subgen. *Bullockia* Bridson floribus hermaphroditibus, a subgen. *Lycioserissa* (Roem. & Schultes) Bridson spinis carentibus, connectivo antherae dorsaliter haud vel vix fuscato, limbis calycis vix lobatis differt. Typus subgeneris: *Canthium lactescens* Hiern

Shrubs or trees, always deciduous; spines absent; brachyblasts sometimes present; lenticels frequently apparent on young stems. Leaves usually confined to new growth at apex of stems, paired, petiolate, papery to subcoriaceous; heterophylly observed in only one species; tertiary nerves finely or less often coarsely reticulate or obscure; domatia present, pubescent or sometimes absent; stipules sheathing, the apical one awned, but soon triangular to ovate, obtuse to apiculate and often becoming corky, sometimes caducous, with tufts of white or less often rust-coloured hair within. Flowers ☿, 4–5-merous, pedicellate or subsessile, borne in pedunculate few–many-flowered cymes; inflorescence-branches present or sometimes very reduced, often with the flowers borne along one side of the ultimate branch; bracteoles obscure or absent. Calyx-tube broadly ellipsoid to ovoid; limb reduced to a rim, sometimes shortly dentate, scarcely equalling disk. Corolla white or greenish yellow, usually rather small, glabrous outside; tube broadly cylindrical ± equal to lobes or a little longer, sparsely hairy to hairy within, often hairy at throat; lobes reflexed, obtuse to acute or subacuminate. Anthers ovate, dorsal face lacking darkened connective or with only a central band. Style slender, shortly exceeding the corolla-tube, or rarely somewhat longer, glabrous; stigmatic knob ± spherical, the point of attachment recessed about ⅓ of the way from the base, 2-lobed; ovary 2-locular. Fruit a 2-seeded drupe, small to moderately sized, strongly dorsiventrally flattened, typically heart-shaped and strongly or rarely scarcely indented at apex; pyrenes ellipsoid, flattened on ventral face, thinly woody, sometimes triangular at point of attachment, with a shallow but distinct crest extending from point of attachment to apex, usually somewhat rugulose. Seeds ± obovoid, with ventral face flattened; apex shortly crested; embryo scarcely curved.

A subgenus of 22 species restricted to eastern and southern tropical Africa (extending as far west as eastern Zaire) and South Africa. Species 2–16.

1. Leaves with 2–4 main pairs of lateral nerves; lateral branches (some at least) reduced to cushion-shoots or almost entirely suppressed so that each node appears to have four leaves; inflorescence seldom more than 6-flowered; stipules usually with fawn or rust-coloured hairs inside; corolla-tube ± 1 mm. long, where known 2
Leaves with (3–)4–10 main pairs of lateral nerves; lateral branches always clearly apparent; inflorescences 1–50-flowered; stipules usually with white hairs inside; corolla-tube (1–)1.75–3 mm. long 4

2. Young stems with stiff upwardly angled hairs; leaf-blades
sessile, not exceeding 2 cm. long (**T** 8) . . . . . . 4. *C. rondoense*
Young stems lacking such characteristic indumentum;
petiole at least 1 mm. long; leaf-blades at least 2.5 cm.
long 3
3. Petioles 1–9 mm. long; stems not remarkably slender;
leaves sometimes ciliate; inflorescences 1–6(–10)-
flowered . . . . 2. *C. pseudoverticillatum*
Petioles ± 1 mm. long; stems slender, the young ones ± 1
mm. wide; leaves never ciliate; flowers solitary (T8,
Kilwa District) . . . . 3. *C. sp. A.*
4. Stipules with marginal area thinner and shiny (obscured
by inrolling when young), readily caducous, seldom
persisting beyond second node; fruit not or only
slightly bilobed and indented at apex; mature leaves
truncate to subcordate at base; lateral nerves acutely
angled, with 3 pairs arising near base 5
Stipules without differentiated margin, the apical ones
herbaceous and acuminate or more often bearing a
filiform to linear lobe, the base frequently becoming
corky then breaking down and eventually caducous;
fruit clearly bilobed and indented at apex; mature
leaves seldom truncate or subcordate at base; lateral
nerves not very acutely angled, clearly spaced towards
base 6
5. Stems and stipules white-grey; petioles 1–1.8 cm. long   15. *C. vollesenii*
Stems and stipules darker grey or reddish; petioles 1–3 mm.
long . . . . 16. *C. burttii*
6. Leaves with 8–10 main pairs of lateral nerves; blades
broadly elliptic to round; stems thick; inflorescences
20–50-flowered with 3–5 orders of branching . . .   9. *C. lactescens*
Leaves with 4–7(–8) main pairs of lateral nerves; blades
occasionally round; stems not markedly thick;
inflorescences up to 25-flowered, usually with 1–2
orders of branching 7
7. Leaf-blades tomentose beneath; tertiary nerves almost
obscured by indumentum . . . .   14. *C. racemulosum* var.
nanguanum

Leaf-blades glabrous or pubescent beneath, but hairs not
so dense as to obscure the tertiary nerves 8
8. Leaf-blades with tertiary nerves obscure, broadly elliptic to
round or rarely elliptic, mature at time of flowering;
domatia absent; stipules acuminate, (2–)3–5 mm. long
altogether . . . .   10. *C. peteri*
Leaf-blades with tertiary nerves apparent to conspicuous,
elliptic to broadly elliptic, rarely round, mature,
immature or absent at time of flowering; domatia
absent or present; stipules usually longer, 3–12 mm.
long altogether, with a distinct (often caducous) apical
lobe or sometimes acuminate 9
9. Stipules with lobe subulate to linear; tertiary nerves
moderately coarsely to coarsely reticulate in mature
leaves 10
Stipules with lobe filiform to subulate; tertiary nerves finely
reticulate in mature leaves 13
10. Domatia absent; petiole 4–10 mm. long; leaves immature at
time of flowering; corolla-tube 3 mm. long; fruit not or
very slightly longer than wide, only slightly tapered
towards base . . . .   12. *C. kilifiense*
Domatia present, often conspicuous; petiole 3–5 mm. long;
leaves mature or immature at time of flowering;
corolla-tube 2 mm. long (where known); fruit heart-
shaped or square to broadly oblong in outline 11

11. Fruit ± square to broadly oblong in outline, scarcely
    tapering towards base; young stipules altogether 3–6
    mm. long . . . .   11. *C. shabanii*
    Fruit ± heart-shaped, usually but not always distinctly
    tapered towards base; young stipules altogether 5–12
    mm. long 12
12. Leaves mature or almost mature at time of flowering,
    frequently with 1–3 pairs present per branch, even in
    flowering stage; corolla slightly pubescent at throat;
    tertiary nerves coarsely reticulate, not conspicuously
    discolorous . . . .   13. *C. keniense*
    Leaves immature or sometimes absent at time of flowering
    (but can be present with immature fruit), rarely more
    than one pair per branch (except occasionally on main
    branch when fruiting); corolla conspicuously
    pubescent at throat; tertiary nerves moderately
    coarsely reticulate, conspicuously discolorous . . .   14. *C. racemulosum*
13. Main and lateral nerves densely pubescent beneath (only
    immature leaves known); T 7,Mbeya District   . . .   8. *C. sp. C.*
    Main and lateral nerves glabrous to sparsely pubescent
    beneath, even when immature 14
14. Leaves large, 11.5–15.5 × 7–9 cm., shortly acuminate at
    apex; domatia present but inconspicuous; fruit ± 1 × 1
    cm. (T 6,Morogoro District) . . . .   6. *C. sp. B*
    Leaves usually somewhat smaller, 5–12 × 3–9 cm., distinctly
    acuminate at apex; domatia usually conspicuous; fruit
    6–10 × 7–10 mm. 15
15. Forest tree, 6–20 m. tall; bark on young stems purplish
    brown but soon turning grey; leaves subcoriaceous,
    shiny above when fully mature, somewhat glaucous
    beneath when younger; flowers 4-merous . . . .   5. *C. siebenlistii*
    Shrub or small tree, 1–7 m. tall, only occasionally found in
    forest; bark on young stems purplish brown, not
    turning grey; leaves chartaceous and dull above when
    fully mature, scarcely discolorous when young, but said
    to be sticky; flowers 4–5-merous   . . . .   7. *C. parasiebenlistii*

2. **C. pseudoverticillatum** *S. Moore* in J.B. 43: 352 (1905); Bullock in K.B. 1932: 377
(1932); K.T.S.: 430 (1961); T.T.C.L.: 489 (1949): Type: Kenya, Kwale District, Shimba Hills,
Kassner 383 (BM, holo.!, K, iso.!)

Shrub or small tree 2–6 m. tall, ?sometimes scandent; young branches sparsely
pubescent to pubescent; bark reddish or light grey or less often white-grey when older.
Leaves paired at apex of main branches or very short lateral branches, often giving
impression of 4 per node due to extreme reduction of branches, mature or sometimes not
fully mature at time of flowering; blades elliptic to broadly elliptic, 2.5–7.5(–10) cm. long,
1–5 cm. wide, acute to subacuminate at apex or sometimes obtuse to rounded, acute to
cuneate at base, papery or chartaceous when mature, glabrous or sometimes ciliate
above, glabrous to pubescent with rather prickle-like hairs on nerves beneath; lateral
nerves in 3–4 main pairs; tertiary nerves obscure or apparent, rather coarsely reticulate;
domatia conspicuous; petioles 1–9 mm. long, glabrous or sparsely pubescent to pubescent
beneath; stipules with a filiform to subulate lobe present at apical node, later with
triangular base 2–3 mm. long which becomes corky and soon breaks down to reveal tufts
of rust-coloured or, less often, fawn or rarely whitish hair within. Flowers (4–)5-merous
borne in small 1–6(–10–20)-flowered shortly pedunculate cymes; peduncles 0.5–3 mm.
long; pedicels 1–5 mm. long, glabrous or with a line of short hairs on either side;
bracteoles inconspicuous. Calyx-tube 1–2 mm. long, glabrous or sparsely pubescent; limb
reduced to a truncate or repand rim, sometimes shortly dentate or with unequal linear
lobes up to 1 mm. long. Corolla yellow or creamy green; tube 1 mm. long, densely but
finely pubescent at throat; lobes ovate or ovate-triangular, (1.25–)1.5–2.25 mm. long,
1–1.75 mm. wide at base, acute. Stigmatic knob ± spherical, ± 1 mm. in diameter, ribbed.
Fruit ± oblong in outline, ± 1 cm. long, ± 0.9 cm. wide, indented at apex, pedicel
lengthening to 0.5–1.3 cm.; pyrene narrowly oblong-ovoid, 9 mm. long, 4 mm. wide,
rugulose. Fig. 154, p. 867.

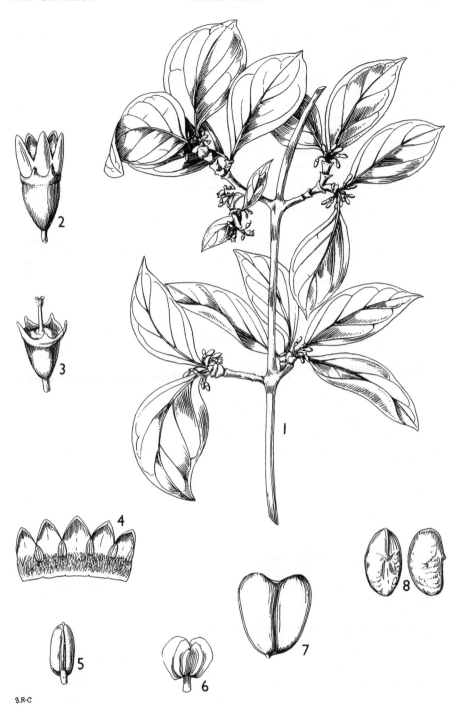

FIG. 154. *CANTHIUM PSEUDOVERTICILLATUM* — **1**, flowering branch, × ²⁄₃; **2**, flower, × 5⅓; **3**, flower with corolla removed, × 5⅓; **4**, corolla opened out, × 6; **5**, dorsal view of stamen, × 20; **6**, stigmatic knob, × 20; **7**, fruit, × 2; **8**, pyrene (2 views), × 2. 1–5, from *R.M. Graham* in *F.D.* 2341; 6, from *Greenway & Kanuri* 12088; 7, from *Archbold* 3107; 8, from *Faden* 71/713. Drawn by Stella Ross-Craig, with 6–8 by Sally Dawson.

subsp. **pseudoverticillatum**

Kenya.   Kwale District: Shimba Hills, 18 Mar. 1902, *Kassner* 383!; Kilifi District: Sokoke Forest, 28 July 1971, *Faden* 71/713!; Lamu District: Boni Forest Reserve, 7.7 km. W of Mararani [Marereni] & 3 km. N. of road, *Robertson & Luke* 5616!

Tanzania.   Handeni District: Kwamkono, 10 Mar. 1986, *Archbold* 3107!; Tanga District, Sawa Creek, 17 Feb. 1957, *Faulkner* 1959!; Morogoro District: Uluguru Mts., 24 Dec. 1934, *E.M. Bruce*!

Distr.   **K**7; **T**3, 6; Mozambique

Hab.   Bushland; 0–975 m.

Syn.   *Plectronia microterantha* K. Schum. & K. Krause in E.J. 39: 541 (1907). Type: Kenya, Kwale District, Shimba Hills, *Kassner* 383 (B, holo.†, BM, K!, iso.)
    Canthium robynsianum Bullock in K.B. 1932: 377, fig. 2 (1932); K.T.S.: 430 (1961). Type: Kenya, Kilifi District, Malindi, Mida, *R.M. Graham* in *F.D.* 2341 (K, holo.!, BM!, EA, iso.)

Note.   Slight differences have been noted between *C. pseudoverticillatum* (**K** 7 & **T** 6, alt. 455–975 m.) and material formerly kept separate as *C. robynsianum* (**K** 7 & **T** 3, alt. 0–300 m.); *C. robynsianum* tends to have the leaf-blades ciliate with sparsely pubescent to pubescent nerves beneath, petioles 3–9 mm. long and pedicels 2–5 mm. long, while *C. pseudoverticillatum* has the leaves entirely glabrous or only occasionally sparsely pubescent beneath, petioles 1–5 mm. long and pedicels 1–2 mm. long. In my opinion there is too much overlap for recognition of *C. robynsianum* at subspecific level.
    A second subspecies will be recognised from Somalia, from between Mogadishu and Bala; it differs in the entirely glabrous leaf-blades obtuse to rounded at apex and lacking domatia, 7–20-flowered inflorescences, white-grey bark and whitish hair inside the stipules. The specimen from Lamu District approaches this subspecies in its obtuse to rounded leaf apices and white-grey bark but in all other respects fits subsp. *pseudoverticillatum*.

3. **C. sp. A**

Shrub; very young branches slender, sparsely pubescent, later covered with flaky dark brown bark. Leaves paired at apex of extremely reduced lateral branches; blades elliptic, 2.8–3.5 cm. long, 1.4–2 cm. wide, subacuminate at apex, acute at base, thinly papery, glabrous; lateral nerves in 3–4 main pairs; tertiary nerves apparent, rather coarsely reticulate; domatia conspicuous; petiole not exceeding 1 mm. long; stipule with a subulate limb and triangular base, 3–4 mm. long altogether, caducous, with a few rusty coloured hairs inside. Flowers solitary (in only known specimen); pedicel 1.5–3 mm. long, sparsely pubescent; bracteoles inconspicuous. Calyx-tube ± 1 mm. long; limb reduced to a rim. Corolla not known. Fruit broadly oblong in outline, 9 mm. long, 10 mm. wide, scarcely tapered towards base, indented at apex; pedicel reaching 6–8 mm. long; pyrene narrowly oblong-ovoid, 9 mm. long, 5 mm. wide, with a distinct crest around apex, acute at point of attachment, rugulose.

Tanzania.   Kilwa District: Selous Game Reserve, Libungani, Mitanga, 23 Feb. 1971, *Ludanga* in *M.R.C.* 1266!

Distr.   **T**8; not known elsewhere

Hab.   Bushland; ± 500 m.

Syn.   [*C. pseudoverticillatum* sensu Vollesen in Opera Bot. 59: 67 (1980), *non* S. Moore]

Note.   This poorly known taxon is very close to an unnamed species from eastern Zimbabwe and the neighbouring part of Mozambique and may be distinguished by the leaves on the young branchlets which are unlike those on the lateral spurs, strongly attenuate at both ends, the clearly apparent hairs inside the stipules, 1–5-flowered inflorescences, the more strongly tapered fruit borne on pedicels 8–11 mm. long and the less strongly crested pyrene with an obtuse point of attachment.

4. **C. rondoense** *Bridson* sp. nov. affinis *C. pseudoverticillati* S. Moore sed foliis sessilibus usque 2 × 1.2 cm., pilis in ramunculorum sursum subpatentibus differt. Typus: Tanzania, Lindi District, S. face of Rondo Escarpment, Mchinjiri, *Eggeling* 6428 (K, holo.!)

Shrub or dwarf spruce-like tree 1–1.5 m. tall; young stems slender, covered with rather stiff upwardly angled hairs; bark reddish. Leaves paired on extremely reduced lateral branches, apparently mature at time of flowering, sessile; blades elliptic to broadly elliptic, 1–2 cm. long, 0.5–1.2 cm. wide, obtuse at apex and base, papery, sparsely pubescent on both faces, ciliate; lateral nerves in 2–3 main pairs, tertiary nerves apparent, coarsely reticulate; domatia absent; stipules triangular, up to 3 mm. long, acute to apiculate, pubescent outside with a tuft of rusty hairs inside. Flowers 4–5-merous, borne in reduced sessile 1–2-flowered fascicles; pedicels ± 1 mm. long. Calyx glabrous or with short stiff hairs; tube 1 mm. long; limb repand or more often dentate, 0.75 mm. long. Corolla white; tube 1 mm. long, pubescent at throat; lobes ovate, 1.5 mm. long, 1.25 mm. wide, acute. Style slightly exceeding corolla-tube; stigmatic knob ± spherical, ± 0.6 mm. across. Disk glabrous. Fruit drying with nerves dark, ± square in outline, 8 mm. long, 8–9 mm. wide, somewhat indented at apex, only slightly narrowed towards base; pedicels lengthening to 2 mm.; pyrene narrowly ovoid with ventral face flattened, 8.5 mm. long, 5 mm. wide, with point of attachment ¼ way down from apex with shallow crest extending around apex, somewhat rugulose.

TANZANIA. Lindi District: S. face of Rondo Escarpment, Mchinjiri, Dec. 1951, *Eggeling* 6428! & Rondo Forest Reserve, 9 Feb. 1991, *Bidgood et al.* 1433A! & 14 Feb. 1991, *Bidgood et al.* 1572!
DISTR. T 8; not known elsewhere
HAB. Semi-evergreen forest with *Dialium, Albizia, Milicia* and *Pteleopsis*, on sand; 750–775 m.

NOTE. Although the small leaves and unusual indumentum give the impression of a juvenile form, the presence of flowers and fruit indicate otherwise.

5. **C. siebenlistii** (*K. Krause*) *Bullock* in K.B. 1932: 379 (1932); T.T.C.L.: 489 (1949). Type: Tanzania, Lushoto District, E. Usambara Mts., Monga, *Grote in Herb. Amani* 3755 (EA, lecto.!, chosen here)

Glabrous tree 6–20 m. tall, sometimes with a fluted trunk; bark on young branches dull red-brown turning to dark grey, lenticellate. Leaves with 1(–2) pairs per branch; blades elliptic to broadly elliptic, 6–8.5 cm. long, 3–6.2 cm. wide, distinctly acuminate at apex, acute at base, herbaceous to subcoriaceous, shiny when fully mature, discolorous, glabrous or sometimes pubescent on midrib beneath; lateral nerves conspicuous; domatia with conspicuous hairy tufts; petioles 4–7 mm. long; stipules (at apical and second nodes) with a slender filiform to subulate lobe, 3–5 mm. long altogether, later developing a triangular rather corky base, eventually caducous, not conspicuously hairy within. Flowers borne on naked branches or with immature leaves. 4-merous, borne in 7–22-flowered pedunculate cymes; peduncles 2–10 mm. long in flowering material or 7–14 mm. long in fruiting material, with lines of pubescence; pedicels up to 3 mm. long, pubescent; bracteoles inconspicuous. Calyx-tube ± 1 mm. long; limb reduced to a rim. Corolla greenish yellow; tube 2–2.5 mm. long, pubescent inside; lobes 2.5 mm. long, 1.5–1.75 mm. wide, acute. Style twice as long as corolla-tube; stigmatic knob ± spherical, 0.5–0.75 mm. in diameter, ribbed. Fruit heart-shaped, 8–10 mm. long and wide, distinctly tapering towards base and well-indented at apex; pyrene oblong-ellipsoid, 9 mm. long, 5 mm. wide, with shallow irregular crest around apex, rugulose.

TANZANIA. Lushoto District: W. Usambara Mts., Shume-Magamba Forest Reserve, 17 Mar. 1984, *Iversen et al.* 84961!; Iringa District: Uzungwa Mts., Mwanihana Forest, Sanje, logger's camp, 30 Dec. 1980, *Rodgers* 391B!; & Nyumbenito, Dec. 1981, *Rodgers & Hall* 1946!
DISTR. T3, 6/7, 7; not known elsewhere
HAB. Forest or elfin forest; 1000–2300 m.

SYN. *Plectronia siebenlistii* K. Krause in E.J. 57: 35 (1920)

NOTE. K. Krause cited *Siebenlist* s.n. and *Grote in Herb. Amani* 3755 as syntypes of this species, but both were destroyed at B. A duplicate of the latter, preserved in EA, has been chosen as the lectotype, it is a rather poor specimen but clearly representative of the taxon. Bullock (cit. supra) cited *Holtz* 1846, which although collected in 1909 was not cited by Krause; a fragment of this fruiting material is preserved at K.
    Bullock makes the following statement "this is a small savannah tree which, like the two following species (i.e. *C. lactescens* and *C. crassum*), develops a thick bark which serves as a protection against fire." I have found no basis for this supposition. All recent gatherings are from forest and not subject to fire.
    See also note after *C. parasiebenlistii*.

6. **C. sp. B**

Tree up to 10 m. tall, glabrous; bark on young stems red-brown, lenticellate. Leaves with 1 pair per branch; blades broadly elliptic, 11.5–15.5 cm. long, 7–9 cm. wide, shortly acuminate at apex, acute and slightly attenuated at base, herbaceous, not discolorous (when dry), entirely glabrous; lateral nerves in 7–8 main pairs; tertiary nerves apparent but not conspicuous, not very finely reticulate; domatia present as tufts of hair but not eye-catching; stipules at apical node with a subulate lobe, 0.9–1.2 cm. altogether, later with base becoming rather corky, eventually caducous, slightly hairy within. Inflorescences 7–14-flowered pedunculate cymes; peduncles 0.8–2 cm. long, with lines of pubescence; pedicels 0.8–1.2 cm. long, puberulous; bracteoles inconspicuous. Flowers not known.

Fruit heart-shaped, ± 1 cm. long and wide, tapered towards base and indented at apex. Pyrene oblong-ellipsoid, 9 mm. long, 5 mm. wide, rugulose.

TANZANIA.   Morogoro District: Shikurufumi [Chigurufumi] Forest Reserve, Mar. 1955, *Semsei* 2032! DISTR.   T 6; known only from above specimen
HAB.   Forest; 500–800 m.

NOTE.   This poorly known taxon is close to *C. siebenlistii* but the leaves and stipules are much larger, the fruit slightly bigger and the domatia less conspicuous.

7. **C. parasiebenlistii** *Bridson*, sp. nov. affinis *C. siebenlistii* (K. Krause) Bullock sed habitu frutescenti vel arborescenti, 1–7 m. alta, ramunculis semper ferrugineis, foliis papyraceis differt. Typus: N. Zambia near Mbala [Abercorn], Sunzu Hill, *Robson* 507 (K, holo.!, BM, LISC, iso.!)

Shrub or small tree 1–5(–8–12) m. tall; lateral shoots well developed or short, with nodes close together; young stems glabrous, covered with purple-brown or red-brown lenticellate bark. Leaves in 1(–2) pairs at apex of main stem and lateral branches, usually immature and often sticky at time of flowering; blades broadly elliptic to round or sometimes oblong-elliptic, 5–12(–14.5) cm. long, 3–9 cm. wide, acuminate at apex, acute to obtuse then somewhat attenuate or sometimes rounded or unequal at base, glabrous or sometimes with a few hairs on the nerves above, glabrous or sparsely pubescent beneath, papery; lateral nerves in 5–7(–8) main pairs; tertiary nerves finely reticulate; domatia present as conspicuous tufts of hair; petioles (3–)5–10 mm. long, glabrous or sparsely pubescent; stipules triangular-acuminate, apical one with a filiform to subulate lobe, 3–12 mm. long altogether, with a few white hairs inside, eventually deciduous. Flowers 4–5-merous; inflorescences restricted to uppermost node or sometimes upper two nodes of mature stem, pedunculate, (1–)3–13-flowered cymes; peduncles 2–15 mm. long, glabrous or with lines of pubescence; pedicels 1–3 mm. long, sparsely pubescent to pubescent; bracteoles inconspicuous. Calyx-tube 1 mm. long, glabrous or sparsely puberulous; limb not exceeding disk. Corolla greenish yellow; tube 2 mm. long, pubescent at throat; lobes triangular-ovate, 1.75–2.5 mm. long, acute and thickened at apex. Style distinctly longer than corolla-tube; stigmatic knob ± 0.5 mm. long and wide, ribbed. Fruit broadly oblong to heart-shaped, 6–9 mm. long, 7–9 mm. wide, indented at apex, tapering to base; pedicel lengthening to 5–7 mm.; pyrene narrowly obovoid or oblong-ellipsoid, 5–8(–9) mm. long, 3.5–4(–5) mm. wide, with a shallow crest around apex, rugulose.

TANZANIA.   Kigoma District: Ititie, 26 Dec. 1963, *Azuma* 1028!; Iringa District: 20 km. W. of Mafinga on Madibira road by Ndembera R., Penny Penn's farm, 25 Nov. 1986, *Brummitt et al.* 18165! DISTR.   T 4, 7; Zaire, Malawi and Zambia
HAB.   River bank in evergreen fringe of *Brachystegia-Uapaca* woodland or riverine forest; 1675 m.

SYN.   *Canthium sp. 3* of F.F.N.R.: 404 (1962)

NOTE.   *C. parasiebenlistii* is very close to *C. siebenlistii* but the two have been kept apart because they occupy different phytochoria, *C. siebenlistii* in the forests along the eastern arc mountains of Tanzania, whereas *C. parasiebenlistii* occurs in riverine forest and woodland of the Zambezian region.
A broad view of *C. parasiebenlistii* has been taken because of the lack of correlated collections at different stages from any one geographical area and habitat. Future revision may be necessary when better data are available, however, some variation noted now is worth recording. The specimen from Kigoma differs in the leaves having oblong-elliptic blades, obtuse to rounded, sometimes unequal at base, with 7–8 main pairs of lateral nerves and shorter petioles, 3–5 mm. long. Most of the specimens from southern Malawi tend to have a longer stipule-lobe, 5–12 mm. long, and slightly bigger fruit than specimens from Zambia. The specimens seen from Zambia were all recorded as being collected on rocky places or near boulders; those from the north have reduced internodes while those from the south and Zaire (Shaba) have longer internodes.

subsp. **A**

Bark not obviously red-brown; domatia present but not conspicuous; petioles 7–12 mm. long.

TANZANIA.   Lindi District: Rondo Plateau, slope below St. Cyprian's College, 5 Feb. 1991, *Bidgood et al.* 1351! & 6 km. N. of Rondo Forest Station, 12 Feb. 1991, *Bidgood et al.* 1516! & *Bidgood et al.* 1524! DISTR.   T 8, known only from above specimens
HAB.   Regenerating coastal forest after cultivation with some patches of undisturbed forest, or regenerating thicket; 500–800 m.

NOTE.   The above specimens, all in fruit, most probably belong to the *C. siebenlistii/C. parasiebenlistii* complex. They differ from *C. siebenlistii* in the smaller habit (shrubs or small trees 3–5 m. tall), longer petioles and stipules, the less conspicuous domatia and in the fruit-shape (somewhat wider than long, 7.5–8 × 10 mm.). Within *C. parasiebenlistii* they come closest to the specimens from southern Malawi (noted above), as suggested by geographical evidence. However, they differ in the greyish bark and smaller domatia.

8. C. sp. C

Small tree; immature stems sparsely pubescent, older stems covered with rusty-red lenticellate bark. Leaves restricted to new growth, immature at time of flowering; immature blades elliptic, up to 4 cm. long, up to 2 cm. wide, acuminate at apex, ± acute at base, glabrous above, densely pubescent on nerves beneath; lateral nerves in ± 5 main pairs; tertiary nerves apparent, pubescent; domatia not apparent; petiole up to 4 mm. long, glabrescent to pubescent; mature stipules with a triangular bark-like base, ± 4 mm. long, bearing a linear subulate lobe 4 mm. long, pubescent within. Flowers 5-merous, borne in 5–15-flowered pedunculate cymes; peduncles 0.6–1.4 cm. long, puberulous; pedicels 1–2 mm. long, puberulous; bracts and bracteoles inconspicuous. Calyx-tube 1 mm. long, puberulous; limb reduced to a rim shorter than disk. Corolla-tube 2 mm. long, pubescent at throat; lobes oblong, 2.5 mm. long, 1 mm. wide, acute. Anthers spreading. Style distinctly longer than corolla-tube; stigmatic knob ± 0.75 mm. long. Fruit not known.

TANZANIA. Mbeya District: Izumbwe, 11 Nov. 1976, *Leedal* 3965!
DISTR. T 7; known only from above specimen
HAB. Not recorded; 1525 m.

NOTE. This poorly known taxon is possibly related to *C. parasiebenlistii*, but the dense indumentum on the leaf-nerves beneath gives it a characteristic appearance.

9. C. lactescens *Hiern*, Cat. Afr. Pl. Welw. 1: 511 (1898); Bullock in K.B. 1932: 379 (1932); T.T.C.L: 489 (1949); F.P.S. 2: 430 (1952); I.T.U., ed. 2: 339 (1952); K.T.S.: 429 (1961); F.F.N.R.: 404 (1962); Fl. Pl. Lign. Rwanda: 548, fig. 184.1 (1983); Fl. Rwanda 3: 148, fig. 43.1 (1985). Type: Angola, Huila, *Welwitsch* 3157 (LISU, holo., BM, COI, K, iso.!)

Tree 2.5–9 m. tall; stems thick, often with short internodes, covered with dark grey to reddish bark, glabrous or occasionally pubescent; lenticels usually apparent. Leaves restricted to new growth at apex of main shoots and short lateral branches, usually only one pair on each branch; blades broadly elliptic to round, 7–18 cm. long, 5–13 cm. wide, acute to subacuminate at apex, acute to obtuse or truncate at base, glabrous or occasionally pubescent beneath and sparsely pubescent above; lateral nerves in 8–10 main pairs, acutely angled, often turning black when dry; tertiary nerves apparent but not conspicuous; domatia usually present, but not markedly conspicuous; petioles 0.5–2 cm. long; stipules broadly triangular to ovate, often long-acuminate when young, 0.6–1.2 cm. long, resembling bark when mature, not conspicuously hairy within. Flowers (4–)5-merous, restricted to leafless nodes of previous season's growth, borne in 20–50-flowered pedunculate cymes, branched 3–5 times, but ultimate segments with flowers arranged on one side; peduncle 1–3.5 cm. long, pubescent on either side; pedicels 0.5–3 mm. long; bracteoles inconspicuous. Calyx-tube 1.5 mm. long, glabrous or sometimes sparsely pubescent; limb reduced to a rim. Corolla cream to yellowish; tube 2–2.5 mm. long, finely pubescent at throat, rather wider than calyx-tube; lobes triangular-ovate, 1.5–3 mm. long, 1–2 mm. wide at base, acute, thickened towards apex and sometimes appendaged inside, spreading or reflexed. Style shortly exceeding corolla-tube; stigmatic knob ± hemispherical, strongly ribbed below, ± 1 mm. wide. Fruit yellowish, edible, ± square in outline, straight sided or tapering at base, strongly bilobed and indented at apex, 0.7–1.2 cm. long, 0.8–1.2 cm. wide; pyrene ellipsoid to ovoid, 0.8–1 cm. long, slightly rugulose. Figs. 131/1, 13, p. 752 & 155, p. 872.

UGANDA. Karamoja District: Mt. Moroto, Lodoketeminit, 8 Apr. 1959, *Kerfoot* 928!; Ankole District: Ruampara, near Gayaza, Oct. 1932, *Eggeling* 626!; Mengo District: 13 km. W. of Lwampanga, 9 July 1956, *Langdale-Brown* 2183!
KENYA. Northern Frontier Province: Leroki Plateau, *Rammell* in F.D. 3321!; Uasin Gishu District: Ol Dane Sapuk–Kapsaret [Kaposoret], 29 May 1951, *G.R. Williams* 211!; Masai District: 17.5 km. on Narok–Nairobi road, Siyabei R., 16 July 1962, *Glover & Samuel* 3218!
TANZANIA. Biharamulo District: Nyabugombe, 28 Nov. 1966, *Tanner* 5580A!; Dodoma District: Manyoni, 11 km. E. of Itigi Station, 8 Apr. 1964, *Greenway & Polhill* 11456!; Iringa District: Mufindi, Sao Hill, Feb. 1966, *Procter* 3105!
DISTR. U 1, 2, 4; K 1–4, 6; T 1, 2, 4, 5, 7; Zaire, Rwanda, Burundi, Sudan, Ethiopia, Zambia, Zimbabwe and Angola

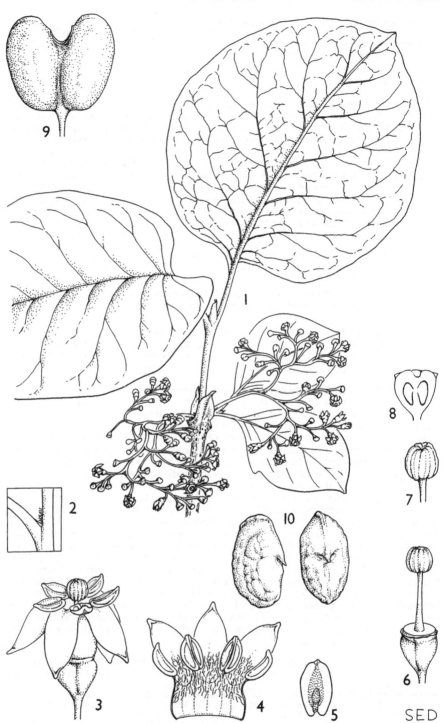

FIG. 155. *CANTHIUM LACTESCENS* — **1**, flowering branch, × ⅔; **2**, portion of leaf showing domatium, × 3; **3**, flower, × 6; **4**, part of corolla opened out, × 6; **5**, dorsal view of anther, × 8; **6**, calyx with style and stigmatic knob, × 6; **7**, stigmatic knob, × 8; **8**, section through ovary, × 6; **9**, fruit, × 3; **10**, pyrene (2 views), × 3. 1–8, from *Gereau & Lovett* 2784; 9, 10, from *Tanner* 5900. Drawn by Sally Dawson.

Hab. On rocky hillsides or in thickets or riverine woodlands; 1000–2300 m.

Syn. [*C. crassum* sensu Hiern in F.T.A. 3: 145 (1877), pro parte, quoad *Schweinfurth* 1695, non Hiern
sensu stricto]
C. umbrosum Hiern, Cat. Afr. Pl. Welw. 1: 479 (1898). Type: Angola, Huila, near Lopollo,
Welwitsch 2576 (LISU, holo., BM, COI, K, iso.!)
Plectronia lactescens (Hiern) K. Schum. in Just's Bot. Jahresb. 1898 (1): 393 (1900)
P. umbrosa (Hiern) K. Schum. in Just's Bot. Jahresb. 1898 (1): 393 (1900)
P. psychotrioides De Wild. in Ann. Mus. Congo., Bot. sér. 4, 1: 228 (1903). Types: Zaire, Shaba,
Lukafu, *Verdick* 312 & 338 (BR, syn.!)
Canthium lactescens Hiern var. *grandifolium* S. Moore in J.L.S. 37: 161 (1905). Type: Uganda,
Ankole District, near Mulema, *Bagshawe* 329 (BM, holo.!)
C. randii S. Moore in J.B. 49: 152 (1911). Type: Zimbabwe, near Harare [Salisbury], *Rand* 1393
(BM, holo.!)
Plectronia randii (S. Moore) Eyles in Trans. Roy. Soc. S. Afr. 5: 493 (1916)

10. **C. peteri** *Bridson*, sp. nov. affinis *C. kilifiensis* Bridson sed per anthesin foliis maturis,
basi acutis vel leviter attenuatis; venis tertiariis obscuris differt. Typus: Tanzania,
Uzaramo District, Msua to Bagala [Bagalla], *Peter* 56193 (K, holo.!, B, iso.)

Erect or scandent shrub, 1–4 m. tall, glabrous; young and older stems with pale grey
shiny bark which flakes off in tissue-like fragments. Leaves paired at apex of main and
short lateral branches or occasionally with two pairs of leaves on main branch, mature, or
almost so, at time of flowering; blades broadly elliptic to round or sometimes elliptic,
4.5–8.5 cm. long, 2.5–5 cm. wide, rounded, obtuse or acute at apex, acute or somewhat
attenuated at base, papery, glabrous; lateral nerves in (4–)5 main pairs; tertiary nerves
obscure; domatia absent; petiole 5–8 mm. long; stipules triangular, long-acuminate,
(2–)3–5 mm. long, with a few hairs inside. Flowers 5-merous, borne in 3–17-flowered
pedunculate cymes, mostly restricted to the top 1–3 nodes of previous season's growth;
peduncle 2–5 mm. long, with pubescent lines; pedicels up to 2.5 mm. long; bracteoles
inconspicuous. Calyx-tube ± 1 mm. long; limb reduced to a repand or shortly dentate rim.
Corolla yellow; tube up to 2 mm. long, finely pubescent at throat; lobes triangular-ovate, 2
mm. long, 1.5 mm. wide at base, acute. Style slightly longer than corolla-tube; stigmatic
knob ± spherical, ± 0.75 mm. in diameter, ridged. Fruit ± heart-shaped, 1.1–1.3 cm. long,
1–1.1 cm. wide at top; pyrene oblong-ellipsoid, 1.1 cm. long, 5 mm. wide, rugulose.

Kenya. Tana River District: Tana River National Primate Reserve, 2 km. SE. of Main Gate, 19 Mar.
1990, *Luke et al.* TPR 659!
Tanzania. Pangani District: Karoti, Mkwaja, 19 June 1956, *Tanner* 2939!; Uzaramo District: Msua–
Bagalla, 5 Nov. 1925, *Peter* 31896!; Kilwa District: Selous Game Reserve, Kingupira Forest, 18 Dec.
1975, *Vollesen* in *M.R.C.* 3106!
Distr. **K** 7; **T** 3, 6, 8; not known elsewhere
Hab. Forest,· or riverine thicket; 30–450 m.

Syn. *C. sp. aff. keniense* Bullock sensu Vollesen in Opera Bot. 59: 68 (1980)

11. **C. shabanii** *Bridson*, sp. nov. affinis *C. siebenlistii* (K. Krause) Bullock sed venis
tertiariis foliorum grossioribus, fructibus latioribus ad basim vix decrescentibus differt.
Typus: Tanzania, Lushoto District, Muheto–Mazinde, *Shabani* 342 (K, holo.!, EA, SRGH!,
iso.)

Shrub or small tree 1.3–8 m. tall, glabrous; green twigs with prominent lenticels; young
branches covered with fawn to light brown bark. Leaves usually restricted to current
season's growth, up to 4 pairs per branch; blades elliptic to broadly elliptic, 6–11 cm. long,
3.5–7 cm. wide, acute to subacuminate at apex, acute or obtuse to rounded at base, papery
to subcoriaceous, lateral nerves in 4–5(–6) main pairs; tertiary nerves rather coarsely
reticulate, apparent or occasionally conspicuous; domatia with conspicuous tufts of hair;
petioles 3–5 mm. long, often drying black together with midrib; stipules 3–6 mm. long
altogether, triangular, acuminate, ± linear on apical nodes, caducous, sparsely pubescent
within. Inflorescence 3–12-flowered pedunculate cymes; peduncles 3–12 mm. long, with
lines of pubescence; pedicels 1–2 mm. long; bracteoles inconspicuous. Calyx-tube 1 mm.
long; limb a repand rim ± equal to disk. Corolla-bud 3 mm. long, obtuse; tube pubescent
within. Fruit square to broadly oblong in outline, scarcely tapering towards base, strongly
indented at apex, 0.9–1.1 cm. long, 1.1–1.5 cm. wide; pedicels reaching 3–6 mm. long;
pyrene obovoid-oblong, 1.1 cm. long, 6 mm. wide, rugulose.

Tanzania. Lushoto District: W. Usambara Mts., Vuruni [Veruni] valley, *Koritschoner* 497! &

Mazumbai [Mazumbei]–Baga, 13 Apr. 1916, *Peter* 55882! & Muheto–Mazinde footpath, 18 Mar. 1969, *Shabani* 342!

Distr.   **T** 3; not known elsewhere
Hab.   Forest; ± 1320–2020 m.

Note.   The data from *Koritschoner* 497 states that the roots and fruits are used as a poison, but this may not be reliable.

12. **C. kilifiense** *Bridson*, sp. nov. affinis *C. keniensis* Bullock sed petiolis 4–10 mm. longis, domatiis nullis, tubo corollae ± 3 mm. longo, fructu leviter longiore quam latiore, ad basim leviter decrescenti differt. Typus: Kilifi, Roka, *Dale* 3840 (K, holo.!, EA, iso.)

Shrub or small tree up to 7 m. tall, glabrous; young woody stems with fawn shiny bark which flakes off in tissue-like fragments; older stems with light grey bark. Leaves immature at time of flowering, paired at apex of main and lateral branches or with two pairs of leaves on main branch; blades broadly elliptic or occasionally elliptic, 6.3–10.5 cm. long, 3–7 cm. wide, obtuse to acute or occasionally rounded at apex, obtuse to rounded or occasionally acute at base, subcoriaceous; lateral nerves in 5–6(–7) main pairs; tertiary nerves rather coarsely reticulate; domatia absent; petiole 4–10 mm. long; stipules with a rather thick subulate arista present at apical node, 4–9 mm. long, later with a triangular base which becomes corky and soon breaks down, slightly but not conspicuously hairy inside. Flowers 5-merous, restricted to upper 1–3 nodes of previous season's growth, borne in pedunculate 1–13-flowered cymes; peduncle 3–6 mm. long, glabrous; inflorescence-branches usually 2, with the flowers borne to one side, with lines of pubescence on upper side; pedicels up to 1.5 mm. long; bracteoles inconspicuous. Calyx-tube 1.25 mm. long; limb reduced to a truncate limb. Corolla (known from well-developed bud only); tube ± 3 mm. long, densely covered with fine hairs at throat; lobes ± 2 mm. long. Stigmatic knob ± spherical, 1 mm. in diameter, ridged. Fruit ± oblong-square in outline, very slightly longer than wide and only slightly narrowed towards base, indented at apex, 0.9–1.2 cm. long and wide; pyrene oblong-ellipsoid, 1 cm. long, 0.5 cm. wide, rugulose. Figs. 131/5, p. 752 & 132/19, p. 754.

Kenya.   Kilifi District: Arabuko, *R.M. Graham* in *F.D.* 1711! & Arabuko-Sokoke Forest, N. of Sokoke Forest Station, 9 June 1973, *Musyoki & Hansen* 1012! & Malindi, May 1960, *Rawlins* 856!
Distr.   **K** 7; not known elsewhere
Hab.   Dry lowland forest or *Brachystegia* woodland; 30–130 m.
Syn.   [*C. keniense* sensu Bullock in K.B. 1932: 377 (1932); K.T.S.:429 (1961), pro parte, quoad *R.M. Graham* in *F.D.* 1711, *non* Bullock]
    [*C. siebenlistii* sensu K.T.S.: 432 (1961), *non* (K. Krause) Bullock]

13. **C. keniense** *Bullock* in K.B. 1932: 377 (1932); K.T.S.: 429 (1961), pro parte excl. *R.M. Graham* in *F.D.* 1711. Type: Kenya, Nairobi–Kikuyu, *Battiscombe* in *F.D.* 872 (K, holo.!)

Shrub or tree 2.5–8.5 m. tall, glabrous; young stems with fawn to greyish bark; lenticels obscure. Leaves restricted to new growth at apex of stems and lateral shoots; blades elliptic, 6–14 cm. long, 2.8–6.5 cm. wide, acute to acuminate at apex, acute to obtuse at base; midrib and 5–6 main pairs of lateral nerves frequently turning black when dry; tertiary nerves apparent, rather coarsely reticulate; domatia present as conspicuous tufts of whitish hair; petiole 3–5 mm. long; stipules triangular, acuminate or somewhat lobed at apex, 0.5–1 cm. long, resembling bark when mature, without conspicuous hairs inside. Flowers 5-merous, restricted to top 1–4 nodes of woody stem, borne in pedunculate 5–20-flowered cymes, arranged to one side on each of the two main branches, or ± subumbellate in few-flowered inflorescences; peduncle 0.3–1.5 cm. long, usually with lines of pubescence; pedicels 0.5–1 mm. long; bracteoles inconspicuous. Calyx-tube ± 1.5 mm. long; limb reduced to a truncate rim ± 0.5 mm. long, sometimes obscurely dentate. Corolla yellowish green; tube 1.75–2 mm. long, finely pubescent at throat; lobes ovate, 1.75–2 mm. long, 1.5 mm. wide at base, acute and thickened at apex, spreading or reflexed. Style slightly longer than corolla-tube; stigmatic knob ± hemispherical, ribbed below with stigmatic surface clearly apparent on upper face. Fruit heart-shaped, 1.3–1.4 cm. long and wide; pedicels thickening and lengthening to 3–4 mm. long; pyrene narrowly obovoid, 1.4 cm. long, 6.5 mm. wide, coarsely rugulose.

Kenya.   Northern Frontier Province: Ndoto Mts., 9 Dec. 1960, *Kerfoot* 2454!; Nairobi District: French Mission, 3 Dec. 1932, *Napier* 2358!; Masai District: Ngong, *V.G.L. van Someren* in *C.M.* 9196!
Distr.   **K** 1, 4, 6; not known elsewhere
Hab.   In forest clumps; 1675–1800 m.

14. **C. racemulosum** *S. Moore* in J.L.S. 40: 87 (1911). Type: Mozambique, Mandanda Forest, *Swynnerton* 541 (BM, holo.!, K, iso.!)

Shrub or small tree, 1–7 m. tall; green twigs glabrous or densely pubescent; bark flaking in small scales, mid- to dark grey. Leaves paired at apex of main stem and short lateral branches, immature at time of flowering; blades elliptic to broadly elliptic or less often round, 3.5–10 cm. long, 1–11.5 cm. wide, obtuse to acute or sometimes subacuminate at apex, acute to cuneate or sometimes obtuse at base, chartaceous, entirely glabrous or pubescent above and tomentose beneath; lateral nerves in 6–7 main pairs; tertiary nerves moderately coarsely reticulate, often discolorous; domatia inconspicuous; petioles 2–5 mm. long, glabrous to tomentose; stipules at apical nodes of green shoots with well-developed linear lobe, 0.8–1.2 cm. long, at woody nodes ovate-triangular, 0.5–1.1 cm. long, glabrous or pubescent when young, becoming corky and resembling bark when mature, with tufts of white hair inside, eventually deciduous. Flowers 5-merous, usually restricted to leafless terminal node of previous season, borne in 4–30-flowered pedunculate 2-branched cymes, with flowers arranged on one side; peduncle 0.4–1.5 cm. long, with lines of pubescence or densely pubescent; pedicels 0.5–2 mm. long, glabrous or pubescent; bracteoles not apparent. Calyx-tube 1–2 mm. long, glabrous or densely covered in spreading hairs; limb reduced to a repand rim, equalling or exceeding disk. Corolla pale green or yellow, glabrous or sometimes pubescent outside; tube 2–2.5 mm. long, densely pubescent at throat; lobes triangular-ovate, 2–2.75 mm. long, 1.5–1.75 mm. wide at base, acute. Style distinctly longer than corolla-tube; stigmatic knob 0.75 mm. long, 0.5 mm. wide, grooved. Fruit ± heart-shaped, 1.2 cm. long, 1.1–1.3 cm. wide across the top; pedicels lengthening to 4–5 mm.; pyrenes narrowly oblong-obovoid, 1.1–1.2 cm. long, 5–6 mm. wide, rugulose.

var. **racemulosum**

Leaf-blades glabrous to glabrescent above, glabrous to sparsely pubescent and with midrib pubescent beneath; calyx-tube glabrous or sometimes pubescent; corolla always glabrous outside.

TANZANIA. Lindi District: Namanga, 3 Feb. 1949, *Anderson* 303! & Nachingwea, 31 Jan. 1954, *Anderson* 961! & 62 km. N. of Lindi, 4–6 km. SE. of Ruchungi, 31 Nov. 1977, *Wingfield* 4390!
DISTR. **T** 8; Mozambique, Malawi and Zimbabwe
HAB. Woodland; 40–575 m.

SYN. [*C. siebenlistii* sensu Vollesen in Opera Bot. 59: 67 (1980), *non* (K. Krause) Bullock]

var. **nanguanum** (*Tennant*) *Bridson*, comb. nov. Type: Tanzania, Masasi District, near Nangua on Tunduru–Masasi road, *Eggeling* 6392 (K, holo.!, EA, iso.)

Leaf-blades pubescent to densely pubescent above, densely pubescent to tomentose beneath; calyx-tube tomentose; corolla glabrous or hairy outside.

TANZANIA. Kilwa District: Selous Game Reserve, ± 19 km. SSW. of Kingupira, 14 Dec. 1976, *Vollesen* in M.R.C. 4226!; Masasi, 12 Dec. 1942, *Gillman* 1205!; Lindi District: N. of Nakilala Thicket, 14 Dec. 1975, *Vollesen* in M.R.C. 3080!
DISTR. **T** 8; Mozambique
HAB. Woodland on rocky ground or rock outcrops; 175–350 m.

SYN. *C. burttii* Bullock var. *nanguanum* Tennant in K.B. 22: 440 (1968)
[*C. burttii* sensu Vollesen in Opera Bot. 59: 67 (1980), *non* Bullock]

15. **C. vollesenii** *Bridson*, sp. nov. affinis *C. racemulosi* S. Moore sed foliis late ellipticis vel rotundatis, basi truncatis vel cordatis vel interdum obtusis, cortice incano differt; affinis *C. burttii* Bullock sed petiolis longioribus, cortice incano differt. Typus: Tanzania, Kilwa District, Kingupira Forest, *Vollesen* in M.R.C. 4825 (K, holo.!, C!, DSM, iso.)

Small tree or shrub 4–6 m. high; twigs covered with whitish grey bark; green shoots glabrous or pubescent. Leaves paired at apex of main stem and reduced lateral branches, almost mature at time of flowering; blades broadly elliptic to round, 9.5–13 cm. long, 4.5–10.5 cm. wide, acute to shortly acuminate at apex, obtuse or truncate to cordate at base, papery to chartaceous, glabrous or occasionally pubescent on both faces; lateral nerves in 5–7 main pairs, acutely angled; tertiary nerves apparent; domatia present as small tufts of hair; petiole 1–1.8 cm. long; mature stipules ovate-triangular, 0.5–1 cm. long, at first thinner at margin but becoming similar to bark, caducous. Flowers 5-merous, restricted to leafless terminal node of previous season, borne in (2–)11–17-flowered pedunculate 2-branched cymes, the flowers borne to one side; peduncle 0.3–1.2 cm. long, puberulous in two lines; pedicels 1–5 mm. long; bracteoles not apparent. Calyx-tube 1 mm. long; limb reduced to a rim. Corolla yellowish green; tube 3 mm. long, finely pubescent at throat; lobes triangular-ovate, 2.5 mm. long, 2 mm. wide, acute. Style slightly exceeding the corolla-tube; stigmatic knob resembling a match-head, grooved. Immature fruit ± spherical, 8 mm. in diameter, covered in pale thin corky scales.

TANZANIA.   Ulanga District: Selous Game Reserve, ± 20 km. NW. of Kingupira, 19 Dec. 1976, *Vollesen* in *M.R.C.* 4240!; Kilwa District: Selous Game Reserve, Kingupira Forest, 30 Dec. 1975, *Vollesen* in *M.R.C.* 3169! & 11 Dec. 1977, in *M.R.C.* 4825!
DISTR.   T 6, 8; not known elsewhere
HAB.   Ground-water forest or thicket; ± 100 m.

SYN.   *C. sp. aff. lactescens* Hiern sensu Vollesen in Opera Bot. 59: 68 (1980)

16. **C. burttii** *Bullock* in K.B. 1933: 146(1933); T.T.C.L.: 488 (1949). Type: Tanzania, Dodoma District, Manyoni Kopje, *B. D. Burtt* 3542 (K, holo.!, EA, iso.)

Small tree or shrub 1.5–6.5 m. tall; green twigs glabrous or densely covered with spreading hairs; older branches covered with reddish to dark grey, smooth bark; lenticels apparent but not conspicuous. Leaves paired at apex of main stem and short lateral branches, not fully mature at time of flowering; blades broadly elliptic to round, 5–12 cm. long, 2.8–10.5 cm. wide, acute to shortly acuminate at apex, cuneate to obtuse or occasionally truncate at base, papery, pubescent to velutinous on both faces or sometimes glabrous; lateral nerves in (4–)5–6(–7) main pairs, acutely angled; tertiary nerves finely reticulate; domatia obscure; petioles short, 1–3 mm. long; stipules ovate, 0.5–1.1 cm. long, acuminate, central area pubescent or sometimes glabrous, membranous and shiny at margin, occasionally hyaline at very edge, readily caducous. Flowers 5-merous, restricted to leafless terminal node of previous season, borne in 3–11(–15)-flowered pedunculate cymes, either 2-branched with flowers borne on one side or ± subumbellate; peduncle 0.4–2 cm. long, pubescent or with 2 pubescent lines; pedicels 1–5 mm. long; bracteoles not apparent. Calyx-tube 1–1.25 mm. long, glabrous or densely pubescent; limb reduced to a rim. Corolla greenish yellow, glabrous or with very few hairs outside; tube 2–3 mm. long, pubescent inside; lobes oblong-ovate, 1.75–2.25 mm. long, 1.25 mm. wide, acute. Style equalling or shortly exceeding corolla-tube, ± resembling a match-head, ± 1 mm. long and wide, grooved. Fruit yellow or sometimes orange, broadly oblong-ovoid in outline, 6–8 mm. long and wide, scarcely bilobed, not indented at apex, glabrous to glabrescent; pyrene obovoid, flattened on ventral face, 7 mm. long, 4 mm. wide, rugulose.

subsp. **burttii**

Plants pubescent.

TANZANIA.   Mwanza District: Mbarika, 11 Apr. 1953, *Tanner* 1366!; Tabora, near Bee Keeping Institute, 20 Dec. 1977, *Shabani* 1224!; Iringa District: Kimiramatonge [Kerimatonge] Hill, NW. side of mile 13.3 from Msembi, 15 Dec. 1970, *Greenway & Kanuri* 14800!
DISTR.   T 1,2,4,5,7;? Zambia
HAB.   In thickets on rocky hillsides; 765–1525(–1980) m.

NOTE.   The only specimen seen from Zambia is sterile and somewhat less pubescent than typical.

subsp. **glabrum** *Bridson,* subsp. nov. a subsp. *burttii* planta glabra differt. Typus: Zambia, Pemba, *Trapnell* 1304 (K, holo.!)

Plants glabrous.

TANZANIA.   Ufipa District: Sumbawanga, Kasanga Escarpment, 24 Nov. 1959, *Richards* 11824!
DISTR.   T 4; Zaire (Shaba) and Zambia
HAB.   Roadside in dense vegetation; 780m.

SYN.   [*C. burttii* sensu F.F.N.R.: 404 (1962), pro parte, quoad *Fanshawe* 1815, *non* Bullock]

Subgen. 3. **Lycioserissa** (*Roem. & Schultes*) *Bridson,* comb. et stat. nov. Type of subgenus: *Lycium inerme* L. f. (*Lycioserissa capensis,* nom. illegit.)
                     Lycioserissa Roem. & Schultes, Syst. Veg. 4: 353 (1819)
          [*Plectronia* sensu Sond. in Harv. & Sond., Fl. Cap. 3: 17 (1865) et auctt. div., *non* L.]

Shrubs or trees, sometimes with spines on trunks, young coppice shoots and saplings; spines paired or often ternate, supraxillary but not associated with brachyblasts; lenticels not markedly apparent on young stems. Leaves tending to be confined towards apex of branches but not strictly so, occasionally subtending inflorescences, mature or immature at time of flowering, paired but often ternate on young shoots, heterophylly sometimes present (leaves on young and coppice shoots smaller and broader than main leaves, those on saplings very much narrower), petiolate, papery to subcoriaceous, glabrous, somewhat discolorous; tertiary nerves obscure or sometimes apparent near margin; domatia pit-like, ciliate or densely pubescent; stipules sheathing at base, bearing a linear lobe, occasionally becoming corky with age, pubescent within. Flowers ♂, 5-merous, pedicellate, borne in pedunculate or sessile (1–)3–40-flowered cymes; inflorescence-branches present or reduced; bracteoles inconspicuous. Calyx glabrous; tube broadly ellipsoid to ovoid; limb with tube ± obsolete, lobed almost to base; lobes small, triangular to linear, often rather unequal. Corolla white or yellow-green, glabrous outside; tube broadly cylindrical, subequal to lobes or a little longer, glabrous to sparsely hairy inside; lobes reflexed, shortly apiculate or not, often puberulous at margins. Stamens set at corolla-throat, erect; anthers oblong-ovate, attached shortly above the base, dorsal face covered with darkened connective, except for marginal band. Style slender, shortly exceeding the corolla-tube, glabrous; stigmatic knob ± spherical, point of attachment recessed ± ⅓ of the way from the base, 2–3-lobed; ovary 2–3-locular. Fruit a 2–3-seeded drupe, moderately sized, in 2-seeded species dorsiventrally flattened, bilobed and indented at apex or not, or 3-lobed in 3-seeded species: pyrene ellipsoid, flattened on ventral face, thinly woody, truncate at point of attachment, with a slight or well-developed crest extending from point of attachment to apex, usually rugulose. Seeds narrowly obovoid with ventral face flattened, very shortly crested at apex; embryo curved.

A subgenus of three species restricted to eastern Africa and South Africa extending as far west as eastern Zaire. Species 17.

17. **C. oligocarpum** *Hiern* in F.T.A. 3:138 (1877); E.P.A.:1009 (1965); Fl. Pl. Lign. Rwanda: 558, fig. 183.2 (1982); Fl. Rwanda 3:148, fig. 43.2 (1985). Type: Ethiopia, Begemder, *Schimper* 1127 (K, holo.!, BR, E, iso.!)

Shrubs or trees 1.5–20 m. tall, glabrous; trunks, coppice shoots and young branches usually armed with ternate or less often paired spines; bark greyish. Leaf-blades elliptic, oblong-elliptic, occasionally broadly elliptic or narrowly elliptic (not in Flora area), 3–14.5 cm. long, 1.5–6 cm. wide, acuminate or less often obtuse to acute at apex, rounded or obtuse to acute at base, leaves on coppice or young shoots smaller and rounded, leaves on saplings very much narrower than mature foliage, papery to subcoriaceous, strongly discolorous (when dry); tertiary nerves obscure; domatia pit-like, ciliate to densely hairy; petiole 0.2–1.5 cm. long; stipules sheathing at base, 2–3 mm. long, bearing a linear lobe, 1–3 mm. long, sometimes rather corky when old, pubescent inside. Flowers 5-merous, borne in pedunculate 3–25(–40)-flowered cymes; peduncles 0.4–1.5 cm. long, glabrous or sparsely and unequally pubescent; inflorescence-branches sometimes reduced; pedicels 0.5–4(–7) mm. long, glabrous or sparsely pubescent; limb-tube ± obsolete; lobes triangular or linear, unequal, up to 1.5 mm. long. Corolla white or yellow-green, acute in bud; tube 2–4.25 mm. long, 1–2 mm. wide, glabrous to sparsely hairy inside; lobes narrowly triangular to triangular or oblong-ovate, 1.5–4 mm. long, 1.5–2 mm. wide at base, shortly apiculate, puberulous at margins outside. Style shortly exceeding tube; stigmatic knob as broad as long, ± 1 mm. across. Fruit slightly longer than wide or wider than long, 1.3–2.5 cm. long, 1.4–2 cm. wide, bilobed and indented at apex; pyrene ellipsoid, flattened on ventral face, scarcely or distinctly curved, 1.2–1.4 cm. long, 7–8 mm. wide, truncate at point of attachment with crest running from point of attachment to apex, sometimes not well developed, somewhat rugulose.

KEY TO INFRASPECIFIC VARIANTS

Leaves obtuse to acute or rarely subacuminate at apex, small
    (2.5–5 × 1.3–2.8 cm.); domatia conspicuous densely hairy
    tufts (rarely with the pit perceptible); fruit somewhat wider
    than long or with dimensions ± equal   . . . . . .   d. subsp. **friesiorum**
Leaves subacuminate or acuminate at apex, small or large;
    domatia with small or large, ciliate to pubescent pits or
    rarely with pit obscured by hairs; fruit longer than wide or
    occasionally ± as wide as long:
 Leaves with 3–5(–6) main pairs of lateral nerves, papery to
    stiffly papery, usually dull above:

Young stems usually green (or dark when dry); domatia
   ciliate to pubescent, usually less densely than in
   subsp. *intermedium*; fruit longer than wide;
   inflorescences (2–)7–23(–40)-flowered   . . . .   a. subsp. **oligocarpum**
Young stems usually with all but apical 2–3 nodes covered
   with pale or greyish bark; domatia pubescent,
   occasionally with pits obscured; fruit ± as long as
   wide; inflorescences 3–12-flowered   . . . . .   c. subsp. **intermedium**
Leaves with (5–)6–7 main pairs of lateral nerves, stiffly
   papery to subcoriaceous, moderately shiny
   above   . . . .   b. subsp. **captum**

a. subsp. **oligocarpum**

Bark on young twigs dark (when dry). Leaf-blades large or small, distinctly acuminate at apex, papery, dull above, with 3–5 main pairs of lateral nerves; domatia ciliate to moderately pubescent. Inflorescences (2–)7–23(–40)-flowered. Fruit somewhat longer than wide, 1.3–2.4 cm. long, 1.5–2 cm. wide. Figs. 131/8, 14, p. 752; 132/20, p. 754 & 156.

UGANDA. Karamoja District: Mt. Morongole, Apr. 1959, *J. Wilson* 723!; Toro District: Ruwenzori Mts., Bwamba Pass, July 1940, *Eggeling* 3989!; Mbale District: Elgon, Dec. 1939, *Dale* 88!
KENYA. Kericho District: SW. Mau Escarpment, Timbilil, Nov. 1962, *Kerfoot* 4551!
TANZANIA. Mpanda District: Kungwe Mt., Kasala Peak, 6 Sept. 1959, *Harley* 9530! & head of Ntali R., 7 Sept. 1959, *Harley* 9538!; Ufipa District: Mbizi Forest, Nov. 1958, *Napper* 1041!
DISTR. U 1–3; K 5; T 4; Zaire, Rwanda, Burundi, Sudan and Ethiopia
HAB. Forest; 1800–2600 m.

SYN. *C. ruwenzoriense* Bullock in K.B. 1932: 376 (1932); F.P.N.A. 2: 352 (1947); E.P.A.: 1010 (1965). Type: Uganda/Zaire, W. Ruwenzori Mts., Butahu [Butagu] Valley, *Mildbraed* 2692 (B, holo. †, K, fragment!)
   [*C. captum* sensu Robyns, F.P.N.A. 2: 352 (1947), *non* Bullock]
   C. sidamense Cufod. in Senck. Biol. 46: 100, t. 10/6 (1965) & E.P.A.: 1011 (1965). Type: Ethiopia, Sidamo, Maleko, near Adola, *Haberland* 2047 (FR, holo.!)

b. subsp. **captum** (*Bullock*) *Bridson*, comb. nov. Type: Tanzania, Rungwe District, Kyimbila, *Stolz* 2293 (K, holo.!)

Bark on young twigs dark (when dry). Leaf-blades large or small, distinctly acuminate at apex, chartaceous to subcoriaceous, moderately shiny above, with (5–)6–7 main pairs of lateral nerves, frequently impressed above; domatia ciliate to sparsely pubescent. Inflorescences 3–10(–30)-flowered. Fruit somewhat longer than wide, 1.4–1.7 cm. long, 1.2–1.4 cm. wide.

TANZANIA. Lushoto District: W. Usambara Mts., ridge N. of Matondwe Hill at head of Kwai valley, 12 June 1953, *Drummond & Hemsley* 2906!; Morogoro District: S. Uluguru Forest Reserve, Mar. 1955, *Semsei* 2072!; Iringa District; Mufindi, Livalonge Tea Estate, 27 Aug. 1977, *Perdue & Kibuwa* 11266!
DISTR. T 3, 6, 7; Malawi and Mozambique
HAB. Forest; 1350–2000 m.

SYN. *Canthium captum* Bullock in K.B. 1932: 376 (1932); T.T.C.L.: 488 (1949); Brenan in Mem. N.Y. Bot. Gard. 8: 452 (1954)

c. subsp. **intermedium** *Bridson*, subsp. nov. a subsp. *friesiorum* (Robyns) Bridson foliis subacuminatis vel acuminatis, domatiis minus pubescentibus et a subsp. *capto* (Bullock) Bridson cortice hinnuleo vel cineraceo ramulorum juniorum obtecto vix nitenti, longitudine fructuum latitudinem ± aequanti differt. Typus: Tanzania, Pare District, Tona, *Peter* 55763 (K, holo.!, B, BR, iso.!)

Bark on young twigs fawn to light grey. Leaf-blades 3.5–9.5 cm. long, 1.8–4.7 cm. wide, subacuminate to acuminate at apex, papery to chartaceous, usually dull above, with 4–5(–6) main pairs of lateral nerves; domatia pubescent, occasionally with pits obscured. Inflorescences 3–12-flowered. Fruit ± as wide as long, 1.3–1.7 cm. long, 1.4–1.6 cm. wide.

KENYA. Teita District: Mbololo Hill, 31 Dec. 1971, *Faden & Smeenk* 71/1009! & Mt. Kasigau, route from Rukanga to summit, upper slopes of peak, *Faden et al.* 71/181! & Ngangao Forest, 9 May 1985, *Faden et al.* 85/253!
TANZANIA. Kilimanjaro, N. slopes, Rongai Forest Station. 17–18 Jan. 1990, *Pócs et al.* 90002/B!; Pare District: Kilomeni–Kissangara, 27 Feb. 1915, *Peter* K744!; Morogoro District: Nguru Mts., Mafulumla, 9 Feb. 1933, *Schlieben* 4083!
DISTR. K 7; T 2,3,6; not known elsewhere
HAB. Dry evergreen forest; 1250–1750(–2100) m.

d. subsp. **friesiorum** (*Robyns*) *Bridson*, comb. et stat. nov. Type: Kenya, N. Nyeri District, West Mount Kenya Forest Station, *R.E. & T.C.E. Fries* 956 (K, lecto.!, chosen here)

SED

FIG. 156. *CANTHIUM OLIGOCARPUM* subsp. *OLIGOCARPUM* — **1**, flowering branch, × ⅔; **2**, young spiny shoot, × 1; **3**, domatium, × 8; **4**, stipule, × 2; **5**, flower, × 4; **6**, portion of corolla, × 4; **7**, calyx with style and stigma, × 4; **8**, stigmatic knob, × 8; **9**, dorsal view of anther, × 8; **10**, longitudinal section through ovary, × 8; **11**, fruit, × 2; **12**, pyrene (2 views), × 2. 1, 3, from *Harley* 9530; 2, from *Bridson* 397; 4, 11, from *Eggeling* 3989; 5–10, from *Harley* 9567; 12, from *Maitland* s.n. Drawn by Sally Dawson.

Bark on young twigs fawn to light grey. Leaf-blades always small (2.5–5 cm. long, 1.3–2.8 cm. wide), obtuse to acute at apex, papery, dull above, with 4–5 main pairs of lateral nerves; domatia densely pubescent, pits usually obscured. Inflorescences (5–)7–15-flowered. Fruit somewhat wider than long, 1.3 cm. long, 1.4 cm. wide.

Kenya. Naivasha District: Aberdare Mts., South Kinangop, *Fey* in *E.A.H.* 11703!; S. Nyeri District: Mt. Kenya, Castle Forest Station, 20 Mar. 1958, *Dyson* in *E.A.H.* 82/58!; Masai District: Ngong Hills, 17 Oct. 1965, *Gillett* 16923!
Distr. **K** 3, 4, 6; not known elsewhere
Hab. Evergreen forest; 2000–2500 m.

Syn. *Rytigynia friesiorum* Robyns in N.B.G.B. 10: 614 (1929); K.T.S.: 472 (1961)
[*Canthium siebenlistii* sensu Chiov., Racc. Bot. Miss. Consol. Kenya: 55 (1935), *non* (K. Krause) Bullock]

Note. (on species as a whole). A fifth subspecies (subsp. *angustifolium* Bridson) is present in eastern Zimbabwe and the adjacent part of Mozambique, it differs from the four subspecies in the Flora area by its narrowly elliptic leaves.

Subgen. 4. **Bullockia** *Bridson* in K.B. 42: 630 (1987)

Shrubs, sometimes scandent, or small trees. Leaves not deciduous, petiolate or subsessile; blades drying brown, bronze, greyish or green, chartaceous to stiffly chartaceous, usually dull above, glabrous or pubescent; tertiary nerves obscure; domatia absent or present but rather small; stipules narrowly ovate to ovate or triangular at base, sometimes produced into a lobe, caducous or somewhat persistent, with a few silky hairs and colleters inside. Flowers unisexual, (4–)5–6-merous; ♂ inflorescence 1–20-flowered, fasciculate or occasionally in shortly pedunculate umbels; ♀ flowers 1(–2); small free bracteoles present on pedunculate inflorescence. Calyx glabrous or pubescent; tube ± ovoid in ♀ flowers and very reduced in ♂ flowers; limb lobed almost to base; lobes usually irregular. Corolla small, rapidly caducous; tube provided with a ring of deflexed hairs inside; lobes subequal to tube, erect, thickened towards apex, not apiculate. Anthers ovate, the dorsal face with a central band of darkened connective tissue, not apiculate. Style moderately slender, slightly longer than corolla-tube, slightly tapered at apex; stigmatic knob ± as wide a long, 2-lobed, hollow with style attached well inside; ovary 2-locular, each locule containing one ± pendulous ovule. Fruit yellow to dark red, fleshy, obovate to broadly obovate, laterally flattened, somewhat bilobed, not or scarcely indented at apex; pyrenes thinly woody, narrowly obovoid, flat on ventral face, crested around the apex, eventually dehiscing from point of attachment back along the crest, slightly rugulose. Seeds narrowly obovoid, with ventral face flattened, apex slightly crested; embryo slightly curved.

A subgenus of six species occurring in eastern tropical Africa and South Africa, possibly with additional species in the Seychelles and Madagascar. Species 18–23.

1. Pedicels and calyx-tube glabrous or occasionally sparsely
   covered with fine hairs; leaves glabrous above, drying
   brown, bronze or greyish 2
   Pedicels and calyx-tube sparsely to densely pubescent;
   leaves glabrescent to pubescent or rarely glabrous
   above, drying greenish or less often brown above and
   grey-green beneath 4
2. Stipules ovate to narrowly ovate or sometimes lanceolate;
   petiole 2–20 mm. long . . . . 18. *C. mombazense*
   Stipules triangular at base, linear above; petiole 1–3 mm.
   long 3
3. Stipules 3–5 mm. long; leaf-blades broadly oblong-elliptic,
   1–2.8 cm. long, obtuse to rounded at base; petiole 1–2
   mm. long . . . . 19. *C. dyscriton*
   Stipules 4–10 mm. long; leaf-blades narrowly elliptic to
   elliptic, 2–6 cm. long, acute at base; petiole 2–3 mm.
   long . . . . 20. *C. fadenii*
4. Lateral nerves acutely angled and slightly to strongly
   impressed above; leaves broadly elliptic; stipules 0.5–1
   cm. long . . . . 21. *C. impressinervium*

Lateral nerves more obtusely angled, not impressed above;
leaves elliptic to oblong-elliptic; stipules up to 5 mm.
long 5
5. Young stems, pedicels and calyx-tube covered with straw- to
rust-coloured, rather stiff upwardly directed
hairs . . . . 22. *C. setiflorum*
Young stems, pedicels and calyx-tube covered with whitish
soft crisped or patent hairs . . . . 23. *C. pseudosetiflorum*

18. **C. mombazense** *Baillon* in Adansonia, 12: 188 (1878); K.T.S.: 430 (1961); Bridson in K.B. 42: 631, fig. 1B (1987). Type: Kenya, Mombasa [Mombassa], *Boivin* s.n. (P, holo., K, fragment!)

Compact, or less often straggling, shrub or small tree 2–7 m. tall (or rarely a ? liane); bark pale grey, glabrous or occasionally pubescent. Leaf-blades pale greyish to bronze above when dry, broadly elliptic to round or broadly oblong-elliptic, 2.2–13 cm. long, 1–7.5 cm. wide, obtuse to rounded at apex, obtuse to rounded or sometimes truncate at base, often unequal, glabrous or sometimes pubescent beneath; lateral nerves apparent, in 4 main pairs; tertiary nerves always obscure; domatia absent or inconspicuous; petiole 0.2–1 cm. long; stipules light brown when dry, ovate or sometimes narrowly ovate, 0.4–1.6 cm. long, 0.2–1 cm. wide, acuminate. Flowers 5–6-merous. Functionally ♂ flowers borne in 3–20-flowered, sessile or rarely pedunculate (up to 2 mm.) umbels; pedicels 3–7 mm. long, glabrous or sometimes pubescent; calyx-tube reduced; limb lobed almost to base; lobes somewhat unequal, linear-triangular, occasionally slightly spathulate, 0.75–2 mm. long, often with a few hairs at apex; corolla whitish or yellowish green, readily caducous; tube ± 1.5 mm. long, with a ring of deflexed hairs at throat inside; lobes triangular-ovate, 1.25–1.5 mm. long, 1–1.5 mm. wide at base, acute, often with a few hairs at apex outside, erect; anthers exserted but almost sessile, 0.75 mm. long; style shortly exceeding corolla-tube; stigmatic knob 0.25 mm. long and wide. Functionally ♀ flowers solitary; pedicels 0.4–1 cm. long; calyx-tube 1.75–2.5 mm. long; lobes more strongly spathulate than ♂ and 2–3 mm. long; corolla as ♂ but lobes generally more thickened; anthers ± 0.5 mm. long, with little or no pollen; style caducous. Fruit dark maroon, laterally compressed obovoid, 0.7–1.2 cm. long, 7–9 mm. wide, not indented at apex; calyx-lobes persistent; pedicel lengthening to 2 cm.; pyrene obovoid, flattened on ventral face, 7 mm. long. 3.5 mm. wide, crustaceous, surface ± smooth except for shallow crest around apex.

KENYA. Teita District: Maungu Hills, S. of Maungu Station on Nairobi–Mombasa road, 31 May 1970, *Faden* 70/176!; Kilifi District: Arabuko Forest on Mida–Jilore Forest Station track, within 3 km. of main road, 20 July 1972, *Gillett & Kibuwa* 20023!; Lamu District: Mararani [Marranani], Boni Forest, 18 Sept. 1961, *Gillespie* 374!
TANZANIA. Tanga District: Sawa Creek, 17 Feb. 1957, *Faulkner* 1962!; Uzaramo District: University of Dar es Salaam campus, between Simba & Killimahewa roads, 24 July 1975, *Wingfield* 3155!; Lindi District: Mlinguru, 18 Dec. 1934, *Schlieben* 5742A!
DISTR. **K** 4, 7; **T** 3, 6, 8, **Z**; **P**; Somalia and Mozambique
HAB. Coastal bushland and wooded grassland to forest; 0–1200 m.

SYN. *Plectronia diplodiscus* K. Schum. in P.O.A. C: 385 (1895). Type: Tanzania, Lushoto District, Usambara Mts., Mlalo, *Holst* 585 (B, holo.†)
P. pallida K. Schum. in E.J. 28: 77 (1899). Type: Zanzibar I., *Stuhlmann* 487 (B, holo.†, HBG, iso.!)
Canthium pallidum (K. Schum.) Bullock in K.B. 1932: 387, fig. 4 (1932);T.T.C.L.: 484 (1949); K.T.S.: 430 (1961)
C. diplodiscus (K. Schum.) Bullock in K.B. 1932: 387 (1932); T.T.C.L.: 484 (1949)
C. greenwayi Bullock in K.B. 1932: 337 (1932); T.T.C.L.: 484 (1949). Type: Tanzania, Pare District, Kisiwani [Kisuani], *Greenway* 2179 (EA, holo., K, iso.!)
C. inopinatum Bullock in K.B. 1932: 389 (1932); K.T.S.: 429 (1961). Type: Kenya, Machakos/Kitui District, Ukambani, *Scott Elliot* 6380 (K, holo.!)

NOTE. Bullock did not notice that the plants were dioecious and the number of flowers varied according to sex; this deceived him into recognising the different sexual forms as taxa. It is most probable that all five species comprising his "anomalous series" fall within the range of variation of *C. mombazense*. *C. greenwayi* has a distinctive facies but probably just represents a reduced form from an exposed habitat. No material of the type of *C. diplodiscus* has been traced, but it is highly probable that it just represented a pubescent from of *C. mombazense*.
This species is occasionally polygamous, two specimens have been seen with odd ♀ flowers on otherwise functionally ♂ branches. Field observations would be of interest.

19. **C. dyscriton** *Bullock* in K.B. 1936: 478 (1936); K.T.S.: 428 (1961); Bridson in K.B. 42: 632 (1987). Type: Kenya, Teita Hills, Nyatchi, *Gardner* in *F.D.* 3000 (K, holo.!)

Small often compact shrubs 0.5–3 m. tall; bark grey; young stems covered with yellowish rusty hairs. Leaf-blades brown when dry, broadly oblong-elliptic, 1–2.8 cm. long, 0.5–1.8 cm. wide, obtuse to rounded or, occasionally, emarginate at apex, rounded at base, somewhat coriaceous, glabrous, sparsely hairy on midrib beneath or pubescent overall beneath; lateral nerves in 3–4 main pairs; tertiary nerves obscure; domatia present as small tufts of hair in the nerve-axils; petioles 1–2 mm. long, pubescent or glabrous beneath and pubescent above; stipules triangular at base, gradually or abruptly tapering to a linear decurrent lobe, 3–5 mm. long, glabrous. Flowers 5-merous. Functionally ♂ flowers borne in up to 8-flowered sessile umbels; pedicels 4–6 mm. long, glabrous; calyx-tube reduced; limb divided to base into triangular-acute or subulate lobes ± 0.5 mm. long, glabrous; corolla-tube ± 1.5 mm. long, with a ring of deflexed hairs at throat inside; lobes erect, triangular-ovate, 1.25 mm. long, ± 1 mm. wide at base, acute, thickened at apex; anthers exserted but subsessile; style ± as long as corolla-tube; stigmatic knob subglobose. Functionally ♀ flowers solitary; calyx-tube 0.75 mm. long. Fruit not known.

Kenya. Machakos District: Thyaa, ¼ km. NW. of SDA Church towards Muthesya, Mbooni area, 15 Aug. 1985, *Muasya* 701!; Teita District: Teita Hills, Nyatchi area, *Gardner* in *F.D.* 3000! & Sagala Hill, above Catholic School, near Sagala town, 1 May 1981, *Gilbert* 6102!
Tanzania. Masai District: W. of Naberera, near Oldoro Lolussoi waterhole, Apr. 1966, *Procter* 3256!
Distr. **K** 4, 7; **T** 2; not known elsewhere
Hab. Rocks, roadside banks and grazed eroded areas; 750–1330 m.

20. **C. fadenii** *Bridson* in K.B. 42: 632, fig. 4 (1987). Type: Kenya, Kiambu District, Thika, behind Blue Posts Hotel, *Faden* 68/012 (K, holo.!, EA, iso.)

Shrub 2.5–4.5 m. tall; stems covered with grey bark, glabrous or pubescent. Leaf-blades brownish when dry, elliptic or sometimes narrowly elliptic, 1.7–6 cm. long, 0.7–2.5 cm. wide, acute to obtuse at apex, acute and often unequal at base, glabrous or sometimes pubescent beneath; lateral nerves in 4(–5) main pairs; tertiary nerves always obscure; domatia absent or obscure; petiole 2–3 mm. long; stipules triangular at base, linear above, 0.4–1 cm. long, glabrous or pubescent outside. Flowers 5-merous. Functionally ♂ flowers borne in sessile umbels cf 1–6 flowers; pedicels 2–5 mm. long, glabrous or pubescent; calyx-tube reduced, glabrous or pubescent; limb lobed almost to base with short subulate lobes from a triangular base, up to 0.5 mm. long; corolla white; tube ± 1 mm. long, with a ring of deflexed hairs at throat inside; lobes triangular-ovate, 1.25 mm. long, 1 mm. wide at base, acute, and somewhat thickened towards apex; anthers almost sessile, 0.5 mm. long; style slightly shorter than corolla-tube; stigmatic knob ± 0.25 mm. long and wide. Functionally ♀ flowers solitary; pedicels 4–6 mm. long; calyx-tube 1.5 mm. long; lobes linear to somewhat spathulate, 0.75–1.5 mm. long; anthers and stigmatic knob not known. Fruit flattened, broadly obovate, 7 mm. long, 7 mm. wide, not indented, glabrous or glabrescent; calyx-lobes persistent; pedicel lengthening to 1–1.5 cm. long; pyrene obovoid, flattened on ventral face, 8 mm. long, 4 mm. wide, slightly rugulose, with shallow crest around apex.

Kenya. Kiambu District: Thika, Chania R., 19 Nov. 1967, *Faden* 67/906! & behind Blue Posts Hotel, 12 Jan. 1967, *Faden* 67/25!; Kitui District: Ngomeni Rock [Igomena Hill], 3 Mar. 1960, *Rauh* 818!
Distr. **K** 4; not known elsewhere
Hab. Bushland to dry forest; 1450–1500 m.

21. **C. impressinervium** *Bridson* in K.B. 42: 633, fig. 5 (1987). Type: Tanzania, Lindi District, Noto Plateau, Mtondoli, *Schlieben* 6109 (K, holo.!, EA, HBG, iso.!)

Shrub or small tree 1–4 m. high; twigs with greyish or brown bark, densely covered with upwardly directed or patent hairs when young. Leaf-blades discolorous, drying brownish above and grey-green beneath or light green above and beneath, broadly elliptic, oblong-elliptic or oblong, 0.9–4.6 cm. long, 0.7–2.8 cm. wide, rounded and either somewhat emarginate or apiculate at apex, rounded, truncate or sometimes subcordate at base, scabrid above, sparsely pubescent beneath; lateral nerves in 3–4 main pairs, acutely angled and impressed above; tertiary nerves obscure; domatia absent; petiole 0.5–2 mm. long, densely pubescent; stipules linear or linear-oblong from a shortly triangular base, 0.5–1 cm. long, pubescent outside. Flowers 4–5-merous. Functionally ♂ flowers, borne in subsessile to shortly pedunculate (1–)3–8-flowered umbels; peduncles up to 5 mm. long, pubescent; pedicels 2–8 mm. long, sparsely pubescent; bracteoles apparent on pedunculate inflorescences; calyx-tube reduced, pubescent or sparsely pubescent; limb lobed to base; lobes linear, 0.5–1.25 mm. long, sparsely pubescent; corolla white or yellow; tube 0.75–1.25 mm. long, with a ring of deflexed hairs at throat; lobes erect, triangular, 0.75–1 mm. long, ± 0.75 mm. wide, pubescent outside. Functionally ♀ flowers solitary; peduncles ± absent or up to 5 mm. long, pubescent; pedicels 3–10 mm. long, pubescent; calyx-tube 1.5 mm. long; limb similar to ♂; corolla not known. Fruit orange, broadly obovoid, 9 mm. long, 7 mm. wide, scarcely indented at apex, sparsely pubescent; stalks lengthening to 2.5 cm.; pyrenes not known.

TANZANIA. Lindi District: Rondo Plateau, 16 May 1903, *Busse* 2575! & Noto Plateau, Mtondoli, 9 Mar. 1935, *Schlieben* 6109!; Newala, 6 Mar. 1959, *Hay* 42!
DISTR. T 8; not known elsewhere
HAB. Bushland, forest scrub or *Brachystegia microphylla*, *Albizia* thicket on escarpment; 450–700 m.

22. **C. setiflorum** *Hiern* in F.T.A. 3: 134 (1877); Bridson in K.B. 42: 634 (1987). Type: Mozambique, near Tete, *Kirk* (K, holo.!)

Scandent shrub 1–4 m. tall; young branches covered with dark greyish bark, not conspicuously marked with lenticels, densely covered with upwardly directed straw-coloured, golden or rust-coloured hairs. Leaf-blades greenish when dry, elliptic, oblong-elliptic or sometimes broadly elliptic, 1.2–6.5 cm. long, 0.6–3.4 cm. wide, acute to obtuse and apiculate at apex, obtuse to rounded at base, sparsely pubescent to glabrescent or rarely glabrous save for midrib above, glabrescent to pubescent beneath, with denser hairs on nerves beneath; lateral nerves in (3–)4(–5) main pairs; tertiary nerves obscure; domatia absent or inconspicuous; petiole 1–2 mm. long, pubescent; stipules triangular at base with a somewhat decurrent linear lobe 2–5 mm. long, pubescent outside. Flowers (4–)5(–6)-merous. Functionally ♂ flowers borne in subsessile to pedunculate, 2–12-flowered, subumbellate cymes; peduncles 1–5 mm. long, pubescent; pedicels 1–4 mm. long, pubescent; calyx-tube reduced and rather sparsely pubescent; limb lobed almost to base; lobes linear to spathulate, 0.5–3 mm. long; corolla yellowish green to cream; tube 1.5–2 mm. long, with a ring of deflexed hairs at throat inside; lobes triangular-ovate, 1 mm. long, 0.75–1 mm. wide at base, sparsely pubescent towards apex outside, acute and thickened at apex, erect or spreading; anthers exserted but almost sessile; style ± equalling corolla-tube. Functionally ♀ flowers solitary or occasionally paired; pedicels 1.5–8 mm. long; calyx-tube 1.5–2 mm. long, densely pubescent; lobes 1.5–3 mm. long; corolla as ♂ but tube 1–1.25 mm. long; anthers somewhat smaller; stigmatic knob somewhat bigger, 0.75 mm. long and wide. Fruit yellow, or turning black when mature, flattened, broadly obovoid, 0.7–1.2 cm. long, and wide, scarcely indented at apex, pubescent; calyx-lobes persistent; pedicel lengthening to 0.6–1.2 cm.; pyrene obovoid, flattened on ventral face, 8 mm. long, 4 mm. wide, slightly rugulose; apical crest slightly defined.

subsp. **telidosma** (*K. Schum.*) *Bridson* in K.B. 42: 634 (1987). Type: Tanzania, Uzaramo District, near Madimola, *Stuhlmann* 6690 (B, holo.†, K, fragment!)

Calyx-lobes 0.5–0.75(–1) mm. long in functionally ♂ flowers and 0.75–1.75 mm. long in functionally ♀ flowers. Male inflorescence infrequently pedunculate; peduncle not exceeding 2 mm. Fruit ± 0.7 cm. long and wide.

KENYA. Kwale District: Mrima Hill, 5–6 Mar. 1977, *Faden* 77/675! & Jombo Mt., 26 Sept. 1982, *Polhill & Robertson* 4842!
TANZANIA. Tanga District: 0.8 km. SE. of Ngomeni, 31 July 1953, *Drummond & Hemsley* 3556!; Kilosa District: Mikumi National Park, Vuma Hills, 29 June 1977, *Wingfield, Mhoro & Mwangoma* 4025!; Morogoro District: 6.5 km. N. of Turiani, Lusunguru Forest Reserve, near Mtibwa Sawmill, 31 Mar. 1953, *Drummond & Hemsley* 1928!
DISTR. K 7; T 3, 6; not known elsewhere
HAB. In evergreen or dry forest or thicket; 60–750 m.
SYN. *Plectronia telidosma* K. Schum. in E.J. 23: 460 (1897)
   Canthium telidosma (K. Schum.) S. Moore in J.L.S. 40: 87 (1911); Bullock in K.B. 1932: 366 (1932); T.T.C.L.: 486 (1949)
   [*C. microdon* sensu Bullock in K.B. 1932: 366 (1932); T.T.C.L.: 486 (1949), pro parte, quoad *Zimmermann* in Herb. Amani 6113, *non* S. Moore]
   [*C. setiflorum* sensu Vollesen in Opera Bot. 59: 67 (1980) *non* Hiern sensu stricto]
NOTE. This subspecies has shorter calyx-lobes than subsp. *setiflorum* (0.5–3 mm. long in functionally ♂ flowers and 1.5–3 mm. long in functionally ♀ flowers), which occurs in Mozambique, southern Malawi, Zimbabwe and South Africa (Natal). One specimen from Kenya, Kwale District, Mrima Hill, *Faden* 77/675, is atypical in its leaves which are glabrous above.

23. **C. pseudosetiflorum** *Bridson* in K.B. 42: 635, fig. 2B & 6 (1987). Type: Kenya, Northern Frontier Province, Moyale, *Gillett* 14096 (K, holo.!, EA, iso.)

Shrub 1–3 m. tall; young branches covered with dark grey bark, often with conspicuous lenticels, covered with pale crisped or occasionally patent hairs when young. Leaf-blades paired or rarely in threes, light green when dry, oblong-elliptic or occasionally broadly oblong-elliptic, 0.6–3.3 cm. long, 0.4–2.4 cm. wide, acute to obtuse and sometimes apiculate at apex; obtuse to rounded or occasionally subcordate at base, pubescent or glabrescent with pubescent nerves above, pubescent beneath; lateral nerves in 3–4(–5) main pairs; tertiary nerves obscure; domatia absent or inconspicuous; petiole 1–3 mm. long, pubescent; stipules triangular at base with a linear somewhat decurrent lobe, 2–4 mm. long, pubescent outside. Flowers 5-merous. Functionally ♂ flowers borne in subsessile to pedunculate 1–6-flowered subumbellate cymes; peduncles 1–4(–6) mm., pubescent; pedicels 1–3(–5) mm. long; calyx-tube reduced and sparsely pubescent; limb lobed almost to base; lobes linear to spathulate, 0.5–1.5 mm. long, sparsely pubescent; corolla yellowish or greenish white, readily caducous; tube 1.25–1.75 mm. long, with a ring of deflexed hairs at throat inside; lobes triangular-ovate, 1–1.25 mm. long, ± 1mm. wide at base, sparsely hairy towards apex outside, acute and thickened at apex, typically erect or occasionally patent; anthers exserted but almost sessile; style ± equalling corolla-tube. Functionally ♀ flowers solitary; pedicels 3–9 mm. long; calyx-tube 1.5–2 mm. long, densely pubescent; lobes 1–2 mm. long; anthers very slightly smaller than in ♂; stigmatic knob slightly larger than in ♂, 0.5 mm. long, 0.75 mm. wide. Fruit yellow, laterally

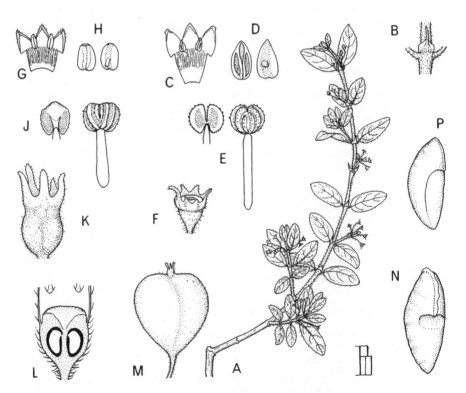

FIG. 157. *CANTHIUM PSEUDOSETIFLORUM*— **A**, flowering branch, ♂, × ⅔; **B**, stipule, × 2; **C**, section through corolla, ♂, × 4; **D**, anthers (2 views), ♂, × 10; **E**, style with stigmatic knob and longitudinal section through stigmatic knob, ♂, × 10; **F**, calyx, ♂, × 6; **G**, section through corolla, ♀, × 4; **H**, anthers (2 views), ♀, × 10; **J**, style with stigmatic knob and section through stigmatic knob, ♀, × 10; **K**, calyx, ♀, × 6; **L**, longitudinal section through ovary, ♀, × 10; **M**, fruit, × 2; **N**, pyrene, × 3; **P**, seed, × 3. A, B, M–P, from *Verdcourt* 766; C–F, from *Dale* 833 & G–L, from *Greenway & Kanuri* 12088. Drawn by Diane Bridson.

flattened, broadly obovoid, 0.8–1.1 cm. long and wide, indented at apex or not, glabrescent to sparsely pubescent; calyx-lobes persistent; pedicel lengthening to 0.6–1.2 cm.; pyrene obovoid, flattened on ventral face, 0.9 mm. long, 0.5 mm. wide, slightly rugulose with an apical shallow crest. Figs. 132/21, p. 754 & 157.

UGANDA. Karamoja District: 19 km. Moroto–Kitale, 5 Oct. 1952, *Verdcourt* 766! & Pian, Lodoketemit [Lodoketeminit], 6 Nov. 1962, *Kerfoot* 4433! & Lokitanyala–Napau Pass, 16 Nov. 1953, *Dale* 833!
KENYA. Northern Frontier Province: Dandu, 3 Apr. 1952, *Gillett* 12680!; Turkana District: Murua Nysigar [Moruassigar], 18 Feb. 1965, *Newbould* 7294!; Baringo District: Kabluk, 3 km. N. of Chebloch Bridge, 12 Mar. 1964, *Brunt* 1524A!
TANZANIA. Mbulu District: Lake Manyara National Park, Msasa Gorge, 1 Dec. 1963, *Greenway & Kirrika* 11104! & 19 Feb. 1964, *Greenway & Kanuri* 11204! & above Maji Moto Springs, 26 May 1965, *Greenway & Kanuri* 12088!
DISTR. U 1; K 1–3, 6; T 2; Ethiopia
HAB. *Acacia, Commiphora* bushland, mixed scrub often amongst granite rocks; 720–2000 m.

SYN. [*C. setiflorum* sensu K.T.S.: 432 (1961); E.P.A. 2: 1011 (1965), *non* Hiern]

NOTE. Specimens from K 6 and T 2 show minor differences compared with material from further north; the fruit is very slightly longer than wide and more distinctly indented at the apex and there is a tendency for the leaves to have more pairs of lateral nerves (3–5 as opposed to 3(–4) in the northern material).

UNKNOWN SPECIES

24. **C. sordidum** (*K.Schum.*) *Bullock* in K.B. 1932: 366 (1932); T.T.C.L.: 486 (1949); K.T.S.:433 (1961). Type: Tanzania, Mwanza District, Bumpeke [Umpeke], *Stuhlmann* 860 (B, holo. †)

SYN. *Plectronia sordida* K. Schum. in P.O.A. C: 386 (1895)

NOTE. Bullock and Greenway (in K.T.S.) both cited *Scheffler* 28 from Kenya, Machakos District, Kibwezi as belonging to *C. sordidum*. This specimen has been redetermined as *Psychotria kirkii* Hiern var. *volkensii* (K. Schum.) Verdc. The type specimen may or may not have belonged to the same taxon.

106. **PYROSTRIA**

A.L. Juss., Gen. Pl.: 206 (1789); Lam., Tab. Encyl., t. 68 (1791) & 1: 289 (1792); DC., Prodr. 4: 464 (1830), pro parte majore; A. Rich., Mém. Fam. Rub.: 136 (1830) & in Mém. Soc. Hist. Nat. Paris 5: 216 (1834); Hook.f. in G.P. 2: 111 (1873); Drake in Bull. Soc. Linn. Paris, nouv. sér. 6: 41 (1898); K. Schum. in E. & P. Pf. IV.4: 94, fig. 33H (1891); Hochr. in Ann. Conserv. Jard. Bot. Genève 11 & 12: 96 (1908); Verdc. in K.B. 37: 563 (1983); Bridson in K.B. 42: 611 (1987); Verdc. in Fl. Masc. 108: 110–118 (1989)

[*Psydrax* sensu DC., Prodr. 4: 476 (Sept. 1830); A. Rich., Mém. Fam. Rub.: 110 (Dec. 1830) & in Mém. Soc. Hist. Nat. Paris 5: 190 (1834), pro parte, *non* Gaertn.]
Canthium Lam. sect. *Psydracium* Baillon, Adansonia 12: 199 (1878) & Hist. Pl. 7: 425 (1880)
Dinocanthium Bremek. in Ann. Transv. Mus. 15: 259 (1933)
Dinocanthium Bremek. sect. *Hypocrateriformes* Robyns in B.J.B.B. 17: 94 (1943)
Dinocanthium Bremek. sect. *Rotatae* Robyns in B.J.B.B. 17: 94(1943), *nom. superfl.*
Pseudopeponidium Arènes in Not. Syst. 16: 19 (1960)
Pyrostria A.L. Juss. sect. *Involucratae* Cavaco in Adansonia, sér. 2, 11: 393 (1971)

Small shrubs to medium-sized trees. Leaves not deciduous, petiolate, less often subsessile or occasionally sessile; blades drying blackish brown, slate-grey or sometimes dull green, typically subcoriaceous, usually glabrous, very infrequently pubescent; tertiary nerves obscure (or rarely apparent); domatia present as glabrous to pubescent cavities or sometimes absent; stipules triangular at base, sometimes produced into a lobe or ovate to linear, persistent or readily caducous, with a few silky hairs and colleters inside. Flowers unisexual or ♂, 4- or 5-merous, borne in pedunculate umbels, sometimes with both peduncle and pedicels somewhat reduced, in dioecious species the ♀ inflorescence contains fewer flowers (often solitary) than the ♂; bracts paired, connate and entirely surrounding inflorescence in bud, persistent, with silky hairs and colleters towards base inside or rarely with dense rusty hairs. Calyx-tube ± ovoid in ♀ and ♂ flowers but very reduced in ♂ flowers; limb reduced to a rim, sometimes shortly dentate or bearing unequal or less often equal lobes or sometimes (in Mauritian species) cupular. Corolla rather fleshy, usually drying dark with the lobes pale above; tube shorter than or longer than the lobes, throat densely congested with crisped or less often straight hairs, lacking a well-defined ring of deflexed hairs inside; lobes spreading, reflexed or sometimes ± erect, thickened towards the apex, sometimes shortly apiculate, upper surface of lobes usually graniculate or very finely colliculate. Stamens set at throat; filaments short, attached near the base of the anther; anthers erect, less often spreading or rarely reflexed, ovate to oblong-ovate, the dorsal face, except for the margin, covered with darkened connective tissue, sometimes extending into a short apiculum. Style slender, slightly longer than corolla-tube, somewhat tapered at apex; stigmatic knob ± as wide as long, 2–several-lobed, lobes only separating to reveal stigmatic surface in ♀ and ♂ flowers, solid, attached to style at base or sometimes slightly recessed; ovary 2(–10 outside Africa)-locular, each locule containing one pendulous ovule. Disk annular or hemispherical; the area between the ovary and disc sometimes elongate (in Madagascan species only). Fruit yellow to red, fleshy; 2-locular fruit subspherical to heart-shaped or less often didymous, bilaterally flattened, often elaborated with extra lobes or wings; multilocular fruit lobed according to number of locules; pyrenes thinly woody to woody, narrowly obovoid to obovoid with ventral face flattened, crested around the apex, eventually dehiscing from the point of attachment back along the crest (fig. 132/22, p. 754). Seeds ± narrowly obovoid with ventral face flattened and apex somewhat crested; endosperm entire; testa finely reticulate; embryo straight or slightly curved with radicle erect; cotyledons small, set perpendicular to ventral face of seed.

A genus of ± 45 species in Africa, Madagascar and the Mascarenes; possibly taxa from Malesia and Indochina should also be included in *Pyrostria*. A total of 14 species are known from Africa.

1. Leaves oblong-elliptic to linear-oblong, 0.8–3.6(–4.5) cm. long, 0.3–1(–1.6) cm. wide; shrub with lateral branches frequently reduced to spurs; leaves with lateral nerves obscure . . . .   8. *P. phyllanthoidea*
   Leaves elliptic to broadly elliptic, larger than above; shrub or small tree with lateral branches seldom reduced to spurs; leaves with lateral nerves apparent 2
2. Leaves with tertiary nerves rather prominent on both faces; apex long-acuminate, often apiculate at tip; young stems with adpressed hairs on either side; immature fruit broader than long; flowers not known . . .   6. *P. sp. C*
   Leaves with tertiary nerves obscure, or if apparent then not prominent; apex obtuse to acuminate; young stems glabrous or sometimes glabrescent on either side 3
3. Stipules triangular or triangular-ovate, usually only present at apical node 4
   Stipules narrowly triangular or with a linear lobe from a triangular base, restricted to apical node or not 5
4. Leaves with (5–)6 main pairs of lateral nerves; corolla-tube 2 mm. long; calyx-limb truncate or bearing unequal lobes . . . .   2. *P. lobulata*
   Leaves with 3–4(–5) main pairs of lateral nerves; corolla-tube at least 3.5 mm. long; calyx-limb divided into equal triangular lobes . . . .   3. *P. sp. A*
5. Small branched shrub 1 m. tall; leaves with 3–4 main pairs of lateral nerves, obtuse to acute at apex; petiole 1–2 mm. long; bracts with conspicuous rusty hairs inside; calyx-lobes ± equal, triangular, 3 mm. long . . .   9. *P. sp. D*
   Taller shrubs; leaves with 4–7 main pairs of lateral nerves; petiole longer but if 1–2 mm. long then leaves acuminate; bracts with white hairs inside or inconspicuously hairy; calyx-lobes often unequal, seldom triangular, 0.5–2(–2.5) mm. long 6
6. Leaves drying brown or brown-black, stiffly chartaceous to subcoriaceous, obtuse to subacuminate at apex; lateral nerves in 4–5 main pairs; inflorescences 4–30-flowered (even when fruiting); flowers ♂ (or mostly so); corolla-tube only slightly longer than lobes . . . . . .   1. *P. bibracteata*

> Leaves drying greenish or black-green, chartaceous or less
> often stiffly chartaceous, subacuminate to acuminate at
> apex; lateral nerves in 5–7 main pairs; inflorescences
> 1–8-flowered (usually 1–2 in fruiting stage); flowers
> unisexual or uncertain; corolla-tube (where known) up
> to 3 times longer than lobes 7
7. Bracts small, 1–3 mm. long; domatia present as small tufts
  of light brown hair . . . .   4. *P. sp. B*
  Bracts larger, 3–5 mm. long; domatia present as glabrous
  or ciliate pits 8
8. Petioles 2–4 mm. long; stipules linear-triangular; flowers
  ? ♂; immature fruit slightly wider than long . . .   5. *P. uzungwaensis*
  Petioles 3–8 mm. long; stipules with a truncate to triangular
  base terminating in a linear keeled decurrent lobe;
  flowers unisexual; fruit slightly longer than wide   7. *P. affinis*

1. **P. bibracteata** (*Bak.*) *Cavaco* in Bull. Mus. Nat. Hist. Nat. Paris, sér. 2, 39: 1015 (1968); Bridson in K.B. 42: 625, figs. 1C & 2C (1987). Type: Seychelles, *Perville* 82 (K, holo.!, P, iso.)

Shrub or small tree 2–10 m. tall, glabrous; young stems covered with pale grey bark. Leaf-blades turning black-brown or brown when dry, elliptic to broadly elliptic or sometimes oblong-elliptic, 4–12.5 cm. long, 1.5–6 cm. wide, acute to obtuse at apex, acute to cuneate or occasionally rounded at base, stiffly papery to subcoriaceous, dull above; lateral nerves in 4–5 main pairs; tertiary nerves obscure; domatia sometimes present as tufts of hair; petiole 2–7 mm. long; stipules triangular at base, lobed above, 0.4–1.4 cm. long altogether, caducous. Flowers ♂ (or mostly so), 4-merous, in 4–30-flowered umbels; peduncles 2–4 mm. long; pedicels 1.5–5 mm. long, usually glabrous; bracts 2–7 mm. long, acuminate. Calyx-tube ± globose, ± 1 mm. in diameter; limb reduced to a rim or unequally lobed, up to 1 mm. long. Corolla yellowish cream; tube 2–3 mm. long, densely congested with hairs at throat; lobes ovate-triangular, 2–2.5 mm. long, 1.25 mm. wide, shortly apiculate. Stigmatic knob 0.75–1 mm. across. Fruit yellow, edible, almost globose or somewhat laterally compressed, 5–8 mm. in diameter, scarcely indented at apex; pyrene obovoid, flattened on ventral face, 5.5–6 mm. long, 2–3 mm. wide, with a shallow crest extending around the apex, slightly rugulose. Fig. 132/22, p. 754.

KENYA. Kwale District: Shimba Hills, Shimba Forest, 14 Mar. 1968, *Magogo & Glover* 294!; Kilifi District: 1.6 km. S. of Jilore, 21 Nov. 1969, *Perdue & Kibuwa* 10022!; Lamu District: 2 km. N. of Hindi, 25 Aug. 1973, *Gillett* 20346!
TANZANIA. Lushoto District: Korogwe, Kabuku Forest, 6 Nov. 1968, *Faulkner* 4166!; Morogoro District: ± 26 km. E. of Morogoro, 26 Nov. 1955, *Milne-Redhead & Taylor* 7385!; Zanzibar I., Mazizini [Massazine], 19 Jan. 1960, *Faulkner* 2462!
DISTR. K 7; T 3, 6, 8; Z; Mozambique, Zimbabwe, Madagascar, Seychelles and ? Aldabra
HAB. In bushland or forest edges; 0–870 m.

SYN. *Plectronia bibracteata* Bak., Fl. Maurit. & Seych.: 146 (1877)
  Canthium bibracteatum (Bak.) Hiern in F.T.A. 3: 145 (1877); Bullock in K.B. 1932: 375 (1932); T.T.C.L.: 487 (1949); K.T.S.: 427 (1961)

NOTE. Material from Aldabra tends to be smaller than material from elsewhere, especially with regard to the fruit (3–5 mm. long). Possibly a distinct taxon is involved, but further study is required for verification.

2. **P. lobulata** *Bridson* in K.B. 42: 627, fig. 1 (1987). Type: Rwanda, Kibungo Prefecture, Rusumo, Akagara Falls, *Bridson* 281 (K, holo., BR, C, COI, EA, LG, UPS, WAG, iso.!)

Shrub or small tree 3–8 m. tall, glabrous; young stems square, very slightly winged, covered with grey or fawn bark. Leaf-blades turning black-brown when dry, broadly elliptic, 2.5–13.5 cm. long, 1–6 cm. wide, acute to obtuse at apex, acute or sometimes rounded and often slightly unequal at base, subcoriaceous, dull above; lateral nerves in (5–)6 main pairs; tertiary nerves obscure; domatia present as inconspicuous tufts of hair in the nerve axils or sometimes absent; petiole 0.2–1 cm. long; stipules triangular to triangular-ovate, 5–10 mm. long, caducous. Flowers unisexual or ? sometimes ♂, 4-merous, borne in pedunculate umbels, functionally ♂ flowers 15–30, functionally ♀ flowers perhaps fewer; peduncles 4–7 mm. long; pedicels 3–4 mm. long, glabrescent to pubescent; bracts paired at apex of peduncle, connate at base, ovate, 5–7 mm. long, acuminate, pubescent near base inside. Calyx-tube in ♂ or ♀ ± globose up to 1 mm. in diameter, possibly ribbed in ♀ judging from very immature fruit; limb a reduced rim, truncate or bearing unequal lobes up to 1 mm. long. Corolla cream or greenish white; tube 2 mm. long, the throat congested with crisped hairs; lobes ovate-triangular, 2–2.5 mm. long, 1–1.25 mm. wide at base, shortly apiculate. Style very slightly longer than corolla-tube; stigmatic knob ± globose, ± 0.5 mm. in diameter. Disk not prominent. Fruits usually borne in umbels of 3–10, distinctly bilobed, with a lateral small lobe on either side of each lobe (tending to disappear on drying), 7–8 mm. long, 0.75–1.3 cm. wide, distinctly indented at apex; pyrene narrowly obovoid, 6 mm. long, 3.25 mm. wide, with a shallow wing around apex, scarcely rugulose. Figs. 131/4, 18, p. 752 & 158.

TANZANIA.   Bukoba/Biharamulo District: Ruiga Forest Reserve, Oct. 1957, *Procter* 741 & Dec. 1958, *Procter* 943!; Ufipa District: Namwele, 24 Feb. 1950, *Bullock* 2589!
DISTR.   **T** 1, 4; Rwanda and Zambia
HAB.   Riverine and hill-top thickets; 1200–1825 m.
SYN.   [*Canthium bibracteatum* sensu F.F.N.R.: 436 (1962); Fl. Pl. Lign. Rwanda: 546, fig. 182.2 (1983) & Fl. Rwanda 3: 148, fig. 42.2 (1985), *non* (Bak.) Hiern]

3. P. sp. A

Small tree 5–18 m. tall, glabrous; young stems square, very slightly winged, covered with grey or fawn bark. Leaf-blades turning black-brown when dry, broadly elliptic, 4–8.8 cm. long, 2–4.6 cm. wide, acute to obtuse at apex, and base, subcoriaceous, dull above; lateral nerves in 3–4(–5) main pairs; tertiary nerves obscure; domatia present as tufts of hair in the nerve axils; petiole 5–8 mm. long; stipules triangular, 5–8 mm. long, caducous. Flowers ? unisexual, 4-merous, borne in pedunculate, 5–7-flowered umbels; pedicels 5–7 mm. long; pedicels 3–5 mm. long, glabrous; bracts paired at apex of peduncle, connate at base, ovate acuminate, somewhat keeled, 6 mm. long. Calyx-tube 0.5–1 mm. long; limb divided into equal triangular lobes, 1 mm. long. Corolla known only from well-developed buds; tube 3.5 mm. long, with few hairs at throat; lobes ovate-triangular, 3.5 mm. long, shortly apiculate. Stigmatic knob broadly ellipsoid, 0.5 mm. in diameter. Fruit not known.

KENYA.   Kitui District: Endau Forest, 2 Jan. 1979, *Owino & Mathege* 182!; Masai District: Emali Hill, 19 Dec. 1971, *Faden et al.* 71/932!; Teita District: Mt. Kasigau, Oct. 1938, *Joana* in *C.M.* 9403!
DISTR.   **K** 4, 6, 7; not known elsewhere
HAB.   Dry evergreen forest; 850–1784 m.
NOTE.   From the limited material it is not possible to be sure if this taxon is unisexual or not.

4. P. sp. B; Bridson in K.B. 42: 628 (1987)

Glabrous shrub; young stems with light brown or greyish bark. Leaf-blades turning blackish when dry, elliptic, 2.7–7 cm. long, 1.2–2.7 cm. wide, acuminate at apex, acute at base, papery; lateral nerves in 5–6 main pairs; tertiary nerves scarcely apparent; domatia present as small tufts of brownish hairs; petiole 1–2 mm. long. Flowers unisexual or ? ♂ (♂ flowers not seen); umbels 2-flowered (? in ♀ inflorescence); peduncles ± 3 mm. long; pedicels 2 mm. long (in young bud); bracts narrowly triangular, ± 2 mm. long. Calyx-limb lobed to the base; lobes linear, ± 0.75 mm. long. Young corolla bud 4 mm. long, apparently not apiculate. Fruit bilobed, broader than long, 8 mm. long, 10 mm. wide, scarcely indented at apex, somewhat narrowed to base; pedicel lengthening to 9 mm.; pyrene not known.

TANZANIA.   Lushoto District: E. Usambara Mts, Longuza, *Zimmermann* in *Herb. Amani* 6107! & s.n.!
DISTR.   **T** 3; not known elsewhere
HAB.   Presumably forest; ± 500 m.
NOTE.   Bullock (in K.B. 1932: 389 (1932)) listed this taxon amongst the species he excluded from *Canthium*, stating "*Canthium didymocarpum* Peters MS in Herb. Amani is better placed in *Rytigynia* (*Zimmermann* in *Herb. Amani* 6107, from Longuza)"

5. P. uzungwaensis *Bridson*, sp. nov. affinis *P. bibracteatae* (Bak.) Cavaco sed stipulis lineari-triangularibus, foliis acuminatis differt; affinis *P. affinis* (Robyns) Bridson sed stipulis lineari-triangularibus, petiolis brevioribus differt. Typus: Tanzania, Iringa District, Uzungwa Mts, Mwanihana Forest Reserve, Sanje, *Lovett* (K, holo.!)

Shrub 3 m. tall; young stems glabrous or with glabrescent lines on either side; older stems with greyish or brown longitudinally ridged bark. Leaf-blades remaining green or

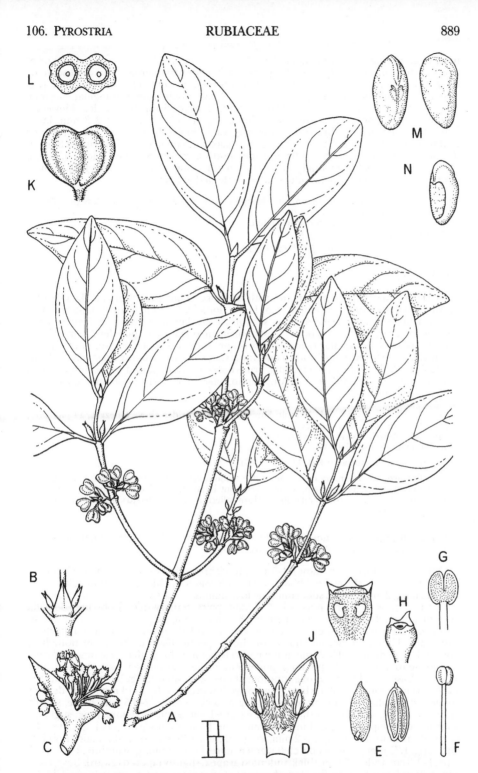

FIG. 158. *PYROSTRIA LOBULATA* — **A**, fruiting branch, × ⅔; **B**, stipule, × 1; **C**, inflorescence, × 2; **D**, half corolla, × 6; **E**, anther (2 views), × 10; **F**, style and stigmatic knob, × 6; **G**, longitudinal section through stigmatic knob, × 10; **H**, calyx, × 6; **J**, longitudinal section through ovary, × 9; **K**, fruit, × 2; **L**, transverse section of fruit, × 2; **M**, pyrene (2 views), × 3; **N**, seed, × 3. A, K–N, from *Bridson* 281; B–J, from *Ford* 729. Drawn by Diane Bridson.

turning greyish or inky when dry, elliptic, 5–10 cm. long, 1.8–2.3 cm. wide, acuminate at apex, acute at base, papery; lateral nerves in ± 5 main pairs; tertiary nerves obscure or drying discolorous on young leaves; domatia present as small glabrous or ciliate pits, obscure or sometimes absent; petioles 2–3 mm. long; stipules linear-triangular, 7–8 mm. long, shortly connate at base, seldom persisting beyond top two nodes. Flowers ? ♂, 4-merous, borne in 2–4-flowered pedunculate umbels; peduncles 2–3 mm. long; pedicels 0.5–1 mm. long; bracts 5–6 mm. long. Calyx-limb lobed to base; lobes linear or slightly triangular at base, 1–2 mm. long. Corolla-tube 5 mm. long, with crisped hairs at throat; lobes ovate, 2 mm. long, 1.5 mm. wide, shortly apiculate. Stigmatic knob ± 0.75 mm. in diameter. Fruits borne singly; pedicels 3–5 mm. long; only immature fruit known, bilobed, ± 1 cm. long, 1.1 cm. wide, scarcely indented at apex.

TANZANIA.   Iringa District: Uzungwa Mts., Mwanihana Forest Reserve, Sanje, 22 Sept. 1984, *D. Thomas* 3942! & 19 July 1986, *Lovett*!
DISTR.   **T** 7; known only from above specimens
HAB.   Ridge-top forest; 650–1250 m.

SYN.   *P. sp. C* sensu Bridson in K.B. 42: 628 (1987)

6. **P. sp. C**

Shrub to 3 m. tall; young branches with adpressed hairs on either side. Leaf-blades turning blackish when dry, elliptic, 6.5–11 cm. long, 2.2–3.8 cm. wide, long-acuminate then apiculate at apex. acute at base, subcoriaceous; lateral nerves in 6–7 main pairs; tertiary nerves prominent on both faces; domatia inconspicuous; petioles 2–3 mm. long; stipules narrowly triangular-lanceolate, 5–6 mm. long, caducous. Flowers not known; peduncles up to 3 mm. long; bracts 5 mm. long. Calyx-limb (in fruiting stage) divided into triangular-subulate lobes, 1.5 mm. long. Fruits borne singly or paired; pedicels up to 1 cm. long; immature fruit broader than long, 8 mm. long, 10 mm. wide, not indented at apex, tapering towards base.

TANZANIA.   Ulanga District: Sali, Muhulu Forest Reserve, on ridge above Mbezi R., 24 Jan. 1979, *Cribb, Grey-Wilson & Mwasumbi* 11177!
DISTR.   **T** 6; known only from above specimen
HAB.   Montane forest; 1350 m.

SYN.   *P. sp. D* sensu Bridson in K.B. 42: 628 (1987)

NOTE.   The prominent tertiary nervation of this taxon is very unusual in *Pyrostria*. However, it is apparently closely allied to the previous species but, in addition to the nervation, the adpressed hairs on the young stems render it distinct. More collections are needed to verify the constancy of these characters.

7. **P. affinis** (*Robyns*) *Bridson* in K.B. 42: 628 (1987). Type: Zaire, Lesse, on the banks of Semliki R., *Bequaert* 3141 (BR, holo.!, K, iso.!)

Shrub or small tree (or ?liane) 3–5 m. tall, glabrous; bark on young branches dark brown, moderately shiny. Leaf-blades drying olive-green, oblong to oblong-elliptic, 4–13.5 cm. long, 2–6 cm. wide, subacuminate to acuminate at apex, cuneate to attenuated at base, stiffly papery; lateral nerves in 5–7 main pairs; tertiary nerves obscure; domatia present as small ciliate pockets; petioles 3–8 mm. long; stipules with a truncate to triangular base terminating in a linear keeled and decurrent lobe, 4–8 mm. long altogether. Flowers 4-merous, borne in shortly pedunculate umbels; functionally ♂ flowers 1–8; peduncles 0.5–3 mm. long; pedicels 2–4 mm. long, glabrous; bracts broadly lanceolate to ovate, 2.5–5 mm. long, pubescent inside; calyx-tube poorly defined; limb lobed to base, or almost so, membranous; lobes somewhat irregular, triangular, 0.5–1.25(–2.5) mm. long; corolla white-green or yellow-green; tube 5–10 mm. long, densely congested with hairs at throat; lobes oblong or oblong-ovate, 2–3.5 mm., acute, shortly apiculate or not, reflexed; anthers partly exserted, subsessile, 1.25 mm. long; stigmatic knob broadly ellipsoid, 1 mm. long, 0.75 mm. wide. Functionally ♀ flowers 1–2; calyx-tube hemispherical to subglobose, 1–2 mm. long; anthers 0.75 mm. long; stigmatic knob 1.25 mm. long, 1.75 mm. wide, deeply cleft when mature. Fruit ± oblong in outline, 1.1–1.4 cm. long, 1–1.3 cm. wide, ± 5 mm. thick, indented at apex, slightly tapered towards base, costa apparent in lower half; pedicel lengthening to 5–8 mm.; pyrene ± narrowly obovoid, 1.1 cm. long, 5 mm. wide, with a very distinct crest at apex, rugulose.

UGANDA.   Kigezi District: S. Maramagambo, halfway along Bileriko road, 11 Oct. 1969, *Synnott* 401!

HAB. Mixed forest; 1200 m.
DISTR. U 2; also in Guinée, Ghana, Cameroon, Zaire and Angola
SYN. [*Coffea melanocarpa* sensu Good in J.B. 64, Suppl. 2: 27 (1926) quoad *Gossweiler* 4453, *non* Hiern]
      Dinocanthium affine Robyns in B.J.B.B. 17: 94 (1943) & F.P.N.A. 2: 350 (1947)
      D. bequaertii Robyns in B.J.B.B. 17:95 (1943) & F.P.N.A. 2: 350, t. 35 (1947). Type: Zaire, Lesse, on
         banks of Semliki R., *Bequaert* 3068 (BR, holo., K, iso.!)
      Canthium affine (Robyns) Hepper in K.B. 16: 338 (1962)
      Rytigynia affinis (Robyns) Hepper in K.B. 17: 171 (1963) & F.W.T.A., ed. 2, 2: 186 (1963)

8. **P. phyllanthoidea** (*Baillon*) *Bridson* in K.B. 42: 629 (1987). Type: Kenya, Mombasa,
Boivin (P, holo., K, fragment!)

Glabrous shrub, 1.8–4.5(–7) m. tall, with lateral branches often short and reduced to
spurs; bark mid- to dark grey, often fissured. Leaves sessile to subsessile; blades drying
grey-green or occasionally blackish, oblong-elliptic to linear-oblong, 0.5–3.6(–4.5) cm.
long, 0.3–1.3(–1.7) cm. wide, rounded at apex, acute at base, with both lateral and tertiary
nerves obscure or occasionally with lateral nerves somewhat apparent, subcoriaceous,
dull above; domatia absent or present as ciliate to pubescent cavities; stipules triangular,
apiculate, 1–3 mm. long. Flowers unisexual or ? ♂, 4-merous; umbels sessile to subsessile.
Functionally ♂ flowers 1–5; pedicels 1.5–3 mm. long; bracts ovate, 2–2.5 mm. long,
acuminate; calyx-tube reduced; limb lobed to base; lobes triangular, 1–1.25 mm. long;
corolla white or yellowish; tube 2–5 mm. long, densely congested with hairs at throat;
lobes ovate, 2–4 mm. long, 1.5–2 mm. wide, shortly apiculate; stigmatic knob ± globose, 1
mm. in diameter. Functionally ♀ (and ♂) flowers 1–2; pedicels usually shorter; calyx-tube
0.75–1 mm. long. Fruit orange-brown, edible, ± heart-shaped, usually slightly wider than
long, 7–9 mm. long, 9–11 mm. wide, somewhat indented at apex; pyrene obovoid,
flattened on ventral face, 7 mm. long, 4 mm. wide with a shallow crest around the apex,
rugulose.

KENYA. Northern Frontier Province: Dandu, 11 Apr. 1952, *Gillett* 12764!; C. Kavirondo District:
      Makoko I., 19 Dec. 1939, *Hornby* S1054!; Kwale District: between Samburu and Mackinnon Road,
      near Taru, 5 Sept. 1953, *Drummond & Hemsley* 4170!
TANZANIA. Mbulu District: Lake Manyara National Park, Msasa R. gorge, 20 Jan. 1965, *Greenway &
      Kanuri* 12058!; Uzaramo District: Kunduchi, 18 km. NNW. of Dar es Salaam, 25 Jan. 1969, *Harris*
      2766!; Iringa District: 96 km. on Iringa–Morogoro road, Nov. 1964, *Procter* 2684!
DISTR. K 1, 3–7; T 1–3, 6, 7; Ethiopia, Somalia and Saudi Arabia
HAB. In *Acacia*, *Commiphora* bushland, often on rocks, or sometimes in *Brachystegia* woodland;
      0–2050 m.
SYN. *Canthium phyllanthoideum* Baillon in Adansonia 12: 220 (1878)
      Plectronia bogosensis Martelli, Fl. Bogos.: 42 (1886). Type: Ethiopia, Keren and Luglia, *Beccari* 155
         (FT, holo.)
      Canthium bogosense (Martelli) Penzig in Atti Cong. Bot. Genova, 1892: 323, t. 344 (1893); Hutch. &
         Bruce in K.B. 1941: 149 (1941); Cufod., E.P.A. 2: 1008 (1965)
      Rytigynia phyllanthoidea (Baillon) Bullock in K.B. 1932: 389 (1932)
      Rytigynia sp. nov. sensu K.T.S.: 471 (1961), quoad *Bally* 5735 & 6607
NOTE. This species occupies two distinct phytochoria, the Somalia-Masai regional centre of
      endemism and the Zanzibar-Inhambane regional mosaic (as recognised by F. White, 1983). At
      present insufficient evidence has been found to justify the recognition of subspecies. The leaves of
      the eastern more coastal element tend to be longer (1.5–4.5 cm.) than the inland element (0.6–2.5
      cm.), but the overlap is too great for use as a key character. Critical field observations concerning
      possible variation between ♂ and ♀ plants in any given area would be of interest, since some
      confusion between such variation and true taxonomic characters could occur.

9. **P. sp. D**

Small branched shrub 1 m. tall, glabrous; young stems covered with thin pale bark
rubbing off to reveal red-brown layer. Leaf-blades elliptic, 4.3–6.5 cm. long, 1.8–3 cm.
wide, obtuse to acute and sometimes apiculate at apex, acute at base, with 3–4 main pairs
of lateral nerves and with tertiary nerves obscure, chartaceous to subcoriaceous, slightly
shiny above; domatia absent; petiole 1–2 mm. long; stipules triangular at base with a lobe
above, 5–6 mm. long, eventually caducous. Flowers unisexual, only functionally ♂ flowers
known, 4-merous; umbels 2–5-flowered; peduncles ± 1 mm. long; pedicels not apparent;
bracts 6–7 mm. long, acuminate with rusty coloured hairs inside. Calyx-tube reduced;
limb lobed almost to base; lobes 4–5, triangular, equal or almost so, ± 3 mm. long. Corolla
cream; tube 3–4 mm. long, densely congested with white hairs at throat; lobes ovate-
triangular, 2.75 mm. long, 1.25 mm. wide, apiculate-acuminate. Stigmatic knob 1 mm.
long, 0.75 mm. wide. Fruit not known.

Tanzania.   Lindi District: SW. end of Rondo Plateau, 15 km. NE. of Mihima, 12 Nov. 1988, *Mackinder & Lock* 36!
Distr.   **T** 8; not known elsewhere
Hab.   Undergrowth of dry forest at escarpment edge; 750 m.

Note.   The well-developed calyx-lobes and reduced habit indicate that this species is most closely related to the South African *P. hystrix* (Bremek.) Bridson.

107. PSYDRAX

Gaertn., Fruct. & Sem. 1: 125, t. 26/2 (1778); DC., Prodr. 4: 476 (1830), pro parte; A. Rich., Mém. Fam. Rub. : 110 (1830), pro parte & in Mém. Soc. Hist. Nat. Paris 5: 190 (1834), pro parte; Bridson in K.B. 40: 687–725 (1985)

Caranda Gaertn., Fruct. & Sem. 2: 17, t. 83/5 (1790)
Mitrastigma Harv., Lond. Journ. Bot. 1: 20 (1842); E.P. Phillips, Gen. S. Afr. Pl.: 587 (1926)
Mesoptera Hook.f. in G.P. 2: 130 (1873) & in Fl. Brit. Ind. 3: 136 (1882); K. Schum. in E. & P. Pf. IV. 4: 92 (1891)

Trees, shrubs, scandent shrubs or sometimes lianes. Leaves mostly evergreen but sometimes deciduous and restricted to apex of branches, paired or rarely ternate, petiolate or sometimes subsessile; blades typically coriaceous but less often chartaceous, glabrous or sometimes pubescent; domatia glabrous or pubescent or sometimes absent; stipules usually with a truncate base bearing a decurrent, slightly to very strongly keeled, sometimes foliaceous lobe or, less often, lanceolate to ovate and soon caducous, never with silky white hairs inside. Flowers 4–5-merous, borne in sessile to pedunculate umbellate cymes, in pedunculate clearly branched cymes or very rarely solitary; bracts and bracteoles typically inconspicuous or occasionally conspicuous or sometimes (not in Africa) entirely enclosing developing inflorescence and eventually rupturing into 2 or more segments. Calyx-tube broadly ellipsoid to almost hemispherical; limb a truncate to dentate rim, occasionally equalling the tube in length but often much shorter. Corolla white to yellow; tube broadly cylindrical, mostly equal to the lobes in length or somewhat longer or shorter, inside usually with a ring of deflexed hairs or sometimes with hairs not restricted to a well-defined ring or absent, often pubescent at throat; lobes reflexed, often thickened towards apex and obtuse to acute, rarely acuminate-apiculate, very rarely apiculate. Stamens set at corolla throat; filaments well developed; anthers lanceolate to narrowly ovate, attached shortly above base, reflexed. Style long, slender, always exceeding corolla-tube, glabrous, sometimes narrowing at apex; stigmatic knob cylindrical, always longer than wide, occasionally flaired or somewhat narrowed at base, hollow to about mid-point, bifid or rarely deeply cleft at apex when mature; ovary 2-locular, each locule containing 1 ovule, attached to the upper ⅓ of the septum. Disk glabrous or pubescent. Fruit a 2-seeded drupe, small (or sometimes large in India) ellipsoid to didymous; pyrenes cartilaginous to woody, ± ellipsoid and flattened on ventral face or laterally flattened-spherical, usually with a very shallow crest extending from point of attachment around apex which eventually splits and with grooves from point of attachment extending to lateral faces, scarcely bullate, rugulose or deeply furrowed. Seeds with endosperm entire; testa very finely reticulate; embryo almost straight or curved to a C-shape (according to the pyrene shape); radicle erect; cotyledons small, parallel to ventral face of seed. Figs. 131/15, 17, 24, p. 752 & 132/23, p. 754.

A large genus of at least a hundred species, occurring throughout the Old World tropics. Combinations for most non-African species have not yet been made; 34 species occur in Africa of which 18 species are represented in the Flora area.

Two subgenera are recognised. Subgen. *Psydrax* includes species 1–16 and is defined by characters given in the first couplet of the key to species below. Subgen. *Phallaria* (Schumach. & Thonn.) Bridson (*Phallaria* Schumach. & Thonn.; *Canthium* sect. *Pleurogaster* DC.) includes species 17 and 18, and is likewise defined below. For further details see Bridson in K.B. 40: 692–725 (1985).

1. Trees, shrubs, or occasionally scandent bushes; corolla-lobes obtuse to acute or very rarely shortly apiculate; corolla-tube inside with a ring of deflexed hairs above mid-point or sometimes without a definite ring of deflexed hairs; stigmatic knob not markedly widened at base (fig. 131/15, p. 752); fruit ovate to didymous or rarely ellipsoid (subgen. *Psydrax*) 2

Lianes or at least scandent shrubs; corolla-lobes
   acuminate-apiculate; corolla-tube inside with a ring of
   deflexed hairs below mid-point; stigmatic knob
   widened at base (fig. 131/17, p. 752); fruit ± didymous
   (subgen. *Phallaria*) 18
2. Leaves deciduous, usually restricted to apex of stem or
   short lateral branches, chartaceous to subcoriaceous,
   not markedly shiny above, pubescent or glabrous 3
   Leaves not (or rarely) deciduous, well spaced along the
   stems (except in some dryland species), typically
   subcoriaceous to coriaceous or occasionally
   chartaceous, usually shiny above, glabrous or very
   occasionally pubescent 5
3. Inflorescence sessile, subumbellate; corolla-tube distinctly
   longer than lobes . . . .   8. *P. robertsoniae*
   Inflorescence a pedunculate compact corymb; corolla-
   tube shorter or slightly longer than lobes 4
4. Leaf-blades 1.5–13 cm. long, obtuse to rounded or
   sometimes acute at base; lateral nerves not impressed
   above; corolla-lobes 1.25–2 mm. long . . . . .   3. *P. livida*
   Leaf-blades 1.3–3 cm. long, rounded or sometimes
   truncate to subcordate at base; lateral nerves impressed
   above; corolla-lobes 2.5–3 mm. long . . . . . .   4. *P. whitei*
5. Inflorescences clearly pedunculate, with branches
   congested to lax 6
   Inflorescences sessile to subsessile or shortly pedunculate,
   with branches absent or rudimentary 8
6. Tall trees with horizontal branches (sometimes said to be
   palm-like), myrmecophilous; leaves chartaceous to
   subcoriaceous, often large with 7–10 main pairs of
   lateral nerves; stipules narrowly ovate to ovate,
   caducous; bracteoles apparent . . . .   1. *P. subcordata*
   Shrubs or trees, but not with habit as above; myrmecophily
   not recorded; leaves subcoriaceous to coriaceous,
   larger or small, with up to 8 main pairs of lateral
   nerves; stipules triangular at base, terminating in
   decurrent lobe; bracteoles inconspicuous 7
7. Disk pubescent; inflorescence-branches lax; leaves larger
   (5.5–15 cm. long); fruit 0.8–1.4 cm. wide . . . .   2. *P. parviflora*
   Disk glabrous; inflorescence-branches moderately
   compact; leaves smaller, 4.5–7.5 cm. long; fruit 0.7–0.8
   cm. wide . . . .   5. *P. faulknerae*
8. Inflorescence 1–2-flowered; leaves with bases rounded or
   occasionally subcordate; petioles 1–3 mm. long;
   corolla-lobes shortly apiculate; fruit obovate in outline 16. *P. recurvifolia*
   Inflorescence 2–50-flowered; leaf-bases and petioles only
   occasionally as above; corolla-lobes never apiculate;
   fruit ± circular in outline or slightly broader 9
9. Leaves subsessile (petioles not exceeding 2 mm. long);
   blades oblong-elliptic, up to 16 cm. long, rounded and
   slightly subcordate at base . . . .   11. *P. kibuwae*
   Leaves not as above; petioles at least 2.5 mm. long; blades
   not exceeding 10.5 cm. long 10
10. Low cushion-shaped shrub with prostrate and erect
   branches to 0.2 m. tall; leaf-blades narrowly oblong-
   elliptic . . . .   13. *P. shuguriensis*
   Erect or sometimes scandent shrubs or small trees, 1.6–10
   m. tall; leaves mostly narrowly elliptic to round 11
11. Leaves broadly elliptic to round, obtuse to truncate at base,
   moderately shiny above; lateral nerves in 2–3 main
   pairs; tertiary nerves almost obscure 12

Leaves narrowly to broadly elliptic or if broadly elliptic to round then acute at base, very shiny above, with lateral nerves in 4–5 main pairs and tertiary nerves clearly apparent 13

12. Leaves small, 0.8–2.8 cm. long; reduced shrubs with numerous lateral spurs set at right-angles to stem; corolla-tube 2.75 mm. long; lobes 4 mm. long   10. *P. polhillii*

Leaves larger, 2.8–3.7 cm. long; shrub to 4 m. tall; branches angled to stem; corolla-tube ± 3 mm. long; lobes ± 3 mm. long . . . . 9. *P. sp. A*

13. Pedicels puberulous to pubescent; tertiary nerves coarsely reticulate, apparent but not conspicuous or sometimes obscure 14

Pedicels glabrous; tertiary nerves finely to coarsely reticulate, usually conspicuous 16

14. Petioles 6–10 mm. long; bark white-grey; corolla-tube distinctly longer than lobes . . . . 8. *P. robertsoniae*

Petioles usually shorter; bark not conspicuously pale; corolla-tube shorter than or equal to lobes 15

15. Leaf-blades narrowly to broadly elliptic or less often obovate, gradually acuminate or sometimes acute at apex; petioles glabrous . . . . 6. *P. schimperiana*

Leaf-blades broadly elliptic to ± round, obtuse or rounded at apex; petioles puberulous . . . . 7. *P. lynesii*

16. Lateral nerves in 5–7 main pairs; blades broadly elliptic, 8–10 cm. long, 4–5 cm. wide, abruptly subacuminate at apex . . . . 15. *P. sp. B*

Lateral nerves in 3–4 main pairs; blades smaller, 3–8 cm. long, 1.3–5 cm. wide, obtuse to acute at apex 17

17. Leaves often ternate; blades elliptic or less often narrowly elliptic; tertiary nerves rather finely reticulate; pedicels 0.6–1.6 cm. long; corolla-tube 3–4 mm. long; lobes 3 mm. long . . . . 14. *P. micans*

Ternate leaves not recorded; blades elliptic to ± round; tertiary nerves fairly coarsely reticulate; pedicels 2–8 mm. long; corolla-tube 4–5 mm. long; lobes 6 mm. long . . . . 12. *P. kaessneri*

18. Leaves stiffly chartaceous, occasionally subcoriaceous, dull or slightly shiny above; lateral nerves in 4–5 main pairs; pedicels 2–3(–4–6) mm. long (in flowering stage); flowers smaller (corolla-tube 3–4 mm. long; lobes 2.5–4 mm. long; stigmatic knob 0.75–1 mm. long); young stems square, distinctly furrowed . . . . 17. *P. acutiflora*

Leaves coriaceous, or occasionally subcoriaceous, shiny above; lateral nerves (4–)5–8 main pairs; pedicels 5–8 mm. long (in flowering stage); flowers larger (corolla-tube 3.75–6 mm. long; lobes 3.5–6 mm. long; stigmatic knob 1–1.75 mm. long); young stems round to square, occasionally furrowed . . . . 18. *P. kraussioides*

---

1. **P. subcordata** (*DC.*) *Bridson* in K.B. 40: 698 (1985); Fl. Rwanda 3: 150, fig. 44.3 (1985). Type: Gambia, near Albreda, *Leprieur & Perrottet* (G, holo.)

Tree 5–15 m. tall, said to have a palm-like habit, associated with ants and often the branches hollow and swollen with access-pores present; young branches distinctly square, glabrous or pubescent. Leaf-blades oblong to ovate or sometimes broadly ovate, 9.5–22 cm. long, 4.5–16.5 cm. wide, shortly acuminate at apex, obtuse, rounded, truncate or sometimes cordate at base, glabrous or sometimes sparsely pubescent above, glabrous or pubescent on nerves or sparsely pubescent beneath, papery to subcoriaceous, dull to moderately shiny above; lateral nerves in 7–10 main pairs; tertiary nerves moderately conspicuous beneath; domatia present as hair-lined cavities in the axils of lateral and tertiary nerves; petioles 1–1.5 cm. long, glabrous or pubescent; stipules narrowly ovate to ovate, 0.4–1 cm. long (or 1.4 cm. on sapling or coppice growth), caducous. Flowers with

S.R.C

FIG. 159. *PSYDRAX SUBCORDATA* var. *SUBCORDATA* — **1**, flowering branch, × ⅔; **2**, flower, × ± 6; **3**, calyx, × ± 10; **4**, longitudinal section through ovary, × ± 16; **5**, fruiting branch, × ⅔; **6**, fruit, × 2; **7**, pyrene (2 views), × 2. 1–5, source unknown; 6, 7, from *Dschang staff in C.N.A.D.* 1779. Drawn by Stella Ross-Craig, with 6, 7 by Sally Dawson.

extremely distasteful smell, 5-merous, borne in pedunculate, 80–120-flowered, dichotomous corymbose cymes; true peduncles rather short, portion above the bract usually much longer, 0.8–2(–4) cm. altogether, pubescent or glabrous; pedicels 1.5–2.5 mm. long, sparsely puberulous to pubescent; bracts connate or sometimes free, 1–4 mm. long; bracteoles free, conspicuous or inconspicuous, up to 3 mm. long. Calyx-tube ± 1 mm. long, glabrous or sometimes pubescent towards base; limb a dentate or truncate rim 0.5–1 mm. long, ciliate, puberulous inside. Corolla white or creamy yellow; tube 2–3.5 mm. long, with a ring of deflexed hairs set above mid-point inside; lobes oblong-ovate, 2.5 mm. long, 1.25–1.5 mm. wide, acute. Anthers not persistent in opened flower. Style 5–7 mm. long; stigmatic knob 1.5–2 mm. long, ribbed. Disk pubescent or sometimes glabrous. Fruit black or greyish black, wider than long, 0.6–0.8 cm. long, 0.9–1.2 cm. wide, usually rather prominent at apex; pyrenes strongly curved, 6–8 mm. long, 4.5–6 mm. wide, 3–4 mm. thick, with a groove on either side from point of attachment to centre of lateral face and with a shallow crest around apex, cartilaginous, very slightly rugulose.

var. **subcordata**

Bracts 1–2 mm. long; bracteoles inconspicuous, sometimes up to 2 mm. long. Calyx-limb 0.5(–1) mm. long. Fig. 159.

UGANDA. Ankole District: Igara, Kalinzu Forest, Feb. 1953, *Osmaston* 2846! & Mar. 1939, *Purseglove* 623!
DISTR.   U 2; W. Africa, Central African Republic, Zaire, Rwanda, Sudan, Zambia and Angola
HAB.   Forest; 1500 m.

SYN.   *Canthium subcordatum* DC., Prodr. 4: 473 (1830); Hiern in F.T.A. 3: 141 (1877); F.W.T.A., ed. 2, 2: 184, fig. 239 (1963), pro parte majore; Fl. Pl. Lign. Rwanda: 550, fig. 182.3 (1983)
    C. glabriflorum Hiern in F.T.A. 3: 140 (1877) & Cat. Afr. Pl. Welw. 1: 474 (1898). Types: S. Tomé, *Mann* 1077 (K, syn.!) & Nigeria, Old Calabar, *Thomson* 112 (K, syn.!)
    C. polycarpum Hiern in F.T.A. 3: 139 (1877); F.P.S. 2: 430 (1952). Type: Sudan, Niamniam Land, Nabambisso, *Schweinfurth* 3051 (K, holo.!, BM, iso.!)
    Plectronia glabriflora (Hiern) K. Schum. in P.O.A. C: 386 (1895); Holland in K.B., Add. Ser. 9: 359 (1915); De Wild., Pl. Bequaert. 3: 185 (1925)
    P. subcordata (DC.) K. Schum. in P.O.A. C: 386 (1895), pro parte
    Canthium welwitschii Hiern, Cat. Afr. Pl. Welw. 1: 475 (1898). Types: Angola, Golungo Alto, Alto Queta, *Welwitsch* 3148 & 3149 (LISU, syn., BM, K, isosyn.!)
    Plectronia welwitschii (Hiern) K. Schum. in Just's Bot. Jahresb. 1898 (1): 393 (1900)
    P. formicarum K. Krause in E.J. 54: 351 (1917). Type: Cameroon, Molundu, Lokomo Bomba – Bangé, *Mildbraed* 4526 (B, holo. †, HBG, iso.!)
    P. laurentii De Wild. var. *katangensis* De Wild., Notes Fl. Kat. 7: 61 (1921). Type: Zaire, Shaba, Biano Plateau (Tshsiuka), *Homblé* 255 (BR, holo.!)
    Canthium sp. 1 sensu F.F.N.R. : 403 (1962)
    [*C. vulgare* sensu F.W.T.A., ed. 2, 2: 184 (1963), pro parte, quoad *Espirito Santo* 2041 & 2052, *non* (K. Schum.) Bullock]

NOTE.   A second variety var. *connata* (De Wild. & Th. Dur.) Bridson with larger bracts, bracteoles and calyx-limb occurs in the Central African Republic and Zaire.

2. **P. parviflora** (*Afzel.*) *Bridson* in K.B. 40: 700 (1985); Fl. Rwanda 3: 152 (1985). Type: Sierra Leone, *Afzelius* 3441a (UPS-THUNB, holo.!, BM, fragment!)

Shrub or tall timber tree, 2–27 m. tall, sometimes with a fluted bole; young stems square or round, glabrous or rarely puberulous. Leaves not restricted to apices of branches; blades drying dull to bright green, brown or blue-black, narrowly elliptic to elliptic, oblong-elliptic or ovate, 5.5–15.5 cm. long, 2–8 cm. wide, distinctly acuminate at apex, acute, obtuse or rounded at base, coriaceous; midrib whitish or red; lateral nerves in 4–8 main pairs; tertiary nerves apparent, sometimes raised above; domatia present as rather prominent blisters in the axils of the lateral nerves, with or without an eruption, glabrous, ciliate or pubescent; petioles 3–10 mm. long; stipules 2–7 mm. long, triangular at base, terminating in a usually decurrent lobe. Flowers 4-merous, borne in pedunculate 20–100-flowered corymbs 2–6 cm. across; peduncles 0.3–2 cm. long, glabrescent to pubescent; pedicels 2–9 mm. long, adpressed pubescent; bracteoles inconspicuous. Calyx-tube 0.75–1 mm. long, glabrous to pubescent; calyx-limb 0.25–1.25 mm. long, shorter or longer than the tube, and often wider, truncate to repand, glabrous to pubescent outside and inside. Corolla whitish; tube 2–3 mm. long, with a ring of deflexed hairs above mid-point inside, and pubescent at throat; lobes oblong-ovate, 1.5–2.5 mm. long, 1.25–1.5 mm. wide, obtuse to acute. Anthers reflexed. Style up to 8 mm. long; stigmatic knob cylindrical, (0.5–)0.75–1 mm. long. Disk pubescent. Fruit black when mature, almost didymous, 5–8 mm. long, 8–14 mm. wide; pyrene ¾-circular in outline, 5–8 mm. wide, with a groove extending from point of attachment, on either side, to lateral face, perpendicular (in 1-seeded fruit) or angled upwards (2-seeded fruit), cartilaginous, slightly bullate.

KEY TO SUBSPECIES

Stipules 6–7 mm. long; leaves turning blue-black when dry; lateral nerves in (6–)7–8 main pairs; tertiary nerves finely reticulate and apparent above; domatia pubescent subsp. c. **melanophengos**

Stipules 2–6 mm. long; leaves remaining green or turning brown when dry, lateral nerves in 4–7 main pairs; tertiary nerves infrequently apparent above, domatia glabrous to ciliate:

 Fruit 0.8–1.1 cm. wide; midribs pale; calyx sparsely pubescent to pubescent; limb longer or shorter than tube; domatia glabrous . . . . subsp. a. **parviflora**

 Fruit 1–1.4 cm. wide; midribs red; calyx glabrous to glabrescent; limb always shorter than tube; domatia ciliate . . . . subsp. b. **rubrocostata**

subsp. a. **parviflora**

Shrub or tree 2–20 m. tall. Leaves drying green; midrib pale; lateral nerves in 4–6(–7) main pairs; tertiary nerves seldom apparent above; domatia glabrous. Calyx sparsely pubescent to pubescent; limb 0.25–1.25 cm. long, shorter or longer than tube. Corolla-tube 2–2.75 mm. long. Fruit 8–11 mm. wide.

UGANDA. Kigezi District: Kinkizi, Kirima, Oct. 1949, *Purseglove* 3132!; Teso District: Kyere Rock, Dec. 1932, *Chandler* 1054!; Masaka District: Lake Kayanja, E. side, 25 Apr. 1969, *Lye & Morrison* 2640!

KENYA. N. Kavirondo District: Kakamega Forest, Kisieni – Kakamega Saw Mill, 2 May 1979, *Bridson* 37!; Kericho District: Cheptuiyet, Belgut, Aug. 1960, *Kerfoot* 2191!; Masai District: Masai-Mara Game Reserve, Endikum (Esoit Oloololo) Escarpment, 20 July 1978, *Msafiri et al.* 411!

TANZANIA. Biharamulo District, Aug. 1952, *Procter* 73!; Mwanza District: Lake Victoria, Maisome I., 15 Mar. 1962, *Carmichael* 867!; Mpanda District: Mahali Mts., Mokoloka, 19 Sept. 1952, *Jefford, Juniper & Newbould* 2456!

DISTR. U 1–4; K 5, 6; T 1, 4; West Africa, Zaire, Rwanda, Sudan, Ethiopia, Zambia and Angola

HAB. Forest, thickets or sometimes bush clumps in grassland; 1070–1675(–1850) m.

SYN. *Pavetta parviflora* Afzel., Remed. Guin.: 47 (1815); DC., Prodr. 4: 492 (1830)
  Canthium afzelianum Hiern in F.T.A. 3: 142 (1877), pro majore parte, excl. specim. *Vogel*. Type: as for *P. parviflora* Afzel., *non Canthium parviflorum* Lam.
  Plectronia vulgaris K. Schum. in P.O.A. C: 386 (1895); K. Krause in Z.A.E. 1907–8, 2: 326 (1911). Type: Tanzania, Mwanza/Kwimba District, shore of Speke Gulf, *B.D. Burtt* 2471 (K, neo.!, chosen by Bridson, 1985)
  Canthium golungense Hiern, Cat. Afr. Pl. Welw. 1: 478 (1898). Types: Angola, Golungo Alto, *Welwitsch* 3153–3156 (LISU, syn., BM, K, isosyn.!)
  Plectronia golungensis (Hiern) K. Schum. in Just's Bot. Jahresb. 1898(1): 393 (1900)
  Canthium golungense Hiern var. *parviflorum* S. Moore in J.L.S. 37: 161 (1905). Type: Uganda, Ankole District, 96 km. up Kagera valley, near Mulema, *Bagshawe* 209 (BM, holo.!)
  Plectronia brieyi De Wild. in F.R. 13: 380 (1914). Type: Zaire, Mayombe, Ganda-Sundi, *Comte de Briey* 226 (BR, holo.!, K, iso.!)
  Canthium vulgare (K. Schum.) Bullock in K.B. 1932: 374 (1932), pro parte, excl. *Holst* 2727 & *Graham* 1823; T.T.C.L.: 488 (1949); I.T.U., ed. 2: 340 (1952); F.P.S. 2: 431 (1952); F.F.N.R.: 403 (1962); Hepper, F.W.T.A., ed. 2, 2: 184 (1963), pro majore parte; Fl. Pl. Lign. Rwanda: 552, fig. 183.4 (1982)
  Canthium giordanii Chiov. in Atti Reale Accad. Ital., Mem. Cl. Sci. Mat. Nat. 11: 35 (1941); E.P.A.: 1009 (1965). Types: Ethiopia, Galla Sidamo, Saio, *Giordano* 2450 & 2539 (FT, syn.!)
  [*C. multiflorum* sensu Hepper, F.W.T.A., ed. 2, 2: 182 (1963), pro parte, quoad syn. *C. afzelianum*, *non* (Schumach. & Thonn.) Hiern]

NOTE. This species is very variable and it is possible that further study, especially using numerical methods, may indicate that several further subspecies should be recognised. Geographical groupings do seem to exist but the disjunctions between them are not clear-cut and the character differences cannot be easily applied to a conventional key. The West African specimens tend to have rather broader, subcoriaceous leaves with 5–6 main pairs of lateral nerves and the calyx-limb is usually longer than the calyx-tube, while in general, the specimens from the Flora area have somewhat narrower coriaceous leaves, often folded, with 4–5(–6) main pairs of lateral nerves and the calyx-limb is shorter than or equal to the calyx-tube. Some specimens from Sudan, Zaire, Rwanda and Tanzania (Biharamulo and Mpanda Districts) are more or less intermediate between the above two groups. One specimen from Ufipa District (*Richards* 13057) has thickly coriaceous leaves with 6–7 main pairs of lateral nerves and seems to resemble specimens from north-western Zambia. These specimens appear yellow-green, when dry, and may possibly represent an ecotype associated with certain metallic deposits. The specimens from Angola (*C. golungense*) could be maintained as a further subspecies.

subsp. b. **rubrocostata** (*Robyns*) *Bridson* in K.B. 40: 702 (1985); Fl. Rwanda 3: 152, fig. 44.1 (1985). Type: Kenya, S. Nyeri District, Kiringa R. – Mukengeria R., *R.E. & T.C.E. Fries* 2107 (UPS, holo., BR, K, iso.!)

Shrub or tree 6–12(–24) m. tall. Leaves often drying brownish; midrib red; lateral nerves in 5–7 main pairs; tertiary nerves seldom conspicuous above; domatia ciliate. Calyx glabrous to glabrescent; limb 0.5–0.75 mm. long, always shorter than tube. Corolla-tube 2–3 mm. long. Fruit 1–1.4 cm. wide.

UGANDA. W. Nile District: Nyagak R., near Niapea [Neapea], May 1936, *Eggeling* 3011!; Mbale District: Elgon, Oct. 1945, *St. Clair-Thompson* T 1798!
KENYA. Baringo District: Katimok Forest, Kamasia, Oct. 1930, *Dale* in *F.D.* 1063!; Kiambu District: Gatamayu [Katamayu], 14 Oct. 1934, *V.G.L. van Someren* in *C.M.* 6720!; Teita Hills, SE. end of Ngangao Mt., 9 Feb. 1966, *Gillett & Osborn* 17115!
TANZANIA. Arusha National Park, Bilo R., 17 Apr. 1968, *Greenway & Kanuri* 13471!; Morogoro District: Uluguru Mts., Mgeta R. above Bunduki, 1 Jan. 1975, *Polhill & Wingfield* 4629!; Iringa District: Dabaga Highlands, Kibengu, 15 Feb. 1962, *Polhill & Paulo* 1496A!
DISTR. U 1, 3; K 3–5, 7; T 2, 3, 6, 7; Sudan and Malawi
HAB. Forest; (1375–)1700–2750 m.

SYN. *Canthium rubrocostatum* Robyns in N.B.G.B. 10: 616 (1929); Bullock in K.B. 1932: 373 (1932); T.T.C.L.: 488 (1949); F.P.S. 2: 431 (1952); I.T.U., ed. 2: 340 (1952); K.T.S.: 432 (1961)
[*C. melanophengos* sensu Bullock in K.B. 1932 : 375 (1932), pro parte, quoad *Rammell in F.D.* 1054 (corrected in the corrigenda in K.B. 1933: 48 (1933)); Chiov., Racc. Bot. Miss. Consol. Kenya : 55 (1935), quoad *Balbo* 90, *non* Robyns]

subsp. c. **melanophengos** (*Bullock*) *Bridson* in K.B. 40: 703 (1985). Type: Rwanda, Rukarara, Rugege Forest, *Mildbraed* 1015 (B, holo. †, K, fragment!)

Tree 10–20 m. tall. Leaves drying blue-black; midrib pale; lateral nerves in (6–)7–8 main pairs; tertiary nerves finely reticulate, apparent above; domatia pubescent. Calyx glabrous; limb 0.75–1 mm. long, always shorter than tube. Corolla-tube 3 mm. long. Fruit 1.4 cm. wide.

UGANDA. Kigezi District, *St. Clair-Thompson* 2532!
DISTR. U 2; Rwanda
HAB. ? Forest; 1525–1850 m.

SYN. *Canthium melanophengos* Bullock in K.B. 1932: 375 (1932), pro majore parte; Fl. Pl. Lign. Rwanda: 548, fig. 185.1 (1982)

3. **P. livida** (*Hiern*) *Bridson* in K.B. 40: 705 (1985). Type: Mozambique, Moramballa, *Kirk* (K, holo.!)

Shrub or small tree 0.8–8 m. tall; green twigs glabrescent to densely pubescent, older stems covered with pale tissue-like bark which flakes off to reveal darker underlayer. Leaves restricted to green stems or occasionally on older stems, not always fully mature at time of flowering; blades narrowly ovate to ovate or occasionally broadly ovate, 1.5–13 cm. long, 1–7 cm. wide, acute to subacuminate, sometimes acuminate or obtuse to rounded at apex, obtuse to rounded or sometimes acute at base, herbaceous, dull to slightly shiny above, glabrous to densely pubescent on both faces; lateral nerves in 4–5 main pairs; tertiary nerves apparent; domatia present as membranous pockets, ciliate; petioles 2–5(– 10) mm. long; stipules triangular-acuminate, 4–7 mm. long, with lobe decurrent when mature. Flowers scented or not, 4-merous, borne in 6–70-flowered pedunculate compact corymbs; peduncle 0.3–1.5 cm. long, pubescent; inflorescence-branches fairly short, rather acutely angled; pedicels 0.3–1(–1.5) cm. long, pubescent; bracteoles inconspicuous. Calyx-tube 1 mm. long, broader than long, pubescent or glabrous save for a few hairs towards base; limb 0.5 mm. long, irregularly dentate, ciliate. Corolla greenish white (and/or yellow); tube 1.25–3 mm. long, pubescent at throat but lacking definite ring of deflexed hairs; lobes oblong, 1.25–2 mm. long, 1–1.5 mm. wide, obtuse to rounded. Anthers reflexed. Style up to 6 mm. long; stigmatic knob 1 mm. long, obscurely ribbed; disk glabrous. Fruit black when mature, broadly oblong in outline, 5–6 mm. long, 7.5–8 mm. wide, not or scarcely indented; pyrene ± ellipsoid, 5–6 mm. long, 3.5 mm. wide, rugulose.

KENYA. Machakos District: near Tawa on Tawa–Ayani road, 10 Mar. 1955, *Nicholson* 50!; Masai District: Emali Valley, 7 Mar. 1940, *V.G.L. van Someren* 155!; Teita District: 3 km. E. of Bura railway station, 17 Jan. 1972, *Gillett* 19572!

Tanzania. Biharamulo District: Lusahunga, 10 Oct. 1960, *Tanner* 5335!; Singida District: 32 km. from Issuna on the Singida–Manyoni road, 13 Apr. 1964, *Greenway & Polhill* 11548!; Lindi District: Lake Lutamba, 1 Nov. 1934, *Schlieben* 5573!

Distr. **K** 4, 6, 7; **T** 1, 3–8; Zaire, Burundi, Mozambique, Malawi, Zambia, Zimbabwe, Botswana, Angola and South Africa

Hab. In thickets and scrub, often on rocky hillsides; 240–1525 m.

Syn. *Canthium lividum* Hiern in F.T.A. 3: 144 (1877)
  Plectronia livida (Hiern) K. Schum. in P.O.A. C: 386 (1895); Eyles in Trans. Roy. Soc. S. Afr. 5: 493 (1916)
  P. syringodora K. Schum. in P.O.A. C: 386 (1895). Types: Tanzania, Mpanda District, Ugalla R., *Boehm* 23a (B, syn.†, K, fragment!) & Karema, *Boehm* 68 (B, syn.†, K, fragment!) & Tabora District, Wala [Walla] R., *Boehm* 92a (B, syn.†)
  Canthium huillense Hiern, Cat. Welw. Afr. Pl. 1: 476 (1898); Bullock in K.B. 1932: 370 (1932); T.T.C.L.: 487 (1949); K.T.S.: 429 (1961); F.F.N.R.: 404 (1962). Types: Angola, Huila, Lake Ivantâla, *Welwitsch* 3145 (LISU, syn., BM, isosyn.!) & Lopollo and Moninho, *Welwitsch* 3146 (LISU, syn., BM, K, isosyn.!)
  Plectronia huillensis (Hiern) K. Schum. in Just's Bot. Jahresb. 1898(1): 393 (1900)
  P. heliotropiodora K. Schum. & K. Krause in E.J. 39: 540 (1907). Types: Tanzania, Lake Tanganyika, *Boehm* 68 (B, syn. †, K, fragment!) and Songea District, Ungoni, Ruanda, *Busse* 875 (B, syn.†, EA, K, isosyn.!)
  P. junodii Burtt Davy in K.B. 1921: 192 (1921). Type: South Africa, Transvaal, Pietersburg, Shilouvane, *Junod* 720 (K, holo.!)
  Canthium clityophilum Bullock in K.B. 1932: 382 (1932); T.T.C.L.: 485 (1949). Type: Tanzania, Morogoro District, Uluguru Mts., Tawa [Tana], *Stuhlmann* 8928 (B, holo.†, K, fragment!)
  C. syringodorum (K. Schum.) Bullock in K.B. 1932: 382 (1932); T.T.C.L.: 488 (1949)
  C. junodii (Burtt Davy) Burtt Davy in K.B. 1935: 568 (1935)
  [*C. vulgare* sensu Chiov., Racc. Bot. Miss. Consol. Kenya: 55 (1935), *non* (K. Schum.) Bullock]
  Plectronia wildii Suesseng. in Trans. Rhod. Sci. Assoc. 43: 59 (1951). Type: Zimbabwe, Marandellas, *Dehn* 590 (SRGH, holo., M, iso.!)
  Canthium gymnosporioides Launert in Mitt. Bot. Staats. München 16: 314 (1957). Type: Caprivi Strip, Okavango, Pupa [Popa] Falls, *Volk* (M, holo.!)
  C. wildii (Suesseng.) Codd in Kirkia 1: 109 (1961)

Note. The inclusion of *Canthium clityophilum* in synonymy here is rather tentative. Bullock places it in his fasciculate series, which suggests that the peduncle is very reduced, also he describes the flower-buds as being 5-lobed. The fragment at Kew consists of detached leaves and fruit only; it is an almost identical match with the distinctly pedunculate specimen *Wallace* 368 from the same area.
  P. livida is a very plastic species, the leaves varying from glabrous to densely pubescent and the flowers in size. Two specimens from Mpwapwa (*Lindeman* 249 & 252), although definable as this species, show similarities to *P. whitei*. Although the leaf-size falls within the range of *P. whitei* the leaves may not have reached their maximum size.

4. **P. whitei** *Bridson* in K.B. 40: 706, fig. 4 (1985). Type: Malawi, Nyika Plateau, *Richards* 22527 (K, holo.!)

Shrub or tree 1.5–9 m. tall (also described as a dwarf shrub 0.3 m. tall); very young stems pubescent; older stems with greyish bark. Leaves restricted to green stems or sometimes on mature stems; blades ovate or sometimes elliptic or round, 1.3–3 cm. long, 0.9–2.7 cm. wide, obtuse or sometimes rounded at apex, rounded to truncate or occasionally subcordate at base, herbaceous to subcoriaceous, somewhat shiny and glabrous above, sparsely pubescent on midrib beneath; lateral nerves in 3–4 main pairs, impressed above; tertiary nerves apparent; domatia present as membranous ciliate pockets; petioles not exceeding 2 mm. long; stipules triangular-acuminate, 3–4 mm. long, keeled when mature. Flowers scented (described as sweet or foetid), 4-merous, borne in (5–)12–70-flowered, compact, pedunculate corymbs; peduncle 0.3–1.5 cm. long, pubescent; inflorescence-branches acutely angled; pedicels 0.2–1.3 cm. long, pubescent; bracteoles inconspicuous. Calyx-tube 1 mm. long, broader than long, glabrous or with a few hairs towards base; limb 0.5–0.75 mm. long, irregularly dentate. Corolla greenish cream; tube 1.5–2 mm. long, pubescent at throat but lacking definite ring of deflexed hairs; lobes oblong, 2.5–3 mm. long, 1.5 mm. wide, obtuse. Anthers reflexed. Style 6–8 mm. long; stigmatic knob 1 mm. long, obscurely ribbed. Disk glabrous. Fruit black when mature, broadly oblong in outline, 6 mm. long, 7 mm. wide, not indented; pyrenes ± ellipsoid with ventral face flattened, 7 mm. long, 4 mm. wide, 3 mm. thick, rugulose. Fig. 160.

Tanzania. Iringa District: Sao Hill, Ipogoro–M'kawa track, 12 Dec. 1961, *Richards* 15559!; Njombe District: Mlangali–Njombe road, 3 Feb. 1961, *Richards* 14214!; Songea District: Matengo Hills, Lupembe Hill, 29 Feb. 1956, *Milne-Redhead & Taylor* 8918!

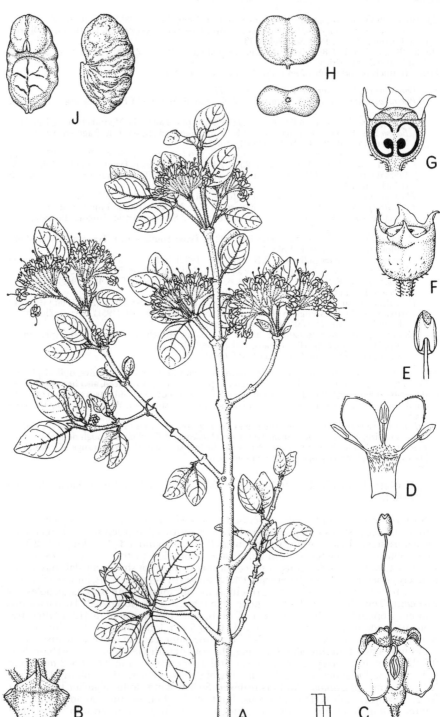

FIG. 160. *PSYDRAX WHITEI* — **A**, flowering branch, × ⅔; **B**, stipule, × 4; **C**, flower, × 6; **D**, section through corolla, × 6; **E**, longitudinal section through stigmatic knob, × 9; **F**, calyx, × 9; **G**, section through ovary, × 9; **H**, fruit (2 views), × 1½; **J**, pyrene (2 views), × 3. A–G, from *Richards* 22527; H–J, from *Pawek* 13763. Drawn by Diane Bridson.

DISTR.   **T** 7, 8; Malawi and Zambia
HAB.   Mixed evergreen forest or thickets, often on rocky ground; 1500–2750 m.

SYN.   *Canthium sp. 2* sensu F.F.N.R.: 404 (1962)
       C. whitei (Bridson) F. White in B.J.B.B. 60: 109 (1990)

5. **P. faulknerae** *Bridson* in K.B. 40: 707, fig. 3 M–T (1985). Type: Tanzania, Tanga District, Mwarongo [Morongo], *Faulkner* 2105 (K, holo.!)

Shrub or tree, or occasionally a scandent shrub, 2–7 m. tall; young branches glabrous or occasionally glabrescent, covered with grey bark when older. Leaves not restricted to apices of branches; blades drying brownish or pale green, often folded, elliptic, 4.5–7.5 cm. long, 2.3–4.3 cm. wide, shortly acuminate to acuminate at apex, acute at base, coriaceous, shiny above, glabrous; lateral nerves in 3–4 main pairs; tertiary nerves mostly inconspicuous; domatia present as erupted or unerupted blisters shortly above the axils of the lateral nerves and frequently in the axils of the tertiary nerves; petioles 2–4 mm. long; stipules with a truncate to broadly triangular base, 1–2 mm. long and a keeled decurrent lobe 2.5–4 mm. long. Flowers 4-merous, borne in 10–40-flowered pedunculate compact corymbose cymes 1.4–2.5 cm. across; peduncle 2–8 mm. long, adpressed pubescent; pedicels 2.5–6 mm. long, rather densely adpressed pubescent; bracteoles inconspicuous. Calyx glabrous; tube ± 1 mm. long; limb shortly toothed, 0.5–0.75 mm. long, sparsely ciliate to ciliate. Corolla whitish; tube 1.5–2.75 mm. long, pubescent at throat usually with some hairs deflexed but not forming a definite ring; lobes oblong, 1.75–2 mm. long, obtuse. Anthers reflexed. Style 5.75–6.5 mm. long; stigmatic knob 1 mm. long, ribbed. Disk glabrous. Fruit black when mature, broadly oblong in outline, 5–5.5 mm. long, 7–8 mm. wide, slightly indented to indented at apex; pyrene broadly ellipsoid, 4–5 mm. long, 3–3.5 mm. wide, 2.5 mm. thick, with ventral face flattened, crested at apex, rugulose.

KENYA.   Kwale District: Shimba Hills, Marere area, near pumping station, 3 Apr. 1968, *Magogo & Glover* 726!; Mombasa District: Mwawesa [Mowesa], *R.M. Graham* in *F.D.* 1823!; Kilifi, 24 Jan. 1937, *Moggridge* 332!
TANZANIA.   Lushoto District: Daluni–Mashewa, 26 Oct. 1935, *Greenway* 4128!; Bagamoyo District: Kikoka Forest Reserve, Apr. 1964, *Semsei* 3818!; Lindi District: Selous Game Reserve, Mpangampanga [Mpangapang] Summit, 14 Feb. 1970, *Rodgers* in *M.R.C.* 938!
DISTR.   **K** 7; **T** 3, 6, 8; not known elsewhere
HAB.   Coastal bush, thicket or *Brachystegia* woodland; 0–725 m.

SYN.   [*Canthium vulgare* sensu Bullock in K.B. 1932 : 374 (1932), pro parte, quoad *R.M. Graham* in *F.D.* 1823 & *Holst* 2717; K.T.S.: 433 (1961), *non* (K. Schum.) Bullock]
       [*C. huillense* sensu Vollesen in Opera Bot. 59: 67 (1980), *non* Hiern]

NOTE.   This species, when juvenile, exhibits heterophylly. One specimen from **T** 6, Uzaramo District, Pugu Hills (*Hawthorne* 1558), has both leaf-types present, and intermediate forms are said to be present on a second duplicate (not seen by author). The young leaves are small and the stems are covered with short prickle-like hairs. These hairs are still present on the semi-mature shoots and on the midribs beneath, but have never been observed on fully mature specimens. The following specimens from **K** 7, Kilifi District, *Langridge* 27 & 124 and Lamu District, *Kuchar* in *E.A.H.* 12842, are probably this species in juvenile foliage.

6. **P. schimperiana** (*A. Rich.*) *Bridson* in K.B. 40: 714 (1985) & in Fl. Rwanda 3: 150, fig. 44.3 (1985). Types: Ethiopia, near Maye-Gouagoua, *Quartin Dillon* (P, syn.) & Tigre, Mt. Seleuda, *Quartin Dillon & Schimper* I. 328 (P, syn.)

Shrub or tree, 2–10 m. tall; young branches square, often somewhat winged at corners, glabrous or rarely puberulous; older branches with fawn or greyish bark. Leaf-blades narrowly to broadly elliptic or less often obovate, 3–10.5 cm. long, 1.3–5 cm. wide, acute or more often gradually acuminate at apex, acute or occasionally obtuse at base, coriaceous, very shiny above, glabrous; lateral nerves in 5 main pairs; tertiary nerves obscure to apparent, coarsely reticulate; domatia absent or inconspicuous; petioles 2–3 mm. long, glabrous; stipules 3–7(–10) mm. long, triangular at base, terminating in a linear keeled decurrent lobe. Flowers both 4- and 5-merous, borne in sessile to subsessile 7–50-flowered umbellate cymes; peduncle, if present, not exceeding 3 mm. long; pedicels 2–7 mm. long, pubescent or rarely glabrescent; bracteoles inconspicuous. Calyx-tube 1–2 mm. long, glabrous or puberulous; limb reduced to a shortly toothed rim. Corolla whitish; tube 2–2.5 mm. long, with a ring of deflexed hairs above mid-point inside; lobes oblong-lanceolate, 3–3.5 mm. long, 1.5 mm. wide, acute. Anthers narrowly oblong, reflexed. Style

4–6 mm. long; stigmatic knob 1.25–1.5 mm. long. Disk glabrous. Fruit black when mature, somewhat wider than long, 5–6.5 mm. long, 6–7.5 mm. wide, apex slightly prominent to scarcely indented, pyrene ± ellipsoid, 6 mm. long, 4–4.5 mm. wide, with a wide groove beneath point of attachment, rugulose to bullate.

subsp. **schimperiana**

Pedicels 4–7 mm. long; angles of young stems apparent but not prominent. Fig. 132/23, p. 754.

UGANDA. Karamoja District: Matheniko, near Lokitanyala [Lokitaungyala], Sept. 1954, *Philip* 637!; Bunyoro District: Butiaba Flats, Sept. 1935, *Eggeling* 2191!; Masaka District: Kabula, Sept. 1945, *Purseglove* 1808!

KENYA. Northern Frontier Province: Moyale, 7 July 1952, *Gillett* 13523!; N. Nyeri District: Zawadi Estate, ± 7 km. on Nyeri–Kiganjo road, 27 May 1974, *Faden & Evans* 74/612!; Kilifi District: Arabuko, June 1922?, *R.M. Graham* in *F.D.* 1995!

TANZANIA. Musoma District: Bolgonja [Bologonja] R. and Mara R. junction, 15 Oct. 1961, *Greenway & Turner* 10263!; Handeni District: Handeni–Bagamoyo road, Sept. 1950, *Semsei* 553!; Kondoa District: Irangi, Kinyassi Scarp, 2 Jan. 1928, *B.D. Burtt* 931!

DISTR. U 1, 2, 4; K 1–7; T 1, 3–5, 7; Zaire, Rwanda, Burundi, Ethiopia, Somalia, Zambia, Malawi and Yemen

HAB. Forest, thicket or bushland; 15–2500 m.

SYN. [*Canthium lucidum* R. Br. in Salt, Voy. Abyss., App.: 62 (1814), *nomen nudum*]
   [*Phallaria schimperi* Hochst. in Flora 24(1), Intell. 1: 27 (1841), *nomen nudum*]
   Canthium schimperianum A. Rich., Tent. Fl. Abyss. 1: 350 (1848); Hiern in F.T.A. 3: 135 (1877); S. Moore in J.L.S. 37: 161 (1905); Robyns in N.B.G.B. 10: 617 (1929); Bullock in K.B. 1932: 385 (1932), pro parte; T.T.C.L.: 485 (1949), pro parte; Cufod. in Phyton 1: 146 (1949); I.T.U., ed. 2: 340 (1952); K.T.S.: 432, fig. 83 (1961); E.P.A.: 1010 (1965); Fl. Pl. Lign. Rwanda: 550, fig. 184.2 (1982)
   Plectronia schimperiana (A. Rich.) Vatke in Linnaea 40: 195 (1876), pro parte; Engl., Hochgebirgsfl. Trop. Afr.: 399 (1892); K. Schum. in E. & P. Pf. IV. 4: 92, fig. 33 E–F (1891); T.S.K.: 106 (1926); Chiov., Fl. Somala 2: 245 (1932)
   [*Plectronia nitens* sensu K. Schum. in P.O.A. C: 385 (1895), pro parte, quoad *Holst* 8868, *non* (Hiern) K. Schum. sensu stricto]
   [*Plectronia lamprophylla* sensu K. Schum. in E.J. 34: 335 (1904), pro parte, quoad syn. *Plectronia nitens, non* K. Schum. sensu stricto]
   Canthium myrtifolium S. Moore in J.B. 45: 266 (1907). Type: Uganda, Toro District, near mouth of Mpanga R., *Bagshawe* 1152 (BM, holo.!)
   Plectronia angiensis De Wild., Pl. Bequaert. 3: 176 (1925). Type: Zaire, Kivu, Angi, *Bequaert* 5693 (BR, holo.!)
   [*Canthium euryoides* sensu Bullock in K.B. 1932: 384 (1932), pro parte, quoad specim. cit.; T.T.C.L.: 485 (1949); F.P.S. 2: 429 (1952); I.T.U., ed. 2: 339 (1952), *non* Hutch. & Dalz.]

NOTE. Although *Canthium euryoides* is cited by Andrews, no specimens of *Psydrax schimperiana* have yet been seen from the Sudan. However, it most probably does occur there.
   The second subspecies, subsp. *occidentalis* Bridson, is known from Ghana, Togo, Dahomey, Nigeria and Cameroon. It has shorter pedicels and the young stems are more prominently winged.

7. **P. lynesii** *Bridson* in K.B. 40: 716, fig. 6A–F (1985). Type: Tanzania, Iringa, *Lynes* I.h. 113 (K, holo.!)

A tree up to 8.5 m. tall; young branches square, very slightly winged, puberulous; older stems covered with greyish fissured bark. Leaves not restricted to apices of branches; blades broadly elliptic to round, 2–4.5 cm. long, 1.3–4 cm. wide, obtuse to rounded at apex, acute at base, coriaceous, very shiny above, entirely glabrous or puberulous on midrib beneath; lateral nerves in 4–5 main pairs; tertiary nerves obscure to apparent, coarsely reticulate; domatia usually present as glabrous cavities; petioles 2–5 mm. long, puberulous; stipules 4–5 mm. long, triangular at base, terminating in a linear, somewhat keeled lobe. Flowers 4–5-merous, borne in subsessile to shortly pedunculate 5–10-flowered subumbellate cymes; peduncles 2–4 mm. long; pedicels 3–7 mm. long, puberulous; bracteoles inconspicuous. Calyx-tube 1.5 mm. long, puberulous towards base; limb 0.75 mm. long, shortly toothed. Corolla-tube 2 mm. long, with deflexed hairs near top inside; lobes narrowly oblong, 4.75 mm. long, 1.25–1.5 mm. wide, acute to obtuse. Anthers ± oblong, reflexed. Style 5 mm. long; stigmatic knob 1.5 mm. long. Disk glabrous. Fruit (only known from 1-seeded fruit) ± spherical, 4–5 mm. in diameter; pyrene broadly ellipsoid, 4.5 mm. long, 3mm. wide and thick, rugulose.

TANZANIA. Iringa District: Iringa, 17 Feb. 1932, *Lynes* I.h. 113! & 22 Mar. 1932, I.h. 255! & on Iringa–Sao Hill road, 5 Apr. 1988, *Bidgood & Vollesen* 928!
DISTR. T 7; not known elsewhere

HAB.   Rocky granite hill with degraded *Uapaca, Brachystegia* bushland; 1700–1975 m.

SYN.   [*Canthium schimperianum* sensu Bullock in K.B. 1932: 385 (1932), pro parte; T.T.C.L.:485 (1949), pro parte, quoad *Lynes* 191 & 255, *non* A. Rich.]

8. **P. robertsoniae** *Bridson*, sp. nov. affinis *P. schimperianae* A. Rich. sed petiolis longioribus, cortice cinereo, tubo corollae lobos excedenti differt. Typus: Kenya, Kilifi District, Watamu, Knocker Plot 34, *S.A. Robertson* 6152 (K, holo.!, EA, MO, US, iso.)

Tree or shrub up to 2 m.; very young stems glabrous or puberulous, later covered with whitish bark. Leaves restricted to apices of new branches or not; blades elliptic, 3–8 cm. long, 1.5–3.4 cm. wide, obtuse at apex, acute or obtuse at base, coriaceous, somewhat shiny above, glabrous; lateral nerves in 4–5 main pairs; tertiary nerves coarsely reticulate; domatia present as small erupted blisters; petioles 0.6–1 cm. long, puberulous or glabrous; stipules with triangular base 1.5–2.5 mm. long, apiculate or lobed and ± 3 mm. long when immature. Flowers 4-merous, borne in sessile 10–30-flowered subumbellate cymes; pedicels 6–10 mm. long, glabrous or pubescent; bracteoles inconspicuous. Calyx-tube ± 1 mm. long, glabrescent; limb reduced to a repand rim. Corolla-tube 3.5–4 mm. long, with deflexed hairs at throat; lobes narrowly oblong, 2–3 mm. long, 1–1.5 mm. wide, rounded. Anthers reflexed, linear. Style 8–9 mm. long; stigmatic knob 1–1.25 mm. long. Fruit almost didymous or broadly oblong in outline, 6 mm. long, 10 mm. wide, slightly indented; pyrene almost semicircular in outline, 6 mm. long, cartilaginous, slightly rugulose.

KENYA.   Kilifi District: Kilifi, Jan. 1937, *Moggridge* 334! & Kambe Kaya Forest, 8 Apr. 1981, *Hawthorne* 273! & Watamu, Knocker Plot 34, 24 Apr. 1990, *S.A. Robertson* 6152!
DISTR.   **K** 7; not known elsewhere
HAB.   Thicket and forest; 5m.

SYN.   *P. sp. B* sensu Bridson in K.B. 40: 716 (1985)

NOTE.   This species is easily separated from others with sessile inflorescences by having the corolla-tube longer than the lobes. Only the three cited gatherings are known, the *Moggridge* and *Robertson* specimens are apparently evergreen, have pubescent pedicels and strikingly white-grey bark, while the *Hawthorne* specimen has the flowers borne on stems with immature leaves, glabrous pedicels and larger leaves with glabrous petioles.

9. **P. sp. A**

Shrub 4 m. tall with branches angled at less than 90° to stem; very young stems puberulous, later covered with greyish or fawn bark. Leaves well spaced along the branches; blades broadly elliptic or round, 2.8–3.7 cm. long, 2–3 cm. wide, rounded at apex, obtuse to truncate at base, coriaceous, slightly shiny above, glabrous; lateral nerves in 2–3 main pairs; tertiary nerves coarsely reticulate, inconspicuous to obscure; domatia present as small glabrous pits; petioles 5–7 mm. long, puberulous; stipules truncate and apiculate, ± 1 mm. long. Flowers 4-merous, borne in ± sessile 6–20-flowered subumbellate cymes; pedicels 3.5–6.5 mm. long, pubescent; bracteoles inconspicuous. Calyx-tube ± 1 mm. long, glabrescent; limb reduced to a repand rim. Corolla-tube ± 3 mm. long, with a ring of deflexed hairs inside, ± ⅓ the distance from the throat; lobes oblong, ± 3 mm. long, 1 mm. wide, obtuse. Stigmatic knob ± 1.25 mm. long. Fruit flattened, ± 5.5–6 mm. long, 6–7 mm. wide, scarcely indented; pyrene (? from 1-seeded fruit) ± circular in outline, 4.5 mm. across, somewhat rugulose.

KENYA.   Tana River District: Tana R., Shekiko, 21 Aug. 1988, *Luke & Robertson* 1368!
DISTR.   **K** 7; known only from above specimen
HAB.   Sand-dune thicket, with *Tarenna graveolens, Hyphaene, Garcinia, Flacourtia, Tamarindus, Mimusops* and *Tricalysia ovalifolia*; 5 m.

10. **P. polhillii** *Bridson* in K.B. 40: 716, fig. 6G–K (1985). Type: Kenya, Tana River District, Kurawa, *Polhill & Paulo* 556 (K, holo.!, EA, iso.)

Shrub 1.5–6 m. tall, with numerous perpendicular lateral spurs; very young stems pubescent, later covered with whitish grey bark. Leaves mostly restricted to apices of lateral spurs; blades broadly elliptic to round, 0.8–2.8 cm. long, 0.5–2.3 cm. wide, rounded at apex, obtuse at base, coriaceous, slightly shiny above, glabrous, sometimes sparsely pubescent on the midrib beneath or occasionally sparsely pubescent beneath; lateral nerves in 3 main pairs; tertiary nerves coarsely reticulate, not very prominent; domatia present as small erupted blisters, glabrous; petioles 1.5–4 mm. long, pubescent; stipules with a shortly truncate base, 0.5–1 mm. long, bearing a linear lobe up to 2 mm. long, pubescent. Flowers 4-merous, borne in subsessile to shortly pedunculate 5–20-flowered subumbellate cymes; peduncles up to 2.5 mm. long; pedicels 2–5 mm. long, sparsely pubescent to pubescent; bracteoles inconspicuous. Calyx-tube ± 1 mm. long, glabrescent to sparsely pubescent; limb reduced to a repand rim, ciliate. Corolla whitish or yellowish cream; tube 2.75 mm. long, with deflexed hairs above mid-point inside; lobes narrowly oblong, 4 mm. long, 1 mm. wide, obtuse. Anthers linear-lanceolate, reflexed. Style up to 6 mm. long; stigmatic knob 0.75 mm. long, slightly ribbed. Disk glabrous. Fruit black when mature, broadly oblong in outline, 6–6.5 mm. long, 9–10 mm. wide, scarcely indented at apex; pyrene almost semi-circular in outline, 6 mm. long, rugulose.

KENYA. Northern Frontier Province: 80 km. E. of Bura, 5 Aug. 1973, *Oxtoby* in *E.A.H.* 15391!; Kwale District: near Taru, between Samburu and Mackinnon Road, 5 Sept. 1953, *Drummond & Hemsley* 4163!; Kilifi District: Sokoke Forest, road to Jilore Forest Station, 3.2 km. from turn-off on Kilifi–Malindi road, 28 July 1971, *Faden & Evans* 71/708!
DISTR. **K** 1, 7; not known elsewhere
HAB. Thickets and *Brachystegia* woodland; 15–350 m.

SYN. *Rytigynia sp. nov.* sensu K.T.S.: 471 (1961), quoad *Drummond & Hemsley* 4163

11. **P. kibuwae** *Bridson* in K.B. 40: 717, fig. 6L–P (1985). Type: Tanzania, Tanga District, Sigi R., Kiwanda, *Kibuwa* 5451 (K, holo.!, NHT, iso.)

Small glabrous tree up to 10 m. high; young stems dark brown, shiny, smooth. Leaf-blades yellow-green when dry, oblong-elliptic, 15–16 cm. long, 6.5–7.2 cm. wide, subacuminate at apex, rounded and occasionally slightly cordate at base; midrib light brown and prominent beneath; lateral nerves in 6–7 main pairs; tertiary nerves very coarsely reticulate and only apparent beneath; domatia present as glabrous pits; petioles not exceeding 2 mm. long; stipules broadly triangular, 4–5 mm. long, with a short subulate apiculum. Flowers 5-merous, borne in sessile to subsessile 30–50-flowered umbellate cymes; pedicels 4–12 mm. long, glabrous; bracteoles inconspicuous. Calyx-tube 2 mm. long, glabrous; limb reduced to a very shortly toothed rim. Corolla (known only from well developed buds) white; tube 2.25 mm. long, with ring of deflexed hairs just below throat inside; lobes oblong-lanceolate, 3.75 mm. long, 1.5 mm. wide. Stigmatic knob 1.75 mm. long. Disk glabrous. Fruit not known.

TANZANIA. Tanga District: Sigi R., Kiwanda, 10 Nov. 1981, *Kibuwa* 5451!
DISTR. **T** 3; known only from above specimen
HAB. River edge on hill slopes; 150 m.

12. **P. kaessneri** (*S. Moore*) *Bridson* in K.B. 40: 719 (1985). Type: Kenya, Kwale District, Gadu, *Kassner* 418 (BM, holo.!, K, iso.!)

A shrub, often scandent to 3m. tall, usually with short lateral branches at right-angles to the stem, glabrous; young branches covered with pale greyish bark, often peeling in tissue-like flakes. Leaves not restricted to apices of branches; blades broadly elliptic to round or sometimes obovate, 3–8 cm. long, 1.4–6 cm. wide, rounded or sometimes obtuse at apex, acute or occasionally obtuse at base, subcoriaceous to coriaceous, moderately shiny above; lateral nerves in 3–4 main pairs; tertiary nerves prominent and coarsely reticulate on both faces; domatia present as small punctured blisters; petioles 2–4 mm. long, glabrous; stipules 4–5 mm. long, with a short truncate to triangular base bearing a linear, strongly keeled lobe. Flowers 5-merous, borne in subsessile 5–30-flowered, subumbellate cymes; peduncles up to 2 mm. long; pedicels 2–8 mm. long (but tending to be accrescent in fruit), glabrous; bracteoles inconspicuous. Calyx-tube 1.25 mm. long, glabrous; limb reduced to a dentate rim, ciliate. Corolla whitish; tube 4–5 mm. long, with a ring of deflexed hairs above mid-point inside; lobes oblong, 5–6 mm. long, 2.5 mm. wide, acute or obtuse. Anthers reflexed, linear-oblong. Style 9 mm. long; stigmatic knob (1.75–)2–2.5 mm. long. Disk glabrous. Fruit black when mature, slightly wider than long, 7–9 mm. long, 9–11 mm. wide, with prominent apex; pyrene almost semicircular in outline, acute and crested at top, rounded at base, rugulose.

KENYA. Tana River District: E. side of Tana R., N. of Garsen, 3 km. N. of Wema, 15 July 1972, *Gillett & Kibuwa* 19927! & Tana River National Primate Reserve, Mwazini, South forest, June 1988, *Medley* 414!; Lamu District: Lunghi Forest Reserve, 23 km. E. of Bodhei, 1 Dec. 1988, *Luke & Robertson* 1532!

TANZANIA. Lushoto District: Korogwe, Lwengera [Luengera] valley, 7 Feb. 1960, *Semsei* 2986!; Bagamoyo District: Kikoka Forest Reserve, 28 Mar. 1964, *Semsei* 3735!; Kilwa District: 3 km. N of Kingupira, 19 Apr. 1975, *Vollesen* in *M.R.C.* 2253!
DISTR. **K** 7; **T** 3, 6, 8; Somalia and Mozambique
HAB. Riverine thickets and forest edges; (20–)90–300 m.

SYN. *Canthium kaessneri* S. Moore in J.B. 43: 351 (1905); Bullock in K.B. 1932 : 382, fig. 3 (1932); T.T.C.L.: 485 (1949); K.T.S.: 429 (1961); Vollesen in Opera Bot. 59: 67 (1980)
   Plectronia longistaminea K. Schum. & K. Krause in E.J. 39: 542 (1907). Type based on same gathering as *C. kaessneri* (B, holo.†, BM, K, iso.!)

13. **P. shuguriensis** *Bridson* in K.B. 40: 721, fig. 7A, B (1985). Type: Tanzania, Kilwa District, Selous Game Reserve, Shuguri, *Vollesen* in *M.R.C.* 4206 (K, holo.!, C, iso.)

Low cushion-shaped shrub with prostrate or short erect branches to ± 0.2 m. tall, glabrous, young branches with fawnish bark. Leaves not restricted to apices of branches; blades narrowly oblong-elliptic, 2.8–4.5 cm. long, 0.9–1.5 cm. wide, obtuse at apex, acute to obtuse at base, subcoriaceous, slightly shiny above; lateral nerves in 4 main pairs; tertiary nerves apparent beneath, rather coarsely reticulate, domatia present as small glabrous pits; petioles 3 mm. long; stipules 4 mm. long, broadly triangular at base, with a linear lobe above. Flowers 5-merous, borne in shortly pedunculate 5–15-flowered subumbellate cymes; peduncles 2–3 mm. long; pedicels 2–3 mm. long, glabrous; bracteoles inconspicuous. Calyx-tube 1 mm. long, glabrous; limb ± 1 mm. long, dentate, ciliate. Corolla white; tube 3.75 mm. long, with deflexed hairs above mid-point inside; lobes oblong, 3.5 mm. long, 1.5 mm. wide, acute. Anthers reflexed, oblong-ovate. Style 6 mm. long; stigmatic knob 1.5 mm. long. Disk glabrous. Mature fruit not known.

TANZANIA. Kilwa District: Ulanga R., Shuguri, 9 Dec. 1976, *Vollesen* in *M.R.C.* 4206!
DISTR. **T** 8; not known elsewhere
HAB. Between rocks in the river; ± 200 m.

SYN. *Canthium sp. nov. aff. kaessneri* S. Moore sensu Vollesen in Opera Bot. 59: 68 (1980)

14. **P. micans** (*Bullock*) *Bridson* in K.B. 40: 721 (1985). Types: Tanzania, Uzaramo District, Dar es Salaam, *Engler* 2127 (B, syn. †, K, fragment!) & 2187 (B, syn.†)

Shrub or small tree 3–4 m. tall, or occasionally a liane up to 10 m. with spines on old stems (fide *Mwasumbi* 11629); young stems glabrous, older stems covered with greyish bark. Leaves not restricted to apices of branches, very often ternate; blades elliptic or less often narrowly elliptic, 3.5–7.5 cm. long, 1.3–3.7 cm. wide, obtuse or acute at apex, acute at base, coriaceous, distinctly shiny above, glabrous; lateral nerves in 3–4 main pairs; tertiary nerves moderately coarsely reticulate, prominent on both faces; domatia present as small blisters with a pin-prick-like puncture; petioles 3–4 mm. long, glabrous; stipules ± 3 mm. long, with a shallow truncate base and linear lobe. Flowers 5-merous, borne in subsessile 2–10-flowered subumbellate cymes; peduncles not exceeding 2 mm. long; pedicels 0.6–1.6(–2.5 fide Bullock) cm. long, glabrous; bracteoles inconspicuous. Calyx-tube 1.5 mm. long, glabrous; limb reduced to a toothed rim, ciliate. Corolla greenish white; tube 3–4 mm. long, with a ring of deflexed hairs above mid-point inside; lobes oblong-lanceolate, 3 mm. long, 1.25–1.5 mm. wide, acute. Anthers reflexed, ± oblong. Style 7–8 mm. long; stigmatic knob 1.25 mm. long. Disk glabrous. Immature fruit 7 mm. long, 6 mm. wide; pedicels accrescent, up to 2.2 cm. long; mature pyrenes not known.

TANZANIA. Rufiji District: Utete, Kibiti, 18 Dec. 1968, *Shabani* 248!; Kilwa District: Nunga Thicket, 18 Jan. 1977, *Vollesen* in *M.R.C.* 4340!; Lindi District: Rondo Plateau, Mchinjiri, Nov. 1951, *Eggeling* 6405!
DISTR. **T** 6, 8; Mozambique
HAB. Thicket or forest; 120–700 m.

SYN. [*Plectronia lamprophylla* K. Schum. in E.J. 34: 335 (1904), excl. syn., *non Canthium lamprophyllum* F. Muell.]
   Canthium micans Bullock in K.B. 1932 : 382 (1932), pro parte; T.T.C.L. : 484 (1949); Vollesen in Opera Bot. 59: 67 (1980). Type: as for *P. lamprophylla* K. Schum.

15. **P. sp. B**

Shrub 2m. tall, glabrous; young branches round, covered with pale fawn bark. Leaves paired, not restricted to apices of branches; blades broadly elliptic, 8–10.5 cm. long, 4–5 cm. wide, subacuminate at apex, acute at base, coriaceous, shiny above; lateral nerves in

5–7 main pairs; tertiary nerves finely reticulate, prominent on both faces, domatia obscure; petioles 3–5 mm. long; complete stipules not observed. Flowers known from immature buds, 5-merous, borne in sessile 5–15-flowered umbellate cymes; pedicels 8–10 mm. long (perhaps not maximum length) glabrous. Calyx glabrous; limb reduced to a toothed rim. Corolla blunt in bud. Disk glabrous.

TANZANIA.   Lushoto District: E. Usambara Mts., Longuza [Longusa], Sigi, 18 Nov. 1917, *Peter* 56091!
DISTR.   **T** 3, known only from above specimen
HAB.   Forest; 300 m.

SYN.   *P. sp. C* sensu Bridson in K.B. 40: 721 (1985)

16. **P. recurvifolia** (*Bullock*) *Bridson* in K.B. 40: 722 (1985). Type: Tanzania, Pemba I., *Greenway* 2781 (K, holo.!, EA, iso.)

Scandent shrub 3–5 m. tall; branchlets inserted at right-angles, pubescent when young, later covered with greyish white bark. Leaves not restricted to apices of branches; blades elliptic to ovate, 1–6 cm. long, 0.7–3.5 cm. wide, obtuse to rounded at apex, obtuse to rounded or occasionally subcordate at base, coriaceous, shiny above, margins recurved; lateral nerves in 3–4 main pairs; tertiary nerves faintly apparent beneath, coarsely reticulate; domatia present, glabrous or pubescent; petioles 1–3 mm. long; stipules with a shallow base; lobe caducous, narrowly oblong, sometimes somewhat spathulate, 3–5 mm. long, at right-angles to stem. Flowers 5-merous, borne in 1–2-flowered sessile fascicles, often at apical node; pedicels 3–7 mm. long in flowering stage, glabrous; bracteoles inconspicuous. Calyx-tube ± 1.5 mm. long, glabrous; limb dentate, 1 mm. long, ciliate. Corolla white; tube 3–4 mm. long, with a ring of deflexed hairs above the middle inside; lobes oblong, 3 mm. long, 1.25 mm. wide, with an apiculum up to 0.75 mm. long, pubescent towards apex outside. Anthers reflexed. Style 6.5 mm. long; stigmatic knob 1.25 mm. long, ribbed. Disk glabrous. Fruit only known from 1-seeded examples, obovate in outline, 7–8 mm. long; pedicel accrescent, 1.7 cm. long.

KENYA.   Mombasa District: Mwawesa [Mowesa], *R.M. Graham* in *F.D.* 1751!; Kilifi District: Kilifi, 17 July 1939, *Moggridge* 534! & Kaya Kivara, 25 May 1987, *S.A. Robertson* 4652!
TANZANIA.   Uzaramo District: Bongoyo I., 14 km. N. of Dar es Salaam, 12 Jan. 1970, *Harris & Gillham* 3935! & Sinda I. near Dar es Salaam, 5 Jan. 1969, *Harris* 2689!; Pemba I., Ras Mkumbuu, 12 Dec. 1930, *Greenway* 2775!
DISTR.   **K** 7; **T** 6; **P**; not known elsewhere
HAB.   In sand above high-water mark, in forest or at swamp edge or in mixed woodland on raised limestone reef; 0–10 m.

SYN.   *Canthium recurvifolium* Bullock in K.B. 1932: 385 (1932); K.T.S.: 430 (1961)

17. **P. acutiflora** (*Hiern*) *Bridson* in K.B. 40: 722 (1985). Types: S. Nigeria, Old Calabar, *Thomson* 97 (K, syn.!) & Cameroon, Mt. Cameroon, *Mann* 1179 (K, syn.!)

Scandent bush or climber, 3.5–7 m. tall, with short lateral branches almost at right-angles to stem, glabrous; young stems square and furrowed. Leaf-blades narrowly to broadly elliptic, 5.3–11 cm. long, 2–6 cm. wide, acuminate at apex, acute at base, stiffly papery to subcoriaceous, dull to shiny above; lateral nerves in 4–5 main pairs; tertiary nerves completely obscure; domatia absent or occasionally present in axils of lateral nerves or tertiary nerves (but not in Flora area); petioles 6–10 mm. long; stipules with a truncate base and linear lobe perpendicular to base, 3–6 mm. long. Flowers 5-merous, borne in sessile to subsessile 5–10-flowered umbellate cymes; peduncle not exceeding 1 mm. long in flowering stage; pedicels 2–3(–4–6) mm. long (in flowering stage), glabrous; bracteoles inconspicuous. Calyx-tube ± 0.75 mm. long, glabrous; limb reduced to a dentate rim. Corolla cream-yellow; tube 3–4 mm. long, with ring of deflexed hairs below mid-point inside; lobes oblong-ovate 2.5–4 mm. long, 1.5 mm. wide with tapering apiculum, 0.75–1.25 mm. long, sparsely pubescent towards base above. Filaments moderately well developed; anthers ovate, less than ⅓ as long as corolla-lobes, reflexed. Style 5.5 mm. long; stigmatic knob cylindrical above, widening at base, 0.75–1 mm. long. Disk glabrous. Fruit ± didymous, 0.75–0.8(–1) cm. long, 1.3–1.5 cm. wide; pedicels somewhat accrescent; pyrenes circular in outline, 0.8 mm. in diameter, on either side with a shallow groove from point of attachment to centre, slightly bullate.

UGANDA.   Mengo District: 13 km. on Kampala–Masaka road, Aug. 1937, *Chandler* 1875! & Entebbe, Lake shore, Oct. 1931, *Eggeling* 39 ! & Kyagwe, near Nansagazi, Nakiza Forest, 40 km. SW. of Jinja, 24 Jan. 1931, *Dawkins* 701!

TANZANIA. Bukoba District: Kaigi, Sept.–Oct. 1935, *Gillman* 406!
DISTR. **U** 4; **T** 1; S. Nigeria, Cameroon, Central African Republic, Zaire, Principe, S. Tomé and Sudan
HAB. Forest edges; 1100–1220 m.

SYN. *Canthium acutiflorum* Hiern in F.T.A. 3: 136 (1877); Hepper in F.W.T.A., ed. 2, 2: 182 (1963), pro parte
  Plectronia henriquesiana K. Schum. in Bol. Soc. Brot. 10: 128 (1892). Type: S. Tomé, Bom Successo, *Moller* 677 (COI, lecto.!, BM, isolecto.)
  P. acutiflora (Hiern) K. Schum. in E.J. 39: 537 (1907)
  P. acarophyta De Wild. in Ann. Mus. Congo., Bot., sér, 5,2: 173 (1907). Type: Zaire, Mondjo, *Pynaert* 317 (BR, holo.!)
  Canthium lacus-victoriae Bullock in K.B. 1932: 384 (1932). Type: Uganda, Sese Is., Sozi I., *Maitland* 424 (K, holo.!)
  C. henriquesianum (K. Schum.) G. Taylor in Exell, Cat. Vasc. Pl. S. Tomé : 210 (1944), pro parte excl. syn.
  [*C. malacocarpum* sensu F.P.S. 2: 429 (1952), *non* (K. Schum. & K. Krause) Bullock]
  C. acarophytum (De Wild.) Evrard in B.J.B.B. 37: 459 (1967)

18. **P. kraussioides** (*Hiern*) *Bridson* in K.B. 40: 723 (1985). Type: Angola, Huila, Morro de Humpata, *Welwitsch* 5353 (LISU, holo., BM, K, iso.!)

Scandent shrub or climber 1.5–7 m. tall, with short lateral branches at right-angles to the stem or somewhat backwardly curved, glabrous; young branches round to square, occasionally furrowed; bark fawn. Leaf-blades elliptic to broadly elliptic or sometimes oblong-elliptic, 4–14 cm. long, 1.5–6.5 cm. wide, acute to acute-acuminate or sometimes abruptly acuminate at apex, rounded, obtuse or acute at base, coriaceous or sometimes subcoriaceous, moderately shiny to shiny above; lateral nerves (4–)5–8 main pairs, often impressed above, and with tertiary nerves completely obscure; domatia absent or present as unerupted or erupted blisters; petioles 5–9 mm. long; stipules with a truncate base and linear to narrowly oblong lobe perpendicular to stem, 4–7 mm. long. Flowers 5-merous, borne in sessile to shortly pedunculate 2–15-flowered subumbellate cymes; peduncles 0–4 mm. long; pedicels 5–8 mm. long in flowering stage, glabrous; bracteoles glabrous. Calyx-tube 1 mm. long, glabrous; limb reduced to a dentate rim, ciliate. Corolla white or cream; tube 3.75–6 mm. long, with a ring of deflexed hairs below mid-point inside; lobes triangular-lanceolate, 3.5–6 mm. long, 1.25–2.5 mm. wide, with a tapering apiculum, 0.75–1.75 mm. long, at apex, sparsely pubescent towards base above. Anthers ovate, less than $\frac{1}{8}$ the length of the corolla-lobes, reflexed. Style 7.5–10 mm. long; stigmatic knob cylindrical above, widening at base, 1–1.75 mm. long. Disk glabrous. Fruit greyish green, almost didymous, 7–7.5 mm. long, 1.4–1.5 cm. wide, often galled; pedicels accrescent; pyrene semicircular to almost circular in outline, 7 mm. in diameter, with shallow groove from point of attachment to centre, cartilaginous, scarcely bullate. Figs. 131/17, p. 752; 132/24, p. 754 & 161, p. 908.

TANZANIA. Mwanza District: North Central Uzinza, 10 June 1937, *B.D. Burtt* 6580!; Morogoro District: Uluguru Mts., Lukwangule, 5 Feb. 1935, *E.M. Bruce* 777!; Iringa District: Itaka, 1 Sept. 1933, *Greenway* 3661!
DISTR. **T** 1, 4, 6, 7; W. Africa, Zaire (Shaba, ? Kivu), Burundi, Mozambique, Malawi, Zambia, Zimbabwe and Angola
HAB. Riverine woodland and forest; 1200–2000 m.

SYN. *Canthium kraussioides* Hiern, Cat. Afr. Pl. Welw. 1: 473 (1898); F.W.T.A. 2: 113 (1931)
  Plectronia kraussioides (Hiern) K. Schum., in Just's Bot. Jahresb. 1898 (1): 393 (1900), as '*P. kranussioides*'
  P. pulchra De Wild. in Ann. Mus. Congo, Bot., sér. 4, 1: 229 (1903); K. Krause in R.E. Fries, Wiss. Ergebn. Schwed. Rhod.-Kongo-Exped. 1911–12, 1 (Nachtr.) : 14 (1921). Type: Zaire, Shaba, Lukafu, *Verdick* 381 (BR, holo.!)
  P. malacocarpa K. Schum. & K. Krause in E.J. 39: 540 (1907). Type: Tanzania, Rungwe District, Kondeland, Ischana, *Stolz* 92 (B, holo.†)
  Canthium egregium Bullock in K.B. 1932: 348 (1932); T.T.C.L.: 485 (1949). Type: Tanzania, Rungwe District, Kyimbila, *Stolz* 1914 (B, holo.†, K, Z, iso.!)
  C. malacocarpum (K. Schum. & K. Krause) Bullock in K.B. 1932: 348 (1932); T.T.C.L.: 485 (1949)
  [*C. henriquesianum* sensu G. Taylor in Exell, Cat. Vasc. Pl. S. Tomé: 210 (1944), pro parte; Hepper in F.W.T.A., ed. 2, 2: 181 (1963), *non* (K. Schum.) G. Taylor]
  [*C. anomocarpum* sensu F.F.N.R.: 402 (1962), *non* DC.]

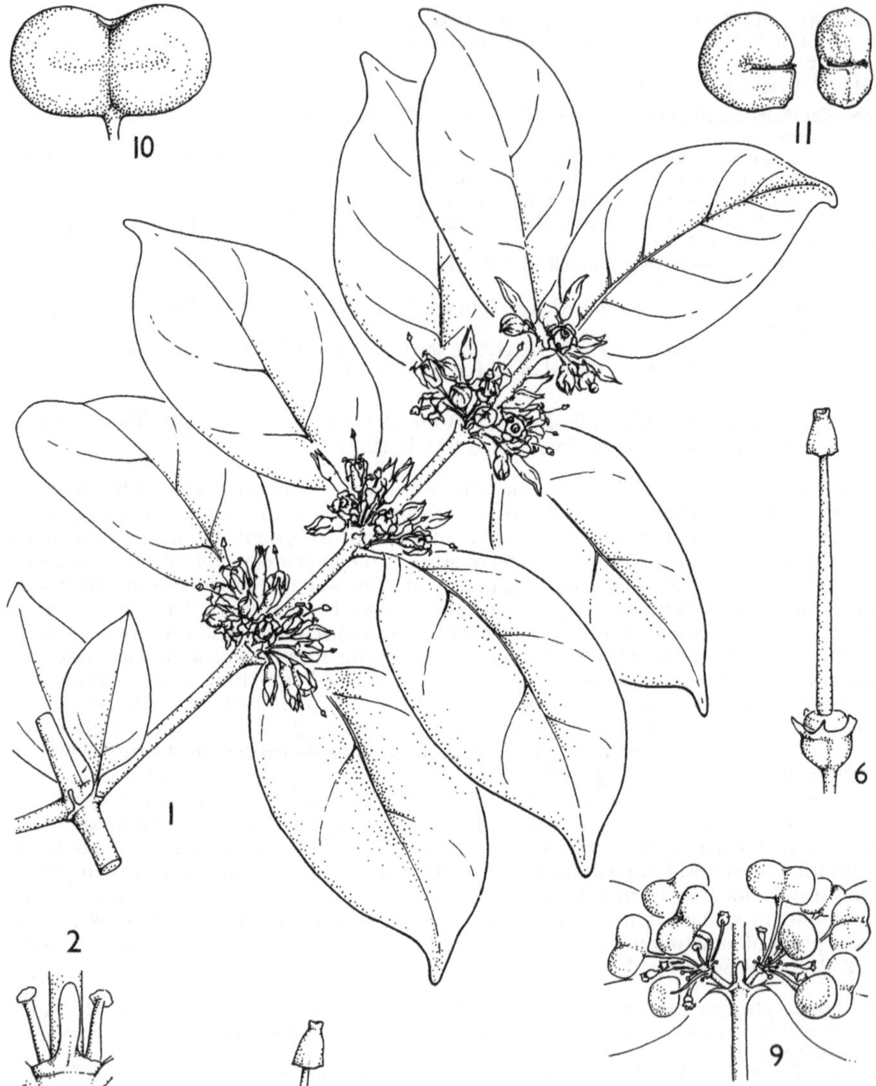

FIG. 161. *PSYDRAX KRAUSSIOIDES* — 1, flowering branch, × ⅔; 2, stipule, × 2; 3, flower, × 4; 4, part of corolla opened out, × 4; 5, anther, dorsal view, × 8; 6, calyx with style and stigmatic knob, × 4; 7, stigmatic knob, × 8; 8, section through ovary, × 8; 9, fruiting node, × ⅔; 10, fruit, × 2; 11, pyrenes, 2 views, × 2. 1, 3–8, from *B.D. Burtt* 6580; 2, from *Richards* 1156; 9–11, from *Pawek* 2901. Drawn by Sally Dawson.

108. **KEETIA**

E. Phillips in Gen. S. Afr. Fl. Pl.: 587 (1926) & in Bothalia 2: 369 (1927); Bridson in K.B. 41: 965–994 (1986)
[*Canthium* sensu Sond. in Harv. & Sond., Fl. Cap. 3: 16 (1865), pro parte; F.T.A. 3: 132–146 (1877), pro parte; Bullock in K.B. 1932: 360–373 (1932), pro parte; F.W.T.A., ed. 2, 2: 181–185 (1963), pro parte, *non* Lam.]

Climbers or scandent shrubs; stems glabrous or frequently pubescent. Leaves paired, petiolate, not restricted to new growth at apex of branches; blades chartaceous or occasionally coriaceous; leaves subtending lateral branches sometimes smaller and broader than main leaves; stipules lanceolate to ovate or triangular at base and acuminate to linear above, never keeled, never with white silky hairs inside. Flowers 4–6-merous, borne in pedunculate, usually distinctly branched cymes; bracts and bracteoles often conspicuous, Calyx-tube ellipsoid to ovoid; limb equalling or sometimes exceeding tube in length, repand to dentate or less often lobed. Corolla white, cream or yellow; tube cylindrical, usually ± equalling lobes, sometimes shorter or rarely longer, typically with a ring of deflexed hairs inside; lobes reflexed, often thickened at apex but never apiculate. Stamens set at throat of corolla; filaments moderately well developed; anthers fully or partly exserted but usually not reflexed, narrowly ovate or oblong. Style long, slender, ± twice as long as the corolla-tube; stigmatic knob cylindrical, distinctly longer than wide, hollow to below apex, bifid at apex when mature (fig.131/16, p. 752); ovary 2-locular, each locule containing 1 ovule attached to upper $\frac{1}{3}$ of the septum. Disk puberulous to pubescent or infrequently glabrous. Fruit a 2-seeded drupe, slightly to strongly bilobed, somewhat laterally flattened, slightly to strongly indented at apex, sometimes 1-seeded by abortion and asymmetrical; pyrenes woody or less often cartilaginous, usually ± ovoid with ventral face flattened, rugulose or somewhat colliculate; point of attachment on ventral face above centre or near apex; lid-like area completely or incompletely defined, either lying along ventral face above point of attachment or across apex, provided with a central crest, eventually dehiscent around circumference (fig. 132/25, 26, p. 754). Seeds ovoid, shaped at apex according to position of lid-like area in pyrene, convoluted; endosperm streaked with tanniniferous areas (resembling a ruminate endosperm, except that the testa is never invaginated) occasionally with tannin granules ± evenly dispersed or less often absent; testa thin, very finely reticulate; embryo straight with erect radicle and small cotyledons lying parallel to ventral face of seed (fig.131/25, p. 752).

A genus of about 40 species, confined to tropical and South Africa.

1. Stipules narrowly ovate to ovate; leaf-blades frequently subcordate to cordate at base 2
   Stipules lanceolate, narrowly triangular or triangular at base then gradually or abruptly narrowing to a linear or subulate lobe, or rarely ± truncate; leaf-blades seldom subcordate to cordate at base 4
2. Leaves not markedly discolorous; bracts linear-lanceolate to lanceolate, 3–6 mm. long; calyx-limb dentate, up to 1.5 mm. long . . . .   1. *K. gueinzii*
   Leaves discolorous, pale beneath; bracts ovate, 0.7–2 cm. long; calyx-limb well developed, at least 1.5 mm. long, usually longer 3
3. Plant with young branches glabrous to hispid and leaves glabrous to sparsely pubescent; calyx-limb truncate to repand, 1.5–3 mm. long . . . .   7. *K. molundensis*
   Plant ± densely covered with rusty coloured hispid indumentum; calyx-limb with long linear lobes 0.7–1.4 cm. long . . . .   8. *K. ferruginea*
4. Sparsely setose on petioles, inflorescence-branches and calyx-teeth; calyx-limb 1.75–2 mm. long; corolla-tube at least 7 mm. long; fruit not at all indented at apex   16. *K. carmichaelii*
   Plants glabrous or with a sparse to dense indumentum of weaker hairs; calyx-limb and corolla-tube never as large; fruit slightly depressed or strongly indented at apex 5

5. Tertiary nerves finely reticulate (occasionally obscured by indumentum in *K. gueinzii*); pyrene with lid-like area lying across apex or occasionally set at an angle of ± 20° to ventral face (fig.132/25, p. 754) 6

   Tertiary nerves obscure or coarsely reticulate, sometimes moderately coarsely reticulate; pyrene with lid-like area set along ventral face or rarely as above (fig. 132/26) 10

6. Leaf-blades narrowly elliptic to oblong-lanceolate; calyx glabrous; corolla-tube 3–4 mm. long; pyrene with lid-like area set at an angle of ± 20° to ventral face — 2. *K. angustifolia*

   Leaf-blades of a broader shape; if corolla-tube as long then calyx usually pubescent; pyrene with area above point of attachment perpendicular to ventral face 7

7. Leaf-blades glabrous to sparsely pubescent above, glabrescent to densely pubescent beneath, cordate or sometimes rounded to truncate at base; stipules lanceolate, gradually acuminate, at least 9 mm. long — 1. *K. gueinzii*

   Leaf-blades glabrous or rarely glabrescent above, entirely glabrous or glabrous save for the nerves beneath, obtuse to rounded at base; stipules triangular at base then gradually or abruptly narrowing to a linear or subulate lobe or, if lanceolate, then up to 5 mm. long 8

8. Tertiary nerves, with the element perpendicular to midrib the more conspicuous, often raised or impressed above; pedicels and calyx-tubes usually densely covered with rusty coloured hairs . . . . — 5. *K. venosa*

   Tertiary nerves evenly reticulate; pedicels and calyx-tubes glabrous or sparsely pubescent (flowers not known in 3, *K. sp. A*) 9

9. Leaf-blades small, narrowly obovate, 2–4 cm. long, 1.2–2.2 cm. wide, acute to acuminate at apex; petiole 2–3 mm. long (**T** 6) . . . . — 3. *K. sp. A*

   Leaves larger, blades elliptic to broadly elliptic, 5.5–11 cm. long, 2.5–5 cm. wide, long-acuminate at apex; petiole 5–10 mm. long . . . . — 4. *K. purseglovei*

10. Young stems densely covered with pale fine soft hair; leaf-blades with pubescent to densely pubescent midribs beneath, drying a characteristic pinkish grey, with 5–6 main pairs of lateral nerves; corolla-tube 3–3.5 mm. long . . . . — 10. *K. purpurascens*

    Young stems glabrous to sparsely pubescent or with yellowish adpressed hairs; leaf-blades with glabrous to sparsely pubescent midribs, mostly drying blackish or brownish; if corolla-tube more than 3 mm. long then with 7–9 pairs of lateral nerves 11

11. Young stems with an indumentum of yellowish adpressed hairs or occasionally glabrous; flowers small; stigmatic knob 0.25–0.5 mm. long; fruit large, strongly lobed, almost didymous, at least 2 cm. wide; domatia obscure — 14. *K. tenuiflora*

    Young stems glabrous or with spreading or crisped hairs; flowers not notably small (where known); stigmatic knob 0.8–1.25 mm. long; fruit bilobed but not so strongly as to be almost didymous, often more laterally compressed; domatia usually present 12

12. Leaf-blades with 4–5 main pairs of lateral nerves; fruit semicircular, flattened above, gradually indented at apex; pyrene with lid-like area lying across apex (flowers not known) (**T** 6) . . . . — 6. *K. sp. B*

    Leaf-blades with 5–9 main pairs of lateral nerves; fruit not shaped as above; pyrene with lid-like area lying along ventral face 13

13. Leaf-blades small, 3.3–7.5 cm. long, 1.7–4.5 cm. wide, obtuse to rounded or sometimes subacuminate; petioles 3–6 mm. long; lateral branches typically short   11. *K. procteri*
     Leaf-blades larger, or if small then usually subacuminate or acuminate at apex; petioles 0.5–1.5 cm. long; lateral branches not short 14
14. Fruit large, either slightly longer than wide (1.7 × 1.6 cm.) or only single-seeded fruit known (1.8 × 1.2 cm.); stipules 2–6 mm. long 15
     Fruit smaller, wider than long (0.8–1.3 × 1.2–1.8 cm.); stipules 0.4–1.3 cm. long 16
15. Stipules linear-lanceolate, 5–6 mm. long; leaf-blades acute to subacuminate at apex; domatia present as inconspicuous tufts of rusty coloured hairs; pyrene not indented on ventral face; calyx-limb dentate (**T** 6)   12. *K. sp. C*
     Stipules ± truncate, 2–4 mm. long; leaf-blades acuminate at apex; domatia present as tufts of white hair; pyrene with conspicuous indentation below point of attachment on ventral face; calyx-limb repand or shortly dentate . . . . 13. *K. lulandensis*
16. Domatia present as pubescent tufts; lateral nerves in 7–9(–10) main pairs . . . . 9. *K. zanzibarica*
     Domatia present as glabrous pits; lateral nerves in 6–7 main pairs . . . . 15. *K. koritschoneri*

1. **K. gueinzii** (*Sond.*) *Bridson* in K.B. 41: 970, fig. 1A–C (1986). Type: South Africa, Durban [Port Natal], *Gueinzius* (S, holo., K, iso.!)

Scandent shrub or liane, 3–25 m. tall; young branches sparsely to densely covered with crisped or spreading golden to rust-coloured hairs. Leaf-blades usually drying brown, bullate or not, oblong-lanceolate to ovate, 5.5–13.5 cm. long, 3.5–6 cm. wide, acuminate at apex, rounded, truncate or more frequently subcordate at base, glabrous to sparsely pubescent above, glabrescent to densely pubescent beneath; lateral nerves in 6–9 main pairs; tertiary nerves finely reticulate; domatia present as tufts of hairs; petioles 3–7 mm. long, sparsely to densely covered with crisped or patent hairs; stipules lanceolate to ovate, 0.9–1.3 cm. long, up to 6 mm. wide at base, gradually acuminate, pubescent outside; leaves subtending lateral branches smaller, circular. Flowers 5-merous, borne in pedunculate 20–50-flowered cymes; peduncles 0.5–1.5 cm. long, sparsely to densely pubescent; pedicels 5–7 mm. long, pubescent to densely pubescent; bracteoles linear-lanceolate to lanceolate, 3–6 mm. long. Calyx-tube 1 mm. long, densely covered with straight or crisped hairs or occasionally glabrescent; limb 1.25–1.5 mm. long, divided into teeth for one-third to half its length, glabrescent to sparsely pubescent, usually ciliate. Corolla creamy white; tube 2.25–4 mm. long, with a ring of deflexed hairs set just below the top inside; lobes oblong-lanceolate to ovate, 2.5–4 mm. long, 1.25–2.25 mm. wide, acute and thickened at apex. Anthers fully exserted, but seldom reflexing. Style 0.5–1 cm. long, glabrous; stigmatic knob 1.25–2.25 mm. long. Disk pubescent. Fruit black when ripe, ± broadly oblong in outline, 7–9 mm. long, 1.1–1.4 cm. wide, slightly indented, glabrous or glabrescent; pyrene obovoid with ventral face flattened to hemispherical, 9–11 mm. long, 6–7 mm. wide, apex with a rhombic area lying perpendicular to the plane face containing a central ridge; point of attachment free. Figs. 131/16, 25, p. 752; 132/25, p. 754 & 162.

UGANDA. Kigezi District: Lake Mutanda, May 1950, *Purseglove* 3391!; Mbale District: Samia Bugwe county, E. boundary of West Bugwe Local Forest Reserve at S. edge of Solo [Nsolo] R., 8 Apr. 1951, *G.H. Wood* 371!; Masaka District: Malabigambo Forest, 3.2 km. SSW. of Katera, 2 Oct. 1953, *Drummond & Hemsley* 4535!
KENYA. Meru District: near base of Kirima Hill, near Nyambeni Hill Tea Estate, 11 Oct. 1960, *Polhill & Verdcourt* 292!; N. Kavirondo District: Kakamega Forest, near water crossing, Kidia R. on road to Kakamega saw mill, 8 Jan. 1968, *Perdue & Kibuwa* 9497!; Kwale District: Shimba Hills, Mwele Mdogo Forest, 6 Feb. 1953, *Drummond & Hemsley* 1138!
TANZANIA. Moshi District: Marangu, Rawuya, 10 Jan. 1940, *Bally* 538!; Lushoto District: Shagayu Forest Reserve, 19 Apr. 1962, *Semsei* 3461!; Iringa District: Sao Hill, Ipogoro–M'kawe track, 12 Dec. 1961, *Richards* 15557!
DISTR. U 1–4; K 2/3, 3–5, 7; T 1–8; Z; Cameroon, Central African Republic, Zaire, Rwanda, Burundi, Sudan, Ethiopia, Malawi, Zambia, Zimbabwe, Angola and South Africa
HAB. Forest and woodland, often on swampy ground; 90–2450 m.

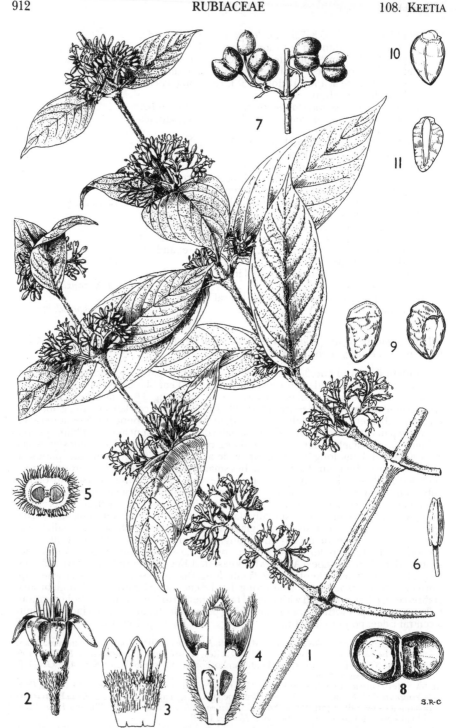

FIG. 162. *KEETIA GUEINZII* — **1**, flowering branch, × ⅔; **2**, flower, × 4; **3**, section of corolla opened out, × 4; **4**, longitudinal section through calyx and ovary, × 12; **5**, transverse section through ovary, × 12; **6**, stigmatic knob, × 8; **7**, infructescences, × ⅔; **8**, transverse section through fruit, × 2; **9**, pyrene, 2 views, × 2; **10**, seed, × 2; **11**, longitudinal section of seed, × 2. 1–8, source not known; 9–11, from *White* 3196. Drawn by Stella Ross-Craig, with 9–11 by Sally Dawson.

SYN. *Canthium gueinzii* Sond. in Linnaea 23: 54 (1850) & in Fl. Cap. 3: 16 (1865); S. Moore in J.L.S. 40: 89 (1911); Bullock in Hook., Ic. Pl. 32, t. 3170 (1932) & in K.B. 1932: 368 (1932); Chiov., Racc. Bot. Miss. Consol. Kenya: 55 (1935); T.T.C.L.: 487 (1949); K.T.S.: 428 (1961); F.F.N.R.: 403, fig. 68I (1962); Fl. Pl. Lign. Rwanda: 548, fig. 183.1 (1982); Fl. Rwanda 3: 148, fig. 45.1 (1985)
  [*C. hispidum* sensu Hiern in F.T.A. 3: 140 (1877), pro parte, quoad *Soyaux* 196 & in Cat. Afr. Pl. Welw. 1: 476 (1898); Robyns in N.B.G.B. 10: 616 (1929); Bullock in K.B. 1932: 369 (1932); T.T.C.L.: 486 (1949); K.T.S.: 429 (1961); Vollesen in Opera Bot. 59: 67 (1980), *non* Benth.]
  [*Plectronia hispida* sensu K. Schum. in P.O.A. C: 386 (1895), pro parte; Z.A.E. 1907–8, 2: 326 (1911); T.S.K.: 106 (1926); E.P.A.: 1009 (1965), *non* (Benth.) K. Schum.]
  P. gueinzii (Sond.) Sim, For. Fl. Cap. Col.: 241 (1907); Bews, Fl. Natal & Zululand: 198 (1921)
  P. charadrophila K. Krause in E.J. 57: 36 (1920). Type: Tanzania, Rungwe District, Bundali Mts., *Stolz* 124 (B, holo.†, K, photo., iso.!)
  P. subcordatifolia De Wild., Pl. Bequaert. 3: 199 (1925). Type: Zaire, Rutshuru, *Bequaert* 6092 (BR, holo.!)
  Keetia transvaalensis E. Phillips in Bothalia 2: 369 (1927). Type: South Africa, Transvaal, Barberton, *Galpin* 519 (K, lecto.!, PRE, isolecto.)
  Canthium scabrosum Bullock in K.B. 1932: 367 (1932); T.T.C.L.: 486 (1949). Type: Uganda, Mengo District, Entebbe, *Fyffe* (K, holo.!)
  C. charadrophilum (K. Krause) Bullock in K.B. 1932: 369 (1932)
  [*C. sylvaticum* sensu Bullock in K.B. 1932: 369 (1932), pro parte, quoad *Grote* 5070, *Haarer* 189 & 1030, *Maitland ex Liebenberg* 1075, *Moon* 584 & *Volkens* 1105a; T.T.C.L.: 486 (1949), pro parte, quoad *Volkens* 1105a, *non* Hiern]

NOTE. *K. gueinzii* is a very plastic species and the indumentum is particularly variable, from sparse to dense and the hairs may be straight or crisped. Although frequently confused with other species, on account of its variability, the lanceolate to ovate stipules and the rounded, or more frequently subcordate to cordate leaf-bases usually make recognition easy.
  Canthium gueinzii has been recorded from West Africa (F.W.T.A., ed. 2, 2: 184 (1963)), but does not appear to extend west of Cameroon. *Canthium hispidum* frequently confused with *K. gueinzii* has not yet been recorded from the Flora area.
  Two species from Zaire, *Plectronia ealaensis* De Wild. and *P. pynaertii* De Wild. are very close to *K. gueinzii*; they may possibly be worth maintaining, at either specific or infraspecific level.

2. **K. angustifolia** *Bridson* in K.B. 41: 971, fig. 2A–L (1986). Type: Rwanda, Cyangugu Prefecture, Pindura–Ibigugu road, near km. 88 on Butare–Cyangugu road, *Bridson* 164 (K, holo.!, BR, C, COI, EA, HNR, LG, WAG, iso.!)

Liane or shrub, 2.5–20 m. tall; young branches glabrous or sparsely pubescent. Leaf-blades narrowly elliptic to oblong-lanceolate, 4–9(–11) cm. long, 1–2.8(–3.2) cm. wide, long-acuminate at apex, obtuse to rounded at base, entirely glabrous or sparsely pubescent beneath and shiny above, often somewhat undulate; lateral nerves in 6–7 main pairs; tertiary nerves rather finely reticulate; domatia present as small hair-lined cavities; petioles 4–8 mm. long, pubescent or sometimes glabrous; stipules triangular at base, abruptly or gradually narrowing to a linear apex, 5–10 mm. long; leaves subtending lateral branches usually absent or not differing in shape. Flowers 4–5-merous, borne in pedunculate 10–40-flowered cymes; peduncles 2–6 mm. long, sparsely pubescent to pubescent; pedicels 1–4 mm. long, glabrous to sparsely pubescent; bracteoles narrowly ovate, 1–2.5 mm. long. Calyx-tube 1–1.5 mm. long, glabrous; limb 0.75–1 mm. long, with well-spaced short teeth near top, sometimes ciliate. Corolla creamy yellow; tube 3–5 mm. long, with a ring of deflexed hairs set at throat and extending to base inside; lobes broadly lanceolate to narrowly ovate, 2.75–3.25 mm. long, 1.5–1.75 mm. wide, acute and thickened at apex. Anthers fully exserted but seldom reflexing. Style 7.5–8 mm. long, glabrous; stigmatic knob 1.5–2 mm. long. Disk pubescent. Fruit slightly broader than long, 1.1 cm. long, 1.3 mm. wide, somewhat indented at apex and narrowing at base; pyrene broadly ellipsoid with ventral face flattened, 1.1 cm. long, 0.9 cm. wide; area above point of attachment subcircular with a central crest, set at an angle of ± 20° to ventral face.

UGANDA. Kigezi District: Impenetrable Forest, Mar. 1947, *Purseglove* 2371! & Luhizha [Luhiza], June 1948, *Purseglove* 2700!
DISTR. U 2; Rwanda and Burundi
HAB. Forest; 2300 m.

SYN. *Canthium sp. A* sensu Fl. Pl. Lign. Rwanda: 554, fig. 183.3 (1983); Fl. Rwanda 3: 152, fig. 45.2 (1985)

3. **K. sp. A**; *Bridson* in K.B. 41: 972 (1986)

Small tree 3 m. high; young branches rusty pubescent. Leaf-blades narrowly obovate, 2–4 cm. long, 1.2–2.2 cm. wide, acute to acuminate at apex, rounded at base; glabrous save for the midrib beneath; lateral nerves in 6–7 main pairs, impressed above; tertiary nerves finely reticulate; domatia present as glabrous pits; petiole 2–3 mm. long, sparsely pubescent; stipules lanceolate, 5 mm. long, pubescent outside, caducous. Cymes pedunculate, known only in fruiting stage; peduncles 0.7–1 cm. long, sparsely pubescent; pedicels ± 6 mm. long, pubescent; bracteoles inconspicuous. Fruit broadly oblong in outline, 1 cm. long, 1.2 cm. wide, slightly indented at apex and somewhat tapered at base; pyrene ellipsoid with ventral face flattened, 1 cm. long, 7 mm. wide, with point of attachment at apex; circular area lying perpendicular to ventral face containing a central crest.

TANZANIA. Kilosa District: Ukaguru Mts., Mamiwa Forest Reserve, summit of Mamiwa, 16 Aug. 1972, *Mabberley* 1496!
DISTR. **T** 6; known only from above specimen
HAB. Mist-forest; 2310 m.

NOTE. The above fruiting specimen has small leaves; the pyrenes are similar to those of *K. gueinzii*. It could possibly represent a reduced form of *K. gueinzii* rather than a distinct taxon. Additional collections are needed to verify its status.

4. **K. purseglovei** *Bridson* in K.B. 41: 972, fig. 2 M–U (1986). Type: Uganda, Kigezi District, Ishasha Gorge, *Purseglove* 3313 (K, holo.!)

Climber to 9 m. tall (? or tree); young branches glabrous or sometimes sparsely pubescent. Leaf-blades elliptic to broadly elliptic, 5.5–11 cm. long, 2.5–5 cm. wide, long-acuminate or sometimes acuminate at apex, acute to rounded and sometimes unequal at base, entirely glabrous; lateral nerves in 5–6 main pairs; tertiary nerves finely reticulate, apparent on lower surface only; domatia present as small glabrous or ciliate pits; petiole 5–10 mm. long, glabrous to pubescent; stipules broadly triangular at base, subulate above, 5–8 mm. long, glabrous; leaves subtending lateral branches similar in shape but smaller. Flowers 4–5-merous, borne in pedunculate 10–50-flowered cymes; peduncles 0.5–1.7 cm. long, sparsely pubescent to pubescent; pedicels 3–7 mm. long, glabrous or sparsely pubescent towards base; bracteoles inconspicuous. Calyx-tube 1 mm. long, glabrous; limb 0.25–0.5 mm. long, repand. Corolla white; tube 2.5–3 mm. long, with a ring of deflexed hairs a short distance below the top inside; lobes oblong-lanceolate, 2 mm. long, 1.25–1.5 mm. wide. Anthers fully exserted but not reflexing. Style 5.75–6 mm. long; stigmatic knob 1.1 mm. long. Disk pubescent. Fruit only known from 1-seeded examples, ± 1 cm. long, 0.8 cm. wide; pyrene ellipsoid, 1 cm. long, 7 mm. wide, rounded at base; lid-like area rather small, scarcely inclining from adaxial face with prominent crest continuing around apex. Seeds with outer ⅔ of endosperm densely and evenly grained with tannins.

UGANDA. Bunyoro District: Budongo Forest, 28 Nov. 1938, *Loveridge* 130a!; Ankole District: Kalinzu, 9 July 1969, *Synnott* 361!; Kigezi District: Ishasha Gorge, Feb. 1950, *Purseglove* 3313!
DISTR. **U** 2; Cameroon
HAB. Forest; 1220–1350 m.

NOTE. This species is perhaps related to the poorly known *K. venosissima* (Hutch. & Dalz.) Bridson from Ghana and Cameroon, but lacks the dense tomentose indumentum on the petioles and midrib characteristic of the latter.

5. **K. venosa** (*Oliv.*) *Bridson* in K.B. 41: 974 (1986). Type: Uganda, W. Nile District, Madi, *Grant* (K, holo.!)

Scandent shrub or climber, 2–7 m. tall; young branches sparsely to densely covered with rusty coloured hairs. Leaf-blades oblong-elliptic or narrowly to broadly elliptic, rarely round, 4.5–14 cm. long, 1.5–6(–7.5–8) cm. wide, acuminate at apex, obtuse to rounded or rarely subcordate at base, glabrous or rarely glabrescent and shiny above, glabrous save for the sparsely to densely pubescent nerves or occasionally sparsely pubescent beneath, papery to subcoriaceous; lateral nerves in 5–9 main pairs; tertiary nerves finely reticulate always with element perpendicular to midrib more conspicuous and often raised or impressed above; domatia present as inconspicuous tufts of hair; petioles 0.5–1.5 cm. long, pubescent; stipules triangular at base rather abruptly narrowing to a linear lobe or acuminate apex, 0.5–1.4(–1.7) cm. long, up to 0.5 cm. wide at base, pubescent outside; leaves subtending lateral branches smaller, ± circular. Flowers 4–5(–6)-merous, borne in pedunculate 20–70-flowered cymes; peduncles 0.5–1.7 cm. long, pubescent; pedicels 2–5 mm. long, pubescent; bracteoles linear-lanceolate, 2.5–6 mm. long. Calyx-tube 0.75–1 mm. long, glabrous or pubescent; limb 0.75–1.25 mm. long, dentate but with teeth usually less than half the length of the tube, ciliate. Corolla creamy white; tube 2.25–3 mm. long, with a ring of deflexed hairs set just below top inside; lobes lanceolate to ovate, 1.5–2.25 mm. long, 1–1.75 mm. wide, acute. Anthers fully exserted but seldom reflexing. Style 4–6.5 mm. long, glabrous; stigmatic knob 0.75–1.25 mm. long. Disk pubescent. Fruit black when mature, ± broadly oblong in outline, 0.8–1.1 cm. long, 1.1–1.5 cm. wide, slightly indented, glabrous; pyrene hemispherical to suborbicular, 6–7 mm. long, 5–6 mm. wide at apex; area from point of attachment lying perpendicular to ventral face triangular, containing a small crest.

UGANDA.   W. Nile District: Koich R., near Rumogi, *Eggeling* 1860!; Teso District: Serere, July 1932, *Chandler* 822!; Masaka District: Sese [Sesse] Is., Bubembe I., 20 Jan. 1956, *Dawkins* 859!
KENYA.   Kwale District: Shimba Hills, Godoni Forest area, 2–3 km. N. of Kwale Forest Station, 16–23 Sept. 1976, *Spjut* 4582! & near Giriama Point, 18 Sept. 1968, *Gillett* 18704! & 1927, *Gardner* 1435!
TANZANIA.   Lushoto District: W. Usambara Mts., Bumbuli at DC's rest-house, 11 May 1953, *Drummond & Hemsley* 2492!; Mpanda District: Kungwe Mts., Kasiha [Kasieha] R., 20 July 1959, *Harley & Newbould* 4479!; Lindi District: Rondo Plateau, Nandembo, 11 Dec. 1955, *Milne-Redhead & Taylor* 7638!
DISTR.   U 1, 3, 4; K 7; T 1, 3, 4, 6, 8; W. Africa, Cameroon, Central African Republic, Zaire, Rwanda, Burundi, Sudan, Mozambique, Malawi, Zambia, Zimbabwe and Angola
HAB.   Forest edges and scrub; 275–1525 m.

SYN.   *Plectronia venosa* Oliv. in Trans. Linn. Soc. 29: 85, t. 49 (1873); K. Schum. in P.O.A. C: 386 (1895); De Wild., Pl. Bequaert. 3: 200 (1925)
    Canthium barteri Hiern in F.T.A. 3 : 143 (1877). Type: Nigeria, Asaba [Assaba], *Barter* 285 (K, holo.!)
    C. venosum (Oliv.) Hiern, F.T.A. 3: 144 (1877); Bullock in K.B. 1932: 371, fig. 1 (1932); T.T.C.L.: 487 (1949); F.P.S. 2: 431 (1952); K.T.S.: 433 (1961); F.F.N.R.: 403 (1962); F.W.T.A., ed. 2, 2: 184 (1963), pro parte; Fl. Pl. Lign. Rwanda: 552, fig. 184.3 (1982); Fl. Rwanda 3: 150, fig. 45.3 (1985)
    C. venosum (Oliv.) Hiern var. *pubescens* Hiern in F.T.A. 3: 144 (1877). Types: Nigeria, Brass, *Barter* 403 & Onitsha, *Barter* 1800 (K, syn.!)
    [*Plectronia cuspido-stipulata* Engl. in Abh. Preuss. Akad. Wiss.: 53 (1894), nomen based on *Holst* 2426]
    [*P. hispida* sensu K. Schum. in P.O.A. C: 386 (1895), pro parte, quoad syn. *P. cuspido-stipulata*, non (Benth.) K. Schum.]
    Canthium sylvaticum Hiern, Cat. Afr. Pl. Welw. 1: 477 (1898); Bullock in K.B. 1932: 369 (1932), pro parte, quoad *Carpenter* 8 & *Maitland* 393 & 400; T.T.C.L.: 487 (1949), pro parte, quoad *Haarer* 1905. Type: Angola, Pungo Andongo, near base of Pedras Cabondo, *Welwitsch* 3134 (LISU, holo., BM, K, iso.!)
    Plectronia sylvatica (Hiern) K. Schum. in Just's Bot. Jahresb. 1898 (1): 393 (1900), as 'silvatica'
    P. barteri (Hiern) De Wild. in Ann. Mus. Congo, Bot., sér. 2, I, 2: 33 (1900)
    P. stipulata De Wild. in Pl. Nov. Herb. Then. 1: 171, t. 38 (1905). Type: Mozambique, Morrumballa [Murrumbala] Forest, *Luja* 430 (BR, holo.!)
    P. myriantha K. Krause in Z.A.E. 1907–8, 2: 327 (1911), non Schlecht. & K. Krause (1908), nom. illegit. Type: Uganda/Zaire, Virunga Mts., Nyavarongo, *Mildbraed* 684 (B, holo.†, K, fragment!)
    P. dundusanensis De Wild., Pl. Bequaert. 3: 183 (1925). Type: Zaire, Dundusana, *Mortehan* 906 (BR, lecto.!)
    P. reygaerti De Wild., Pl. Bequaert. 3: 195 (1925). Type: Zaire, Mobwasa region, *Reygaert* 553 (BR, lecto.!)
    [*Keetia transvaalensis* sensu E. Phillips in Bothalia 2: 369 (1927), pro parte, quoad *Schlechter* 12290, non sensu stricto]
    [*Canthium gueinzii* sensu F.W.T.A., ed. 2, 2: 184 (1963), pro parte, excl. syn. *C. venosissimum* Hutch. & Dalz., non Sond.]
    [*C. zanzibaricum* sensu F.W.T.A., ed. 2, 2: 184 (1963), pro parte, quoad *F.H.I.* 30363, non Klotzsch]
    C. dundusanense (De Wild.) Evrard in B.J.B.B. 37: 459 (1967)
    Canthium sp. C. sensu Fl. Pl. Lign. Rwanda: 554, fig. 185.2 (1982)

NOTE.   This species, like *K. gueinzii*, is both widespread and very variable; the leaves are especially variable in both shape and texture. However, *K. venosa* can usually be recognised by the tertiary venation which has the element perpendicular to the midrib the more conspicuous, either somewhat prominent or impressed above.
    One species from Zaire, *Plectronia vanderystii* De Wild., is very close to *K. venosa* but differs in the longer peduncles and glabrous inflorescence-branches; it may perhaps be worth recognition at infraspecific level.
    A high-altitude form of *K. venosa* was separated as *Canthium sp. C* in Fl. Pl. Lign. Rwanda: 554 (1982), but the characters used seem to break down when compared with specimens from a wider geographical range. Field observations and additional collections from high-altitude areas of Uganda would be of interest.

6. **K. sp. B**; Bridson in K.B. 41: 976 (1986)

Shrub 2 m. tall; young branches glabrous. Leaf-blades broadly elliptic, 3–5.5 cm. long, 1.5–3.2 cm. wide, acute to subacuminate or obtuse then sometimes apiculate at apex, obtuse at base, with 4–5 main pairs of lateral nerves impressed above, and with tertiary nerves obscure, entirely glabrous; domatia obscure; petioles 4–6 mm. long, glabrous; stipules triangular at base, subulate above, 2.25 mm. long, readily caducous; lateral branches developing on one side only and subtending leaves apparently absent. Inflorescences pedunculate cymes, only known in fruiting stage; peduncle 0.5–1 cm. long, glabrous; pedicels 0.6–1 cm. long, sparsely pubescent; bracteoles inconspicuous. Fruit semicircular in outline, ± flat above, 1 cm. long, 1.5 cm. wide, slightly depressed at apex; pyrene shortly obovoid with ventral face flattened, 9 mm. long, 8 mm. wide at top, with point of attachment at apex and circular area lying perpendicular to ventral face with a central crest.

TANZANIA. Ulanga District: Kwiro Forest Reserve, SW. flank of ridge, 18 Jan. 1979, *Cribb et al.* 11027!
DISTR. **T** 6; known only from above specimen
HAB. Montane forest; 1400 m.

7. **K. molundensis** (*K. Krause*) *Bridson* in K.B. 41: 976 (1986). Type: Cameroon, Molundu, Bumba, *Mildbraed* 4236 (B, holo.†, HBG, iso.!)

Climber with young branches glabrous to densely pilose or sometimes tomentose. Leaf-blades discolorous, drying brown or blackish above and pale beneath, elliptic, oblong-elliptic or sometimes ovate, (5.5–)7.5–13 cm. long, 2.5–5.5 cm. wide, long-acuminate at apex, obtuse to rounded or truncate to subcordate or sometimes cordate at base, glabrous or less often sparsely pubescent above, sparsely pubescent or glabrous save for the nerves beneath, sometimes ciliate; lateral nerves in 5–7 main pairs; tertiary nerves obscure or more often clearly apparent and very coarsely reticulate; domatia present in angles of lateral and tertiary nerves or sometimes absent; petioles 4–12 mm. long, sparsely pilose to setose; stipules triangular-lanceolate, ovate-lanceolate or ovate, 0.7–2 cm. long, 3–13 mm. wide, acuminate, glabrous or velutinous outside; leaves subtending lateral branches not differentiated. Flowers 5-merous, borne in shortly pedunculate 10–40-flowered cymes; peduncles 3–15 mm. long, rusty pubescent; pedicels 3–8 mm. long, setose to pubescent; bracteoles resembling small stipules, up to 7 mm. long. Calyx-tube 1–1.5 mm. long, rusty pubescent; limb 1.5–3 mm. long, distinctly wider than tube, truncate or repand or sometimes distinctly dentate, sparsely covered with hairs or glabrous. Corolla-tube 3.5–4 mm. long, with a ring of deflexed hairs set at mid-point inside, pubescent above; lobes oblong-ovate, 3–3.75 mm. long, 1.5–1.75 mm. wide, acute and thickened at apex. Anthers fully exserted, eventually reflexing, sometimes with cells apparent, connective covering dorsal face and drying blackish. Style 0.8–1 cm. long, glabrous; stigmatic knob 1.25–1.5 mm. long. Disk pubescent. Fruit square to oblong in outline, 1.8–2.2 cm. long, 1.7–1.8 cm. wide, slightly bilobed, not indented at apex, slightly tapered towards base; pyrene obovoid with ventral face flattened, ± 1.5 cm. long, 0.9 cm. wide, 0.8 cm. thick, smooth; circular area perpendicular to point of attachment (rather small in relation to size of pyrene) with a central crest.

SYN. *Plectronia molundensis* K. Krause in E.J. 54: 350 (1917)

var. **macrostipulata** (*De Wild.*) *Bridson* in K.B. 41: 976 (1986). Type: Zaire, Kivu, Walikale, *Bequaert* 6533 (BR, holo.!)

Calyx-limb truncate to repand; petioles 3–5 mm. long. Stipules velutinous outside; young stems sparsely to densely setose.

UGANDA. Kigezi District: Kinkizi, Katete, Feb. 1951, *Purseglove* 3581!
DISTR. **U** 2; Nigeria, Zaire and Cameroon
HAB. Valley forest; 1280 m.

SYN. *Plectronia macrostipulata* De Wild., Pl. Bequaert. 3: 190 (1925)
    P. mortehanii De Wild., Pl. Bequaert. 3: 191 (1925). Type: Zaire, Equateur, Dundusana, *Mortehan* 529 (BR, holo.!)
    Canthium macrostipulatum (De Wild.) Evrard in B.J.B.B. 37: 459 (1967)
    C. mortehanii (De Wild.) Evrard in B.J.B.B. 37: 459 (1967)

NOTE. A second variety, var. *molundensis*, has been recorded from Sierra Leone and Cameroon; stipules glabrous or velutinous only at margins and towards apex outside and young stems glabrous.

8. **K. ferruginea** *Bridson* in K.B. 41: 977, fig. 3 (1986). Type: Tanzania. Mpanda District, 64 km. S. of Uvinza, *Procter* 474 (EA, holo.!, BR, iso.!)

Climber or scandent shrub; young branches densely hispid with rusty coloured hairs. Leaf-blades drying discolorous, brownish above, pale beneath, obovate to broadly elliptic, 4.5–10 cm. long, 2.2–6.5 cm. wide, acute then apiculate at apex, subcordate to cordate at base, hispid on both faces with dense hairs on the nerves beneath; lateral nerves in 7–9 main pairs, often impressed above; tertiary nerves very coarsely reticulate or sometimes obscure; domatia obscure; petiole 2–7 mm. long, densely hispid; stipules ovate, 1.5–2.5 cm. long, 1–1.4 cm. wide, acute, densely pubescent outside. Flowers 5-merous, borne in pedunculate congested cymes; peduncle 3–7 mm. long, densely pubescent; pedicels very reduced, up to 1 mm. long; bracts conspicuous, ovate, up to 2 cm. long, 1 cm. wide, acuminate, apiculate, densely hispid outside, glabrescent inside; bracteoles similar to bracts but smaller. Calyx-tube 2.5 mm. long, pubescent; limb divided almost to base into linear lobes, 0.7–1.4 cm. long, densely hispid. Corolla-tube 4 mm. long, lacking ring of deflexed hairs; lobes ovate, 5 mm. long, acute, pubescent at apex outside. Anthers fully exserted, with the cells not apparent. Style 8.5 mm. long; stigmatic knob 2–2.5 mm. long, strongly ribbed. Disk pubescent. Fruit not known.

TANZANIA. Kigoma District: Kasakati [Kasangati], Aug. 1965, *Suzuki* B44!; Mpanda District: 64 km. S. of Uvinza, July 1956, *Procter* 474! & Mpanda – Uvinza, Kafulu, July 1951, *Eggeling* 6174!
DISTR. **T** 4; not known elsewhere
HAB. Thicket on sandstone; 1370–1525 m.

NOTE. A fruiting specimen (*Zenker* 3340) of what can be assumed to be a very closely allied unnamed species is known from Cameroon. This differs from the above by the leaves which are not pale beneath, bullate and have a linear apiculum up to 8 mm. long. The fruit is oblong in outline, 1.6 cm. long, 2.3 cm. wide; the pyrene has the lid-like area lying along the ventral face.

9. **K. zanzibarica** (*Klotzsch*) *Bridson* in K.B. 41: 979, fig. 1D–F (1986). Type: Zanzibar I., *Peters* (B, holo.†)

Scandent shrubs, small trees or lianes; young stems glabrous to sparsely pubescent or occasionally pubescent; bark pale greyish. Leaf-blades often turning blackish or grey when dry, narrowly elliptic to round or sometimes ovate to broadly ovate, 5–15 cm. long, 2–7.5 cm. wide, acuminate or less often obtuse to acute and often apiculate at apex, frequently unequal and rounded to truncate, obtuse or rarely subcordate at base, entirely glabrous or with the nerves sparsely pubescent or sometimes glabrescent to sparsely pubescent or rarely pubescent beneath; lateral nerves in 7–9(–10) main pairs; tertiary nerves coarsely reticulate; domatia usually present as tufts of hair; petiole 0.5–1.5 cm., glabrous to pubescent; stipules with a triangular base topped by a subulate to linear lobe 0.4–1.2 cm. long; leaves subtending lateral branches smaller but scarcely differing in shape. Flowers 4–5-merous, borne in 30–60-flowered pedunculate cymes; peduncles 0.5–1.5 cm. long, glabrescent to sparsely pubescent; pedicels 2–8 mm. long, glabrescent to densely pubescent; bracteoles up to 2 mm. long. Calyx-tube 0.75–1 mm. long, glabrous to densely pubescent; limb dentate, 0.5–1 mm. long, usually glabrous and ciliate or sometimes pubescent. Corolla white; tube 2–3.25 mm. long, with a ring of deflexed hairs above the middle inside; lobes oblong-lanceolate, 1.5–2.75 mm. long, 1–1.5 mm. wide, acute to obtuse. Anthers fully exserted, erect. Style 5–8 mm. long, glabrous or pubescent; stigmatic knob 0.8–1.25 mm. long. Disk pubescent. Fruit oblong or cordate in outline, 0.8–1.3 cm. long, 1.2–1.7 cm. wide; pyrene broadly ellipsoid with ventral face flattened, 1–1.1 cm. long, 0.75–0.8 cm. wide; area above point of attachment circular on ventral plane with a central crest.

KEY TO SUBSPECIES

Style distinctly pubescent or with just a few hairs; stigmatic
    knob (0.75–)1–1.25 mm. long; calyx glabrous or sometimes
    hairy towards base; leaves ovate to broadly ovate,
    sometimes round or occasionally elliptic, typically rounded
    to truncate at base, dull above . . . .   subsp. a. **zanzibarica**
Style glabrous; stigmatic knob 0.5–1 mm. long; calyx sparsely to
    densely pubescent; leaves oblong-elliptic, oblong-ovate or
    sometimes elliptic or ovate, obtuse to rounded, seldom
    truncate at base, dull to shiny above:

Flowers larger (calyx-limb 0.5–1 mm. long; corolla-tube (2–)
2.75–3.5 mm. long; stigmatic knob 0.75–1 mm. long);
leaves oblong-ovate, sometimes ovate, infrequently
elliptic . . . .   subsp. b. **cornelioides**
Flowers smaller (calyx-limb 0.3–0.5 mm. long; corolla-tube
2–2.5 mm. long; stigmatic knob 0.5–0.75 mm. long);
leaves oblong-elliptic, narrowly ovate or ovate . . .   subsp. c. **gentilii**

subsp. a. **zanzibarica**

Leaf-blades ovate to broadly ovate, sometimes round, occasionally elliptic, rounded to truncate
(rarely subcordate) or occasionally obtuse at base, entirely glabrous or with petioles and nerves very
sparsely hairy beneath or rarely sparsely pubescent, always dull above. Calyx-tube glabrous or
sometimes pubescent towards base; limb 0.6–1 mm. long. Corolla-tube 2.75–3.5 mm. long. Style
distinctly pubescent, with just a few hairs or rarely glabrous; stigmatic knob (0.75–)1–1.25 mm. long.
Fruit 1–1.2 cm. long, 1.4–1.7 cm. wide. Fig. 132/26, p. 754.

Kenya. Kwale District: Shimba Hills National Reserve, 22 Nov. 1971, *Bally & Smith* 14351!; Kilifi
District: Sokoke Forest, 28 Feb. 1945, *Jeffery* K104!; Lamu District: Pangani, 5 Mar. 1977, *Hooper &
Townsend* 1210!
Tanzania. Lushoto District: Korogwe, 28 Oct. 1963, *Archbold* 313!; Uzaramo District: near Dar es
Salaam airport, 31 July 1965, *Harris* 148!; Zanzibar I.: near Chuini, 31 Jan. 1929, *Greenway* 1266!
Distr. **K** 1/7, 7; **T** 3, 6, 8; **Z**; Mozambique
Hab. Coastal bushland, thickets, forest edges, sometimes on marshy ground; 0–500m.

Syn. *Canthium zanzibaricum* Klotzsch in Peters, Reise Mossamb., Bot. 1: 291 (1861); Hiern in F.T.A. 3:
138 (1877), pro parte; Bullock in K.B. 1932: 373 (1932), pro parte; T.T.C.L.: 488 (1949); K.T.S.:
433 (1961)
Plectronia zanzibarica (Klotzsch) Vatke in Oest. Bot. Zeitschr. 25: 231 (1875); Engl. in Abh. Preuss.
Akad. Wiss.: 26 (1894)
[*Canthium zanzibaricum* Klotzsch var. *glabristylum* sensu Hiern in F.T.A. 3: 139 (1877), pro parte,
non Hiern sensu stricto]

Note. Some specimens from Mozambique at altitudes over 450 m. tend towards subsp. *cornelioides*.

subsp. b. **cornelioides** (*De Wild.*) Bridson in K.B. 41: 979 (1986). Types: Zaire, Shaba, Lukafu, *Verdick*
32 & 124 (both BR, syn.!)

Leaf-blades oblong-ovate or sometimes occasionally elliptic, obtuse to rounded or infrequently
truncate at base, entirely glabrous or glabrescent to pubescent. Calyx-tube sparsely to densely
pubescent or occasionally only pubescent towards base; limb 0.5–1 mm. long. Corolla-tube (2–)2.75–
3.5 mm. long. Style glabrous; stigmatic knob 0.75–1 mm. long. Fruit 1.1–1.3 cm. long, 1.5 cm. wide.

Tanzania. Mbeya District; Unyika area, Mlowo, 3 Oct. 1976, *Leedal* 3856!
Distr. **T** 7; Zaire (Shaba), Zambia, Malawi and Mozambique
Hab. Not recorded

Syn. *Canthium zanzibaricum* Klotzsch var. *glabristylum* Hiern in F.T.A. 3: 139 (1877), pro parte. Type:
Malawi, Lake Malawi [Nyasa], entrance of Roangiva (probably Loangwa, which = Dwambasi),
Kirk (K, lecto.!)
Plectronia cornelioides De Wild. in Ann. Mus. Congo, Bot., ser. 4, 1: 159 (1903)
P. hispida (Benth.) K. Schum. var. *glabrescens* K. Krause in R.E. Fries, Wiss. Ergebn. Schwed.
Rhod.-Kongo-Exped. 1911–1912, 1, (Nachtr): 15 (1921), *nomen*, based on *R.E. Fries* 556 & 556a
[*Keetia transvaalensis* sensu E. Phillips in Bothalia 2: 369 (1927), pro parte, quoad *Borle* 293, *non*
E. Phillips sensu stricto]
[*Canthium hispidum* sensu Bullock in K.B. 1932: 369 (1932), pro parte, quoad syn. *Plectronia
hispida* var. *glabrescens*]
[*C. zanzibaricum* sensu Brenan in Mem. N.Y. Bot. Gard. 8: 452 (1954); F.F.N.R.: 403 (1962), *non*
Klotzsch sensu stricto]

subsp. c. **gentilii** (*De Wild.*) Bridson in K.B. 41: 980 (1986). Type: Zaire, Equateur, Djuma valley, *Gillet*
2884 (BR, lecto.!)

Leaf-blades oblong-elliptic, narrowly ovate to ovate, acute to obtuse or rounded at base, entirely
glabrous or glabrescent to sparsely pubescent beneath, dull to somewhat shiny above. Calyx very
sparsely to densely pubescent; limb 0.3–0.5 mm. long. Corolla-tube 2–2.5 mm. long. Style glabrous;
stigmatic knob 0.5–0.75 mm. long. Fruit 0.9 cm. long, 1.3 cm. wide.

Uganda. W. Nile District: East Madi, Jan. 1952, *Leggat* 58!; Bunyoro District: Murchison Falls
National Park, Rabongo Forest, 18 Feb. 1964, *H.E. Brown* 2021!; Masaka District: Buddu county, 6
km. NE. of Lukaya, 12 July 1969, *Lye* 3465!
Tanzania. Bukoba District: Kishoju, near Mwisha R., ? Nov. 1948, *Ford* 731! & Ngono R. bridge,
Kyaka road, Jan. 1458, *Procter* 792!; Mwanza District: Uzinza, Geita, 3 July 1953, *Tanner* 1552!

DISTR. U 1, 2, 4; **T** 1, 4; Central African Republic, Zaire, Rwanda, Sudan and Ethiopia
HAB. Thickets and forest, often near water; (250–)900–1525 m.

SYN. [*Canthium zanzibaricum* sensu Hiern in F.T.A. 3: 183 (1877), pro parte, quoad *Schweinfurth* 2905;
        Bullock in K.B. 1932: 373 (1932), pro parte, quoad *Maitland* 852 & *Snowden* 42; F.P.S. 2: 430
        (1952); Fl. Pl. Lign. Rwanda: 552, fig. 186.1 (1982); Fl. Rwanda 3: 152, fig. 46.1 (1985), *non*
        Klotzsch sensu stricto]
      Plectronia gentilii De Wild. in Ann. Mus. Congo, Bot., sér. 5, 1: 82 (1904); Miss. Laurent.: 294
        (1906) & in Ann. Mus. Congo, Bot. sér. 5, 3: 294 (1910)
      [*Canthium sylvaticum* sensu Bullock in K.B. 1932: 370 (1932), pro parte, quoad *Scott Elliot* 7192,
        non Hiern]
      C. gentilii (De Wild.) Evrard in B.J.B.B. 37: 459 (1967)

NOTE. There is a character overlap between subsp. *cornelioides* and subsp. *gentilii*; some specimens
    from **T** 4 (Kahama District, *Bullock* 3044 & Mpanda District, *F.G. Smith* 1215) could be equally well
    referred to either subspecies. However, I feel the two are best maintained separate for the present.
    Specimens from Sudan and Ethiopia tend to have the pedicels and calyces less densely pubescent
    than the rest of the material. In my opinion Bullock's inclusion of the Angolan types, *Canthium
    gracile* Hiern and *C. tenuiflorum* (here maintained as a distinct species), in his synonymy of
    Canthium zanzibaricum was erroneous.

10. **K. purpurascens** *(Bullock) Bridson* in K.B. 41: 981 (1986). Type: Tanzania, Rufiji,
Musk 67 (K, holo.!, EA, iso.)

Small tree or scandent shrub; young stems densely covered with soft straw-coloured to
rusty hairs, older stems with light-coloured bark. Leaf-blades turning a pinkish brown or
greyish when dry, elliptic to oblong-elliptic, 4–10 cm. long, 2–5.5 cm. wide, acute to obtuse,
subacuminate or sometimes emarginate, then shortly mucronate at apex, rounded or
sometimes obtuse at base, glabrous save for the pubescent midrib above, glabrescent to
sparsely pubescent with densely pubescent or occasionally pubescent nerves beneath;
lateral nerves in 5–6 main pairs; tertiary nerves coarsely reticulate and rather obscure;
domatia absent; petiole 4–8 mm. long, pubescent; stipules with a triangular base
produced into a subulate lobe, 3–5 mm. long altogether, pubescent outside; leaves
subtending lateral branches smaller but scarcely differing in shape. Flowers 5-merous,
borne in 30–50-flowered branched pedunculate cymes; peduncles 1–1.5 cm. long,
pubescent; pedicels 2–4 mm. long, densely pubescent; bracteoles up to 2 mm. long.
Calyx-tube 1 mm. long, pubescent to densely pubescent; limb 1 mm. long, dentate to
mid-point, puberulous, densely ciliate. Corolla white; tube 3–3.5 mm. long, with deflexed
hairs set at mid-point inside; lobes oblong-lanceolate, 2.5–3 mm. long, 1.25–1.5 mm. wide,
acute. Anthers fully exserted but not reflexed. Style 8 mm. long, glabrous; stigmatic knob 1
mm. long. Disk pubescent. Fruit edible, oblong-cordate in outline, 1.1–1.2 cm. long;
1.5–1.6 cm. wide; pyrene ellipsoid with ventral face flattened, 1.2 cm. long, 0.7 cm. wide,
rugulose; area above point of attachment circular with a central crest separating a
depression on either side.

TANZANIA. Ulanga District: Magombera Forest Reserve, 8 Nov. 1961, *Semsei* 3403!; Kilwa District:
    Selous Game Reserve, [Luwegu] Luwego R., 2 Dec. 1970, *Rees* T68! & Kingupira waterhole, 15 July
    1975, *Vollesen* in *M.R.C.* 2584!
DISTR. **T** 6, 8; not known elsewhere
HAB. Usually near rivers or waterholes; 15–300 m.

SYN. *Canthium purpurascens* Bullock in K.B. 1932: 368 (1932); T.T.C.L.: 487 (1949), pro parte; Vollesen
    in Opera Bot. 59: 67 (1980)

NOTE. This species is very close to *K. zanzibarica*, but I have decided to maintain it as a distinct
    species; apart from the obvious differences in indumentum other more subtle differences, such as
    colour of dry leaves, glabrous style, fewer lateral nerves and shorter stipules have been noted.

11. **K. procteri** *Bridson* in K.B. 41: 981, fig. 4A–J (1986). Type: Tanzania,
Tabora/Mpanda District, Ugalla R., Senga, *Procter* 2088 (K, holo.!)

Scandent shrub or liane; young branches sparsely pubescent. Leaf-blades broadly
elliptic to ovate, 3.3–7.5 cm. long, 1.7–4.5 cm. wide, obtuse to rounded or sometimes
subacuminate at apex, obtuse to rounded at base, entirely glabrous or glabrescent
beneath; lateral nerves in 5–6 main pairs; tertiary nerves obscure to apparent, coarsely
reticulate; domatia present, glabrous or pubescent; petiole 3–6 mm. long, glabrous to
sparsely pubescent; stipules triangular, 3–5 mm. long, long-acuminate. Flowers (4–)5-
merous, borne in 20–50-flowered pedunculate cymes; peduncles 0.5–1.2 cm. long,
pubescent; pedicels 3–5 mm. long, pubescent; bracteoles up to 2 mm. long. Calyx

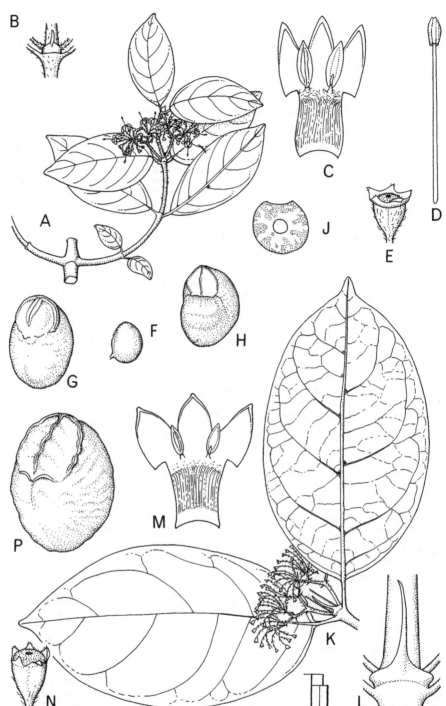

FIG. 163. *KEETIA PROCTERI* — **A**, flowering branch, × ⅔; **B**, stipules, × 2; **C**, section through corolla, × 6; **D**, style with stigmatic knob, × 6; **E**, calyx, × 6; **F**, fruit (single-seeded), × 1; **G**, pyrene, × 3; **H**, seed, × 3; **J**, transverse section of seed, × 3. *K. KORITSCHONERI* — **K**, node with inflorescences, × ⅔; **L**, stipule, × 2; **M**, section through corolla, × 6; **N**, calyx, × 6; **P**, pyrene, × 3. A–E, from *Procter* 2138; F–J, from *Koritschoner* 2227; K–L, from *Koritschoner* 1327; M, N, from *Koritschoner* 1412; P, from *Koritschoner* 669. Drawn by Diane Bridson.

pubescent; tube 1 mm. long; limb dentate, 0.5 mm. long. Corolla-tube 2.75 mm. long, with a ring of deflexed hairs set just below the top inside; lobes oblong-lanceolate, 3 mm. long, 1.25 mm. wide, acute. Anthers fully exserted but seldom reflexing. Style 7.25 mm. long, glabrous; stigmatic knob 1.25 mm. long. Disk pubescent. Fruit, only known from single-seeded examples, 9 mm. long, 7 mm. wide; pyrene ovoid with ventral face flattened, 8.5 mm. long, 6 mm. wide; point of attachment poorly defined, with lid-like area on ventral face surrounding central crest. Fig. 163/A–J.

TANZANIA.   Shinyanga, Nov. 1938, *Koritschoner* 2227!; Tabora District: Wala R., 16 km. S. of Tabora, 31 Oct. 1978, *Lawton* 2091!; Mpanda District: Nyamanzi R., 40 km. N. of Mpanda, Aug. 1962, *Procter* 2138!
DISTR.   T 1, 4; not known elsewhere
HAB.   Riverine forests and thickets; 1100–1210 m.

12. **K. sp. C**; Bridson in K.B. 41: 982 (1986)

Climber, young stems glabrous. Leaf-blades broadly elliptic, 4.5–9 cm. long, 2.5–4.5 cm. wide, acute or subacuminate then apiculate at apex, obtuse to rounded at base, entirely glabrous; lateral nerves in 5–6(–7) main pairs; tertiary nerves obscure; domatia present as inconspicuous rusty coloured tufts of hair; petiole 0.7–1.2 cm. long, slender, glabrous to sparsely pubescent; stipules linear-lanceolate, 5–6 mm. long, readily caducous; leaves subtending lateral branches not observed. Flowers not known. Cymes only known in fruiting stage; peduncles 1–1.3 cm. long, glabrous; pedicels 6–8 mm. long, sparsely pubescent; bracteoles inconspicuous. Fruit oblong-cordate in outline, 1.7 cm. long, 1.6 cm. wide; persistent calyx-limb dentate; pyrene ellipsoid flattened on ventral face, 1.6 cm. long, 0.9 mm. wide; point of attachment not prominent; lid-like area on ventral face, not completely demarcated, central crest not prominent.

TANZANIA.   Morogoro District: Shikurufumi [Chigurufumi] Forest Reserve, Mar. 1955, *Semsei* 2025!
DISTR.   T 6; known only from above specimen
HAB.   Forest; 500–800 m.

13. **K. lulandensis** *Bridson*, sp. nov. affinis *K. zanzibaricae* (Klotzsch) Bridson et *K. koritschoneri* Bridson sed fructu majore stipulisque truncatis vel apiculatis differt. Typus: Tanzania, Iringa District, Mufindi, Lulanda Forest, *Gereau et al.* 2899 (K, holo.!, MO, iso.)

Liane to 6 m. tall ?or small tree, young stems glabrous, lenticellate. Leaf-blades chartaceous, pale beneath, elliptic, or oblong-elliptic, 6–11.5 cm. long, 3–6 cm. wide, acuminate at apex, acute to obtuse and unequal at base, entirely glabrous; lateral nerves in 4–5(–6) main pairs; tertiary nerves rather coarsely reticulate; domatia present as tufts of white hair; petiole 6–10 mm. long, glabrous; stipules ± truncate, 2–4 mm. long, apiculate; leaves subtending lateral branches not observed. Flowers 5-merous, borne in 3–10-flowered branched pedunculate cymes; peduncles 6–11 mm. long, puberulous; pedicels 2–4 mm. long, puberulous; bracteoles up to 1.5 mm. long. Calyx-tube up to 2 mm. long, somewhat puberulous; limb 1.5 mm. long, repand or shortly dentate. Corolla-tube 4 mm. long, with deflexed fused hairs set at ⅓ of the way down inside; lobes ovate, 4.25 mm. long, 1.75–2.25 mm. wide, acute. Anthers fully exserted but not reflexed. Style 6 mm. long, glabrous; stigmatic knob 1.25 mm. long. Disk puberulous at edge. Fruit only known from single-seeded examples, ± reniform, 1.8 cm. long, 1.2 cm. wide; pyrene ellipsoid with ventral face flattened and strongly indented beneath point of attachment, 1.7 cm. long, 1.1 cm. wide; lid-like area on ventral face surrounding central crest.

TANZANIA.   Iringa District: Mufindi District, Lulanda Forest, 19 Apr. 1987, *Lovett & Congdon* 2034!, 25 Jan. 1989, *Gereau et al.* 2874! & 26 Jan. 1989, *Gereau et al.* 2899!
DISTR.   T 7; not known elsewhere
HAB.   Montane forest; 1450–1520 m.

14. **K. tenuiflora** (*Hiern*) *Bridson* in K.B. 41: 982 (1986). Types: Angola, Pungo Andongo, near Quilanga and Quibanga, *Welwitsch* 3143 (LISU, syn., BM, K, isosyn.!) & Barrancos, *Welwitsch* 3144 (LISU, syn., BM, K, isosyn.!)

Climber or scandent shrub 2–3.5 m. tall; lateral branches spreading, upwardly curved, glabrous or covered with yellowish adpressed hairs when young; bark fawn. Leaf-blades sometimes drying blackish, elliptic to broadly elliptic, 5–12 cm. long, 2.5–5 cm. wide, acuminate at apex, rounded or acute at base, glabrous to glabrescent with adpressed hairs on nerves beneath, papery, sometimes wavy at margins; lateral nerves in 5-7 main pairs; tertiary nerves coarsely reticulate, apparent or somewhat obscure; petioles 5-10 mm. long, pubescent with usually adpressed hairs; stipules with a triangular base 1-2 mm. long and apical subulate-linear lobe 4-7 mm. long, adpressed pubescent outside; leaves subtending lateral branches smaller but not differing in shape. Flowers 4-merous, borne in densely flowered pedunculate cymes; peduncles 0.4-1.5 cm. long, adpressed pubescent; pedicels slender, 2-7 mm. long, adpressed pubescent; bracteoles inconspicuous, up to 1 mm. long. Calyx-tube 0.5-0.75 mm. long, glabrescent to sparsely pubescent; limb dentate, 0.5-0.75 mm. long, sparsely pubescent. Corolla white; tube 2-2.5 mm. long, with a ring of deflexed hairs below mid-point and finely pubescent above inside; lobes oblong-obovate, 1.5 mm. long, 0.75-1 mm. wide, obtuse. Anthers small, fully exserted but not reflexed. Style 4.25-6 mm. long, glabrous; stigmatic knob 0.25-0.5 mm. long. Disk pubescent. Fruit strongly 2-lobed, each lobe almost spherical, 1.2-1.5 cm. long, 2 cm. wide; pyrenes almost ovoid-spherical, 9 mm. wide; lid-like area lying along ventral face surrounding central shallow crest.

UGANDA. Toro District: Burahya, Kawyawara, *Struhsaker* 169!; Mengo District: 16 km. on Entebbe road, Feb. 1938, *Chandler* 2165! & Kivuvu, July 1917, *Dummer* 3238!
DISTR. U 2, 4; Sierra Leone, Nigeria, Central African Republic, Cameroon, Gabon, Zaire and Angola
HAB. Forest; 1170-1220 m.

SYN. *Canthium tenuiflorum* Hiern, Cat. Afr. Pl. Welw. 1: 477 (1898)
  Plectronia tenuiflora (Hiern) K. Schum. in Just's Bot. Jahresb. 1898(1): 393 (1900)
  P. angustiflora De Wild. in B.J.B.B. 5: 30 (1915). Type: Zaire, Equateur, Dundusana, *Mortehan* 911 (BR, holo.!)
  P. rutshuruensis De Wild., Pl. Bequaert. 3: 197 (1925). Type: Zaire, Kivu, Rutshuru, *Bequaert* 6289 (BR, holo.!)
  Canthium brownii Bullock in K.B. 1932: 370 (1932). Type: Uganda, Mengo District, Entebbe, *E. Brown* 233 (K, holo.!)
  [*C. zanzibaricum* sensu Bullock in K.B. 1932.: 373 (1932), pro parte, quoad syn. *C. tenuiflorum*; Hepper in F.W.T.A., ed. 2, 2: 184 (1963), pro parte, quoad *Deighton* 3058 and *Small* 471, *non* Klotzsch]

15. **K. koritschoneri** *Bridson* in K.B. 41: 984, fig. 4K-P (1986). Type: Tanzania, Lushoto District, Makuyuni, *Koritschoner* 1412 (EA, holo.!, K, iso.!)

? Scandent shrub or tree 7 m. tall; young branches glabrous. Leaf-blades oblong-elliptic, 10-16 cm. long, 5.8-7.7 cm. wide, shortly acuminate at apex, obtuse to rounded at base, glabrous; lateral nerves in 6-7 main pairs; tertiary nerves rather coarsely reticulate; domatia present as glabrous pits in axils of main and sometimes tertiary nerves; petioles 0.8-1.4 cm. long, glabrous or sparsely pubescent; stipules linear from a triangular base, 0.9-1.3 cm. long, glabrous. Flowers 4-5-merous, borne in pedunculate 30-50-flowered cymes; peduncle 0.8-1.1 cm. long, sparsely pubescent; pedicels 6-8 mm. long, sparsely pubescent; bracteoles inconspicuous. Calyx-tube 0.75 mm. long, glabrous; limb 0.75 mm. long, dentate, ciliate. Corolla-tube 2.25-3 mm. long, with a ring of deflexed hairs a short distance below apex inside; lobes triangular, 2-2.5 mm. long, 1-1.5 mm. wide at base. Anthers fully exserted, erect; stigmatic knob not known. Disk pubescent. Fruit only known from single-seeded examples, 1.3 cm. long, 1.1 cm. wide and thick; pyrene ± broadly ellipsoid, 1.3 cm. long, 1 cm. wide, 0.7 cm. thick, rugulose; area above point of attachment on ventral plane ± circular with a central crest. Fig. 163/K-P, p. 920.

TANZANIA. Lushoto District: Makuyuni District, June 1935, *Koritschoner* 669! & 1412! and Mombo Forest Reserve, 6 Jan. 1956, *Muze* 18!
DISTR. T 3; not known elsewhere
HAB. Forest; 400-1000 m.

16. **K. carmichaelii** *Bridson* in K.B. 41: 984, fig. 5 (1986). Type: Tanzania, Iringa District, Idunduge, *Carmichael* 306 (K, holo.!, EA, iso.!)

Shrub or liane 1.3-20 m. tall; young branches sparsely setose; older branches with fawn-coloured bark. Leaf-blades drying brownish, elliptic to broadly elliptic, 4-7.5 cm. long, 2.4-4 cm. wide, obtuse to subacuminate and often shortly apiculate at apex or sometimes emarginate, obtuse or rounded at base, subcoriaceous, moderately shiny above, sparsely setose on midrib beneath; lateral nerves in 6 main pairs, impressed above; tertiary nerves moderately finely reticulate, rather obscure; domatia present as tufts of brown or rusty coloured hairs; petioles 4–10 mm. long, densely setose; stipules lanceolate, 4–8 mm. long, setose, caducous. Flowers 5–6-merous, borne in pedunculate, 1–20-flowered cymes; peduncles 0.5–2.5 cm. long, sparsely setose; pedicels 3–6 mm. long, sparsely setose; bracteoles inconspicuous, lanceolate. Calyx-tube ellipsoid, 2.5 mm. long, glabrous; calyx-limb well developed, 1.75–2 mm. long, divided almost to base into 5–6 triangular lobes, sparsely setose outside, coarsely reddish setose inside. Corolla yellowish green (only known from mature bud); tube 7 mm. long, with a ring of deflexed hairs at mid-point, pubescent above and below; lobes narrowly oblong, 4 mm. long, 1.75 mm. wide. Stigmatic knob ± 2 mm. long. Disk glabrous. Fruit compressed conical widest at top, 1.1 cm. long, 1.3 cm. wide, 9 mm. thick, apex not at all indented, tapering to base; pyrene (from single-seeded fruit) ± broadly ellipsoid, 10 mm. long, 8 mm. wide, 7 mm. thick, slightly rugulose; circular area lying perpendicular to ventral face containing a central ridge along line of dehiscence; point of attachment not free.

TANZANIA. Iringa District: Idunduge, Nov. 1953, *Carmichael* 306!, Mufindi, Lulanda Forest, 22 Feb. 1988, *Lovett & Congdon* 3151! & Mufindi, Lulanda, 24 May 1989, *Kayombo* 612!
DISTR. **T** 7; known only from above gatherings
HAB. Moist forest; 1500–1825 m.

POSSIBLY VANGUERIEAE

Tree to 9 m.; stems purplish brown, somewhat quadrangular, pubescent with ± ferruginous somewhat bristly hairs. Leaf-blades obovate, ± 13 × 6 cm., acuminate at the apex, cuneate at the base, discolorous, main nerves drying paler above, glabrous save for domatia; petioles very short or obsolete; stipules with base ± 2 mm. long, densely pilose within and filiform appendages up to 1 cm. long, soon falling. Peduncles(?) axillary, one on each side of node. Fruits said to be edible.

TANZANIA. Lindi District: Rondo Plateau, Mchinjiri, Mar. 1952, *Semsei* 704!
DISTR. **T** 8; known only from the above gathering
HAB. Presumably forest

NOTE. We have been unable to identify this with anything described but the material is completely sterile. The Kimwera name is given as 'mterarela'.

Subfamily ANTIRHEOIDEAE (GUETTARDOIDEAE)

Tribe 23. GUETTARDEAE*

109. GUETTARDA

L., Sp. Pl.: 991 (1753) & Gen. Pl., ed. 5: 428 (1754)

Trees or shrubs, the branches in some species armed with spines (not in Old World). Leaves opposite or rarely ternate, petiolate; stipules interpetiolar or intrapetiolar, simple, persistent or soon falling, free or connate. Flowers ♂ or polygamo-dioecious, showing limited heterostyly, (4–)5–6(–9) even 11-merous, usually sessile, mostly secundly arranged in axillary pedunculate apically dichotomously branched inflorescences or rarely solitary or in groups of 2–3; bracteoles present or absent. Calyx shortly tubular, globose or ovoid, the limb truncate or 2–9-toothed. Corolla white, yellowish or red to reddish purple, salver-shaped; throat glabrous or nearly so; lobes imbricate or subvalvate, the margins often wavy or crenulate. Stamens 4–9, inserted in the throat, included; anthers sessile or nearly so, dorsifixed. Ovary 2–9-locular; ovules solitary, pendulous; style slender; stigma ± capitate, slightly bifid. Fruit globose or subglobose, ± fleshy or fibrous with woody or bony endocarp, 2–9-locular, containing 2–9 pyrenes. Seeds with little or no endosperm.

A genus of about 80 species, almost entirely confined to the New World, one widespread in the Old World tropics and 12 or so described from New Caledonia and New Hebrides.
J.B. Gillett reports that *G. uruguensis* Cham. & Schlecht. an E. tropical S. American species has been grown in Nairobi. It has much smaller leaves and flowers than the native species.

---

* By B. Verdcourt.

G. speciosa *L.*, Sp. Pl.: 991 (1753); Hiern in F.T.A. 3: 125 (1877); E. & P. Pf. IV. 4: 96, fig. 34A–D (1891); K. Schum. in P.O.A. C: 387 (1895); T.T.C.L.: 500 (1949); K.T.S.: 445 (1961); Chao in Fl. Taiwan 4: 269, t. 996 (1978); Fosberg & Renvoize, Fl. Aldabra: 152, fig. 23/1, 2 (1980); A.C. Smith & Darwin in Fl. Vit. Nova 4: 148, figs. 60, 61 (1988); Robbrecht, Trop. Woody Rub.: 129, fig. 49 (1988). Type: Java or ? India, *Herb. Linnaeus* 1121.1 (LINN, holo.)*

Shrub or small branched or unbranched tree 1–8(–18) m. tall with trunk up to 15 cm. wide; young shoots velvety tomentose; stems stout, pubescent, at length glabrescent, with obvious leaf scars, lenticellate; bark brown, slightly rough. Leaves crowded at the ends of the twigs; blades obovate, 5–32 cm. long, 3.5–22.2 cm. wide, rounded to obtuse at the apex, rounded to cordate at the base, glabrous or almost so above, glabrous to velvety pubescent beneath; petioles 0.5–3(–12) cm. long; stipules sheathing the terminal bud, broadly elliptic, 0.7–2.3 cm. long, soon falling. Flowers fragrant, in cymose pubescent to glabrous inflorescences from the upper axils; peduncles 0.4–19 cm. long, reddish at apex; secondary peduncles absent or 0.3–3 cm. long; pedicels mostly absent or very short. Calyx-tube green, tinged pink, campanulate, 1–3(–5) mm. long, adpressed pubescent; limb yellowish, 2–5 mm. long, undulate or with lobes 0–2 mm. long. Corolla white, cream or yellowish or greenish white, sometimes slightly pink tinged, adpressed silky pubescent outside; tube narrowly cylindrical, 2–4.6 cm. long; lobes 4–9(–11), oblong, 0.2–1.5 cm. long, 1.5–8(–10) mm. wide. Anthers the same number as the lobes, linear, subsessile in the throat, 3–5 mm. long, included. Ovary 4–9-locular; style slender, 1.8–4.5 cm. long; stigma green, obconic-cylindric, grooved, slightly bifid, reaching to or just beyond the anthers in long-styled flowers and to less than a half the way up the tube in short-styled flowers. Fruit green, streaked brownish red, ovoid or subglobose, 0.8–2.5 cm. long, 1.2–3.5(–4) cm. wide; putamen woody surrounded by cavities filled with pith-like tissue. Seeds 2–5 mm. long, 1.5–5 mm. wide, curved; radicle inferior; stony arilloid tissue present. Fig. 164.

KENYA. Kwale District: Ukunda, 9 May 1953, *Bally* 8906!; Mombasa, 26 May 1934, *Napier* in *C.M.* 3268!; Kilifi District: Vipingo, 16 Dec. 1953, *Verdcourt* 1060!
TANZANIA. Tanga District: Sawa, 6 May 1956, *Faulkner* 1862! & 2 Sept. 1956, *Faulkner* 1911!; Rufiji District: Mafia I., Kanga, 11 Aug. 1937, *Greenway* 5048!; Zanzibar I., Paje, 25 Sept. 1959, *Faulkner* 2371!; Pemba I., *Barraud!*
DISTR. **K** 7; **T** 3, 6, 8; **Z**; **P**; Mozambique, widespread throughout Indian and Pacific Ocean beaches as far north as Pratas I., China and Ryuku Is. in Japan
HAB. On sand and coral rocks usually within reach of the highest tides and normal spray, with *Cordia, Casuarina, Pemphis* etc., also occasionally with mangrove vegetation; sea-level
NOTE. Fosberg claims that this species is distinctly heterostylous and further examination of herbarium material indicates this is correct. My assertion under Guettardoideae, part 1: 4, must be modified. The heterostyly is of the incomplete type where the stigma reaches different heights within the tube but the anthers are in only slightly different positions. The fruit is buoyant and dispersed by the sea.

ADDENDA

Arranged in taxonomic order, generic numbers in bold, species numbers in roman.

Numerous undescribed new taxa and extensions to geographical ranges are not included.

1,1a. For *Psychotria riparia* (K. Schum. & K. Krause) Petit var. *riparia*, use **P. capensis** (*Eckl.*) *Vatke* subsp. **riparia** (*K. Schum. & K. Krause*) *Verdc.* var. **riparia** — see F.Z. 5 (1): 13 (1989).

1,1b. For *Psychotria riparia* (K. Schum. & K. Krause) Petit var. *puberula* Petit, use **P. capensis** (*Eckl.*) *Vatke* subsp. **riparia** (*K. Schum. & K. Krause*) *Verdc.* var. **puberula** (*Petit*) *Verdc.* —see F.Z. 5 (1): 14 (1989).

1,5. For *Psychotria eminiana* (Kuntze) Petit var. *stolzii* (K. Krause) Petit, use **P. eminiana** (*Kuntze*) *Petit* var. **eminiana** — see F.Z. 5 (1): 14 (1989).

* It seems reasonable to select the Linnean Herbarium specimen (excluding fruit) as the type; the Species Plantarum gives a description and no reference but the type locality is Java. The specimen is, however, said to come from India.

FIG. 164. *GUETTARDA SPECIOSA* — **1**, flowering branch, × ⅔; **2**, large leaf, × ⅔; **3**, section through corolla, × 1; **4**, dorsal view of anther, × 6; **5**, longitudinal section through calyx and ovary, × 4; **6**, transverse section through ovary, × 6; **7**, stigma, × 6; **8**, infructescence (front view), × ⅔; **9**, infructescence (dorsal view), × ⅔; **10**, transverse section through fruit at middle, × 1; **11**, transverse section through fruit above base, × 1; **12**, longitudinal section through fruit, × 1; **13**, seed with integument cut to show embryo, × 1. 1, 2 from *Faulkner* 2371; 3–7, from *Verdcourt* 1060; 8–13, from *Drummond & Hemsley* 3277. Drawn by Marie Bywater, with 4, 5, 7, 10–13, by Diane Bridson.

1,8c. For *Psychotria goetzei* (K. Schum.) Petit var. *meridiana* Petit, use 10. **P. zombamontana** *(Kuntze) Petit* — see F.Z. 5 (1): 16 (1989).

1,10. For both *Psychotria meridiano-montana* Petit var. *meridiano-montana* and var. *angustifolia* Petit, use **P. zombamontana** *(Kuntze) Petit* — see F.Z. 5 (1): 16 (1989).

1,34. For *Psychotria sp. F*, use 7. **P. triclada** *Petit* — see Borhidi & Verdcourt in K.B. 45: 705 (1990).

1,35. For *Psychotria sp. G*, use 7/1 **P. verdcourtii** *Borhidi* — Borhidi & Verdcourt in K.B. 45: 706 (1990).

1,39. For additional citations of **Psychotria usambarensis** *Verdc.* from Nguru Mts. — see Borhidi & Verdcourt in K.B. 45: 707 (1990).

1,41. For *Psychotria sp. K*, use **P. pocsii** *Borhidi & Verdc.* subsp. **ferruginea** *Borhidi & Verdc.* —see Borhidi & Verdcourt in K.B. 45: 708 (1990).

1,41a. Additional subspecies — **Psychotria pocsii** *Borhidi & Verdc.* subsp. **pocsii** — see K.B. 45: 707 (1990). Tanzania, Lushoto District, E. Usambara Mts., Amani–Sigi, May 1987, *Borhidi et al.* 87387! & 19 Apr. 1954, *Verdcourt* 1122!

1,43/1. Additional species — **Psychotria lovettii** *Borhidi & Verdc.* — see K.B. 45: 703, fig. 1 (1990). Tanzania, Lushoto District, W. Usambara Mts., Herkulu, 13 Nov. 1984, *Lovett* 242! & Bumba area, Gombelo Forest Reserve, 29 Apr. 1987, *Borhidi et al.* 86056!

1,60h. For *Psychotria kirkii* Hiern var. *diversinodula* Verdc., use 60/1. **P. diversinodula** *(Verdc.) Verdc.* — see F.Z. 5 (1): 29 (1989).
The other varieties of *P. kirkii* are not maintained in F.Z.

1,73. *Grumilea diploneura* K. Schum. Two specimens (*Lovett* 205 & *Lovett et al.* 560) from the Uluguru Mts. apparently match the description of this taxon. A combination in *Psychotria* is needed — **Psychotria diploneura** *(K. Schum.) Bridson & Verdc.*, comb. nov. —basionym *Grumilea diploneura* K. Schum. in E.J. 28: 496 (1900).

4. *Chazaliella abrupta* (Hiern) Petit & Verdc. var. *abrupta* — additional synonym: *Canthium pubipes* S. Moore in J.B. 43: 352 (1905). Type: Kenya, Cha Shimba [Pemba Flats], *Kassner* 393 (BM, holo., K, iso.!)

5,6. For *Chassalia* ? hybrid between *C. albiflora* and *C. zimmermannii* (in note), use 4/1 **C. longiloba** *Borhidi & Verdc.* — see Borhidi & Verdcourt in K.B. 45: 709 (1990).

7,4. The widely distributed Indo-Malesian and Pacific coastal species **Morinda citrifolia** *L.* has now been recorded — Kenya, Kilifi District, Watamu peninsula, 5 Dec. 1978, *Kuchar* 10031!

13,3. *Pentanisia foetida* Verdc. an alternative combination, *Chlorochorion foetidum* (Verdc.) Puff & Robbrecht has been published — see Puff & Robbrecht in Bot. Jahrb. Syst. 110: 547 (1989).

13,4. *Pentanisia monticola* (K. Krause) Verdc. an alternative combination, *Chlorochorion monticola* (K. Krause) Puff & Robbrecht has been published — see Puff & Robbrecht in Bot. Jahrb. Syst. 110: 547 (1989).

14. *Hondbessen* Adans. was incorrectly placed in synonymy with *Paederia*. It is very doubtfully *Canthium* — see Parkinson in Taxon 37: 150 (1988).

22,4. *Kohautia longifolia* Klotzsch — the four varieties are not maintained by D. Mantell in F.Z. 5 (1): 98 (1989).

22,9c. For *Kohautia caespitosa* Schnizl. var. *delagoensis* (Schinz) Bremek., use **K. caespitosa** *Schnizl.* subsp. **brachyloba** *(Sond.) D. Mantell* — see F.Z. 5 (1): 88 (1989).

22,10. For *Kohautia lasiocarpa* Klotzsch var. *subverticillata* (K. Schum.) Bremek., use **K. subverticillata** *(K. Schum.) D. Mantell* subsp. **subverticillata** — see F.Z. 5 (1): 89 (1989).

24A. *Manostachya staelioides* (K. Schum.) Bremek. — additional record, see key to genera on p. 420. Tanzania, Njombe District, 24 km. W. of Njombe on road from Kipengere, 14 Nov. 1966, *Gillett* 17835! — see Verdcourt in K.B. 35: 322 (1980).

33,11. For *Oldenlandia pellucida* Hiern var. *pellucida*, use **O. echinulosa** *K. Schum.* var. **pellucida** (*Hiern*) *Verdc.*, and for *O. pellucida* Hiern var. *echinulosa* (K. Schum.) Verdc. use **O. echinulosa** *K. Schum.* var. **echinulosa** — see Verdcourt in K.B. 33: 608 (1978).

34,8. For *Otiophora parviflora* Verdc. var. *iringensis Verdc.*, use **O. villicaulis** *Mildbr.* var. **iringensis** (*Verdc.*) *Puff* — see F.Z. 5 (1): 149 (1989).

35,1. For *Anthospermum herbaceum* L.f. var. *villosicarpum* Verdc., use **A. villosicarpum** (*Verdc.*) *Puff* — see Puff in Pl. Syst. Evol., Suppl. 5: 292 (1986).

35,3. *Anthospermum randii* S. Moore, of the three cited specimens, *Harley* 9529 and *Richards* 8666 have been redetermined as *A. rosmarinus* K. Schum., while the sterile *B.D. Burtt* 1303 tentatively remains as *A. randii*, now known as **A. ternatum** *Hiern* subsp. **randii** (*S. Moore*) *Puff* — see Puff in Pl. Syst. Evol., Suppl. 3: 285 & 292 (1986).

35,9/1. Additional record — **Anthospermum rigidum** *Eckl. & Zeyh.* subsp. **pumilum** (*Sond.*) *Puff* — F.Z. 5 (1): 158 (1989). Tanzania, Iringa District, Mafinga, 25 Feb. 1982, *Kibuwa* 5514!

36. For *Hydrophylax madagascariensis* Roem. & Schultes, use **Phylohydrax madagascariensis** (*Roem. & Schultes*) *Puff* — see key to genera on p. 422 — see Puff in Pl. Syst. Evol. 154: 343 (1986).

38,1. True *Spermacoce confusa* Gillis does not occur in East Africa, material so called should be referred to **S. tenuior** *L.* var. **tenuior** — see Verdcourt in K.B. 37: 546 (1983).

38,2. Additional variety — **Spermacoce tenuior** *L.* var. **commersonii** *Verdc.* — see K.B. 37: 546 (1983). Tanzania, Zanzibar I., Mwera Swamp, 16 Nov. 1930, *Vaughan* 1682!

38,9/1. Additional record — **Spermacoce filifolia** (*Schumach. & Thonn.*) *Lebrun & Stork.* Tanzania, Kilwa District, ± 19 km. SSW. of Kingupira, 17 Apr. 1976, *Vollesen* in *M.R.C.* 3473.

38,14. For *Spermacoce laevis* Lam., use **S. assurgens** *Ruiz. & Pav.* — see Verdcourt in K.B. 37: 547 (1983).

38,19. For *Spermacoce ocymoides* Burm.f., use **S. mauritiana** *Osia Gideon* — see Verdcourt in K.B. 37: 547 (1983).

38,20. For *Spermacoce hispida* L., use **S. articularis** *L.f.* — see Sivarajan & Nair in Taxon 35: 363–367 (1986).

38,27. *Spermacoce sp. D.*, possibly conspecific with *S. schlechteri* Verdc. — see F.Z. 5 (1): 180 (1989).

39. For *Mitracarpus villosus* (Sw.) DC., use **M. hirtus** (*L.*) *DC.* — see Taxon 26: 573 (1977) & 30: 301 (1981).

59. For *Porterandia penduliflora* (K. Schum.) Keay, use **Aoranthe penduliflora** (*K. Schum.*) *Somers* — see Somers in B.J.B.B. 58: 74 (1988).

89. Heinsenia under synonymy of *H. diervilleoides* K. Schum. subsp. *diervilleoides* for '*H. infundibuliformis* Petit' read *Aulacocalyx infundibuliflora* Petit.

100,6. Further material of *Rytigynia sp.* C (p. 810) has recently been collected which can now be described.

    Rytigynia longipedicellata *Verdc.*, sp. nov. inter congeneres inflorescentiis 1–2-floris, pedunculis obsoletis vel usque 2 mm. longis, pedicellis gracilibus 1.3 cm. longis usque in fructu 2.5–3 cm. longis, ovario 2-loculare differt. Typus: Tanzania, Lindi District, Rondo Plateau, *Bidgood et al.* 1578 (K, holo.!, EA, iso.!)

    Shrub or small tree 1.5–4.5 m. with lenticellate peeling purplish bark. Leaves elliptic to elliptic-oblong, 3–12 cm. long, 1.7–5.5 cm. wide, narrowly obtuse to shortly acuminate, glabrous save for domatia beneath; stipules with dense brownish hairs within. Pedicels ± 1.3 cm. long elongating to 3 cm. in fruit. Corolla cream or with greenish tube; tube ± cupular, 4 mm. long; lobes triangular, ± 3 mm. long.

TANZANIA.   Lindi District: Rondo Plateau, Mchinjiri, Mar. 1952, *Semsei* 700! & Rondo Plateau, 14 Feb. 1991, *Bidgood et al.* 1577!, 1578!
DISTR.   **T** 8; not known elsewhere
HAB.   Forest edge; 700 m.

NOTE.   *Bidgood* 1577 and 1578 were growing adjacent to each other and appeared very different, 1577 having much larger thinner leaves; nevertheless, the three specimens cited above are clearly conspecific. The species will key to couplet 38 of the key where difficulty is evident, but nothing else coming later has such long pedicels.

New names validated by D.M. Bridson and B. Verdcourt in this Part

Canthium subgen. **Afrocanthium** *Bridson*, p. 864
Canthium subgen. **Lycioserissa** (*Roem. & Schultes*) *Bridson*, p. 876
Canthium burttii *Bullock* subsp. **glabrum** *Bridson*, p. 876
Canthium kilifiense *Bridson*, p. 874
Canthium oligocarpum *Hiern* subsp. **captum** (*Bullock*) *Bridson*, p. 878
Canthium oligocarpum *Hiern* subsp. **friesiorum** (*Robyns*) *Bridson*, p. 878
Canthium oligocarpum *Hiern* subsp. **intermedium** *Bridson*, p. 878
Canthium parasiebenlistii *Bridson*, p. 870
Canthium peteri *Bridson*, p. 873
Canthium racemulosum S. Moore var. **nanguanum** (*Tennant*) *Bridson*, p. 875
Canthium rondoense *Bridson*, p. 868
Canthium shabanii *Bridson*, p. 873
Canthium vollesenii *Bridson*, p. 875
Cuviera migeodii *Verdc.*, p. 771
Keetia lulandensis *Bridson*, p. 921
Psychotria diploneura (*K. Schum.*) *Bridson & Verdc.*, p. 926
Psydrax robertsoniae *Bridson*, p. 903
Pyrostria uzungwaensis *Bridson*, p. 888
Rytigynia longipedicellata *Verdc.*, p. 927
Rytigynia longituba *Verdc.*, p. 829

GEOGRAPHICAL DIVISIONS OF THE FLORA

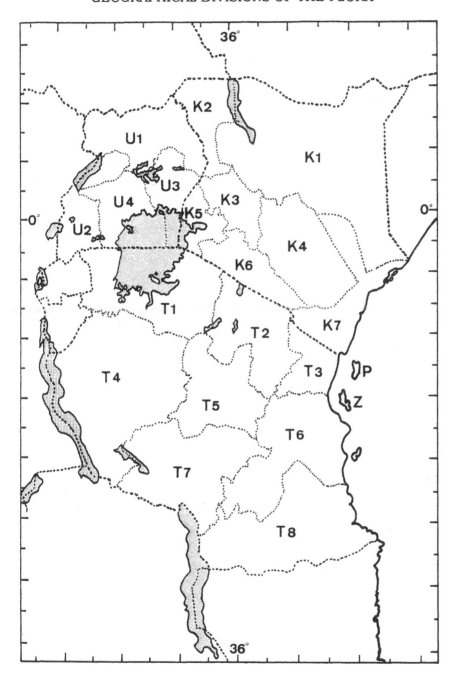

T - #0660 - 101024 - C0 - 244/170/11 - PB - 9789061913573 - Gloss Lamination